Sustainability Sciences in Asia and Africa

Sustainable Aquatic Food Systems

Series Editors
Basanta Kumar Das, ICAR – Central Inland Fisheries Research Institute
Barrackpore, Kolkata, West Bengal, India

Natrah Fatin Mohd Ikhsan, Faculty of Agriculture
Universiti Putra Malaysia, Serdang, Selangor, Malaysia

Sustainability as a scientific endeavor requires broad understanding of the interconnections in our global environment. It is therefore positioned at the crossroads of research in natural sciences, technology and human sciences. This book series supports the global efforts by providing timely coverage of the progress, opportunities, and challenges towards achieving sustainable development goals (SDGs) in Asia and Africa. It fosters the research in transdisciplinary academic fields spanning across agriculture and global food security, health and diseases, access to clean water and affordable energy, climate resilience and effects of global climate change on biodiversity, water and land. This series is planned to narrate the stories of scientific development and progression towards SDGs from the regions of Asia and Africa. The overall aim of the series is to provide a comprehensive publication platform and act as a knowledge engine in the growth of sustainability sciences with a special focus on developing nations. The series constitute of four sub-series each focusing on a specific theme closely linked to a set of SDGs. Each sub-series is leaders and experts in agricultural sciences, environmental sciences, biomedicine and medical sciences respectively. Books in this series are useful for environmentalists, biotechnologists, plant scientists, medical practitioners, conservation biologists, policy makers, NGOs and scientific bodies working in the areas of sustainability sciences in the developing regions.

Aquatic food systems have the potential to nurture human health and support environmental sustainability; however, they are currently threatened by anthropogenic climate change. Providing a growing global population with healthy diets from sustainable aquatic food systems is an immediate challenge. Although global fisheries and aquaculture production surpassed 223 million tonnes, aquaculture contributed to 130 million tonnes. More than 820 million people still have insufficient food, and close to 2 billion people suffer from micronutrient malnutrition, commonly called "hidden hunger."

Aquaculture could be crucial for food and nutritional security with high-quality animal protein (15 %), providing millions of people with livelihood and income generation. It is also expected to contribute towards the Sustainable Development Goals, namely SDG1 – No Poverty, SDG2 – Zero Hunger, and SDG14 – Life Below Water, adopted by all members of the United Nations (UN) in 2015. However, intensive and high-density culture systems could result in suboptimal oxygen levels, accumulation of metabolic waste products and feed leftovers that can create stressful environmental conditions and significantly affect the health status of the aquatic animals. The degraded aquatic environment also leads to decreased growth performance and reduced immune response, which becomes the ideal ground for disease outbreaks. Therefore, technological development to cope with changes in biological characteristics and stabilize production is strongly desired for both capture fisheries and aquaculture sustainability. In this book series, we shift the focus from production to consumption and aim to increase the visibility of aquatic resources as sustainable food that can contribute to "ending poverty, protecting the planet and improving the lives and prospects for everyone, everywhere" (The Sustainable Development Agenda).

Basanta Kumar Das
Piyashi DebRoy • Prateep Kumar Nayak
Editors

Small-scale Inland Fisheries in India

Springer

Editors
Basanta Kumar Das
ICAR – Central Inland Fisheries Research Institute
Barrackpore, Kolkata, West Bengal, India

Prateep Kumar Nayak
School of Environment, Enterprise and Development, Faculty of Environment
University of Waterloo
Waterloo, ON, Canada

Piyashi DebRoy
Economics and Policy Unit
ICAR – Central Inland Fisheries Research Institute
Barrackpore, Kolkata, West Bengal, India

ISSN 2730-6771 ISSN 2730-678X (electronic)
Sustainability Sciences in Asia and Africa
ISSN 3059-4162 ISSN 3059-4170 (electronic)
Sustainable Aquatic Food Systems
ISBN 978-981-95-3848-5 ISBN 978-981-95-3849-2 (eBook)
https://doi.org/10.1007/978-981-95-3849-2

This Springer imprint is published by the registered company Springer Nature Singapore Pte Ltd.
The registered company address is: 152 Beach Road, #21-01/04 Gateway East, Singapore 189721, Singapore

Foreword

It gives me immense pleasure to express that the book on small-scale inland fisheries in India is the first of its kind. The book chapters aptly reflect on the dynamics of the state of the inland fisheries in India along with the factors that govern the multi-dimensionality of the sector. Inland fisheries is a state subject in the Indian system of Governance and Legislation, which divulges that the dynamics, inter-play of actors in the value chains, and policies governing the state of inland fisheries are as vast as the subject itself. Unlike the marine fisheries sector where landings and catch statistics are better regulated, and hence data generation takes place effectively, much less light has been shed into the inland fisheries, particularly in the small-scale sectoral operations. Owing to the diverse and dispersed nature of the inland open water resources, reliable data and catch statistics has been a persistent challenge. Addressing such challenges has now assumed crucial significance because of the huge inland open water resources that the country is endowed with. On this occasion, the publication of the book is very timely which puts forward the different aspects of the small-scale inland fisheries sector of India and enables one to understand the intricacies involved in the management of the fisheries resources. I am confident that this book would serve as an instrument of change in the attitude of the readers regarding the endearing experiences and insightful achievements that have marked the glory of inland fisheries from the past till the recent times. A lot more needs to be attained to streamline the small-scale inland fisheries sector and augment the livelihood of inland fishers for which this book is a highly recommended piece of cognizance and literary insight.

Former Director and Vice Chancellor,
ICAR – Central Institute of Fisheries
Education,
Mumbai, India
5th August 2025

Gopal Krishna

Preface

This book serves as a guide for a clear, concise, and an overarching introduction to the different components of small-scale inland fisheries in India. Small-scale inland fisheries cater to the livelihood of millions of people in developing countries. Small-scale fisheries (SSF) constitute 70% of the total fishing workforce in the country. In the inland fisheries sector, small-scale fishing contributes around 60% of the household income. Inland fisheries in India have gained less importance than the marine sector as it is highly scattered and forms a highly diverse resource group. The book chapters reveal the similarities and variedness in the different inland fisheries management systems within various social-economic-ecological realms. Nutritional security, governance, gender, climate change, policy and regulatory framework, marketing management, and case studies in inland resources in the country have been depicted. This book would help in monitoring the progress made toward the achievement of SDGs in small-scale fisheries in India. Information on global inland small-scale fisheries sector is limited. It is high time that explicit mention of inland SSF in global and national fisheries policies and schemes is necessary by incorporating scientific and field-level inputs. Since more information is now needed on small-scale fisheries, the authors have aimed to get a fuller and more accurate picture of small-scale fisheries in the inland sector of India. In the way to fulfilling this need, the book on Small-Scale Inland Fisheries in India serves as a consolidated knowledge and text for depicting the richness of small-scale inland fisheries that the country inherits and highlights and the priority that should be given to the sub-sector. The book emphasizes on the state of management of different resources of the sub-sector with detailed examples to enable the readers to learn about the theories inherent in the sector and use these resources appropriately during future course of work for holistic management of the fisheries sector of the country. It provides a unified base and a framework to understand the implications of the state of activities in small-scale inland fisheries in light of institutional management, governance, multi-species and multi-gear characteristic nature of the fishery, climate change, Artha-Ganga, and gender equity. It is designed both for attempting to understand that small-scale inland fisheries are "too big to ignore" and for those aiming to conduct research in the subject.

Barrackpore, Kolkata, West Bengal, India — Basanta Kumar Das
Barrackpore, Kolkata, West Bengal, India — Piyashi DebRoy
Waterloo, ON, Canada — Prateep Kumar Nayak

Acknowledgments

This book would not have been completed without the enthusiastic help of Yateesh D. C. and Anurag Singh who have intricately helped in formatting and editing all the chapters and many drafts of the manuscript. We are also thankful to Liton Paul and Gitashree Thengel who have helped immensely in editing and formatting many of the chapters and their drafts. The help of Antara Roy Chowdhury is highly acknowledged who formatting the first of many early drafts of the manuscript.

Contents

About the Editors and Contributors

Editors

Basanta Kumar Das is working as the Director at the Indian Council of Agricultural Research–Central Inland Fisheries Research Institute (ICAR-CIFRI), Kolkata, India. He did his BFSc, MFSc, and PhD at Orissa University of Agriculture and Technology, and post-doctoral work at FRS Marine Lab, Scotland, UK, and University of California, USA. He has been honored with numerous awards and fellowships. He was elected as President of the Section of Animal, Veterinary, and Fishery Sciences for 2023–24 in the 109th Session of the Indian Science Congress (to be held in 2024). He has more than 390 international publications in peer-reviewed journals.

Piyashi DebRoy is currently serving as a Senior scientist in the Economics and Policy Unit at ICAR–Central Inland Fisheries Research Institute, Barrackpore, Kolkata, West Bengal. She has obtained her PhD in fish business management from ICAR–Central Institute of Fisheries Education, Mumbai. She has expertise in the subject of fisheries economics, environmental economics, and gender quity in fisheries and aquaculture. Her research focuses on studying how the socioeconomic dynamics of inland fishers and fish farmers is shaped by the demographic patterns, factors, trends, and drivers in inland open water fisheries in India.

Prateep Kumar Nayak is currently a Professor and Associate Dean of Equity, Diversity, Inclusion and Justice in the Faculty of Environment, University of Waterloo, Canada. He is the Project Director of the Vulnerability to Viability (V2V) Global Partnership for building strong small-scale fisheries. His academic training is in political science, environmental studies, and international development. His research is transdisciplinary with an active focus in combining social and ecological perspectives. He has published over 60 refereed papers, book chapters, and books and over 260 policy, research, review, and technical contributions. He is recognized as a science leader and advocate in the global science-policy-practice forum and serves on international policy, advisory, advocacy, and research committees pertaining to biodiversity, climate, development, and conservation.

Contributors

Dibakar Bhakta ICAR – Central Inland Fisheries Research Institute (ICAR-CIFRI), Barrackpore, Kolkata, India

Shreya Bhattacharya ICAR – Central Inland Fisheries Research Institute (ICAR-CIFRI), Barrackpore, Kolkata, India

Sanjenbam Bidyasagar College of Fisheries, Dr. Rajendra Prasad Central Agricultural University, Pusa, Bihar, India

Simanku Borah ICAR – Central Inland Fisheries Research Institute (ICAR-CIFRI), Barrackpore, Kolkata, India

L. Chakraborty ICAR – Central Inland Fisheries Research Institute (ICAR-CIFRI), Barrackpore, Kolkata, India

Ganesh Chandra ICAR – Central Inland Fisheries Research Institute (ICAR-CIFRI), Barrackpore, Kolkata, India

Basanta Kumar Das ICAR – Central Inland Fisheries Research Institute (ICAR-CIFRI), Barrackpore, Kolkata, India

Piyashi DebRoy ICAR – Central Inland Fisheries Research Institute (ICAR-CIFRI), Barrackpore, Kolkata, India

Anjana Ekka ICAR – Central Inland Fisheries Research Institute (ICAR-CIFRI), Barrackpore, Kolkata, India

Bandana Das Ghosh ICAR – Central Inland Fisheries Research Institute (ICAR-CIFRI), Barrackpore, Kolkata, India

Pranab Gogoi ICAR – Central Inland Fisheries Research Institute (ICAR-CIFRI), Barrackpore, Kolkata, India

Gunjan Karnatak ICAR – Central Inland Fisheries Research Institute (ICAR-CIFRI), Barrackpore, Kolkata, India

Suman Kumari ICAR – Central Inland Fisheries Research Institute (ICAR-CIFRI), Barrackpore, Kolkata, India

Lianthuamluaia ICAR – Central Inland Fisheries Research Institute (ICAR-CIFRI), Barrackpore, Kolkata, India

Sullip Kumar Majhi ICAR – Central Inland Fisheries Research Institute (ICAR-CIFRI), Barrackpore, Kolkata, India

R. K. Manna Reservoir and Estuarine Fisheries Division, ICAR – Central Inland Fisheries Research Institute (ICAR-CIFRI), Barrackpore, Kolkata, India

Mishal P. ICAR – Central Inland Fisheries Research Institute (ICAR-CIFRI), Barrackpore, Kolkata, India

S. K. Mohanty Chilika Development Authority, Bhubaneswar, Odisha, India

Prateep Kumar Nayak School of Environment, Enterprise and Development, Faculty of Environment, University of Waterloo, Waterloo, ON, Canada

Arun Pandit Economics and Policy Unit, ICAR – Central Inland Fisheries Research Institute (ICAR-CIFRI), Barrackpore, Kolkata, India

Pranaya Kumar Parida ICAR – Central Inland Fisheries Research Institute (ICAR-CIFRI), Barrackpore, Kolkata, India

Liton Paul ICAR – Central Inland Fisheries Research Institute (ICAR-CIFRI), Barrackpore, Kolkata, India

Aparna Roy Training and Extension Unit, ICAR – Central Inland Fisheries Research Institute (ICAR-CIFRI), Barrackpore, Kolkata, India

U. K. Sarkar Reservoir and Estuarine Fisheries Division, ICAR – Central Inland Fisheries Research Institute (ICAR-CIFRI), Barrackpore, Kolkata, India

Anurag Singh ICAR – Central Inland Fisheries Research Institute (ICAR-CIFRI), Barrackpore, Kolkata, India

Deependra Singh ICAR – Central Inland Fisheries Research Institute (ICAR-CIFRI), Barrackpore, Kolkata, India

Gitashree Thengal ICAR – Central Inland Fisheries Research Institute (ICAR-CIFRI), Barrackpore, Kolkata, India

Abhilash Wodeyar K. ICAR – Central Inland Fisheries Research Institute (ICAR-CIFRI), Barrackpore, Kolkata, India

D. C. Yateesh ICAR – Central Inland Fisheries Research Institute (ICAR-CIFRI), Barrackpore, Kolkata, India

Small-Scale Inland Fisheries of India: Trends, Transitions, and the Road Ahead

1

Basanta Kumar Das, Piyashi DebRoy, and D. C. Yateesh

Abstract

Small-scale fisheries (SSF) contribute around 50% of global fish catches in developing countries, employing 90% of the world's capture fishers. In India, SSF constitutes 70% of the fishing workforce but harvests only 20% of the potential yield, highlighting operational gaps, inefficiencies, and underutilization. Global SSF studies, mainly marine-focused, lack documentation on inland fisheries, emphasizing the need for comprehensive data on biological, economic, social, and cultural aspects. SSF marketing in India faces inefficiencies, including insufficient infrastructure, cold chain gaps, and logistical shortcomings, highlighting the need for long-term investments. ICAR-CIFRI's interventions in SSF, spanning rivers, estuaries, reservoirs, wetlands, and lagoons, reveal poor socio-economic conditions and emphasize the need for diversification of livelihoods. Future perspectives for Indian SSF demand a holistic approach, addressing systemic challenges, involving stakeholders, empowering women, and enhancing assessment methodologies for a sustainable and inclusive future.

Keywords

Small-scale inland fisheries · Fishers · Livelihood · Small indigenous fish · Policy

Small in name but vast in nature: Small-scale inland fisheries are the arteries and veins for protein fulfilment of rural land-locked regions of India.

B. K. Das (✉) · P. DebRoy · D. C. Yateesh
ICAR – Central Inland Fisheries Research Institute (ICAR-CIFRI), Barrackpore, Kolkata, India

B. K. Das et al. (eds.), *Small-scale Inland Fisheries in India*, Sustainability Sciences in Asia and Africa, https://doi.org/10.1007/978-981-95-3849-2_1

1.1 Introduction

Small-scale fisheries (SSF) dominate much of India's fisheries sector and play a crucial role in sustaining rural livelihoods (Jena and George 2018). India possesses a rich and varied network of inland water bodies, including rivers, canals, ponds, lakes, reservoirs, upland lakes, and floodplain wetlands, which have long supported the livelihoods, food security, and nutrition of millions. The country's freshwater resources include approximately 195,210 km of rivers and canals, 3.54 million hectares of major and minor reservoirs, 2.4 million hectares of ponds and lakes, and nearly 0.8 million hectares of floodplain wetlands and abandoned or derelict water bodies. Inland fisheries in the country play an important role in contributing to rural livelihoods and food availability. As per 2015–16 estimates, the inland fisheries sector accounted for around 1.24% of the national Gross Value Added (GVA) and 7.28% of the GVA in agriculture. According to the Handbook on Fisheries Statistics (2020), approximately 23.1 million individuals are engaged in inland fisheries activities, with women constituting nearly 43.7% of this workforce. The sector recorded a total inland capture fish production of 1.46 million tonnes, with nearly the entire yield being sourced from small-scale fishing activities (FAO 2020).

Fishing in inland areas across India generally involves low-investment methods and equipment, which are more accessible to local communities (Jena and George 2018). The operations are diverse, varying by location and season, and are predominantly carried out by traditional fishers. These communities, often socio-economically marginalized, form a major part of the workforce in artisanal inland fisheries (Sharma and Chandra 2015) (Fig. 1.1).

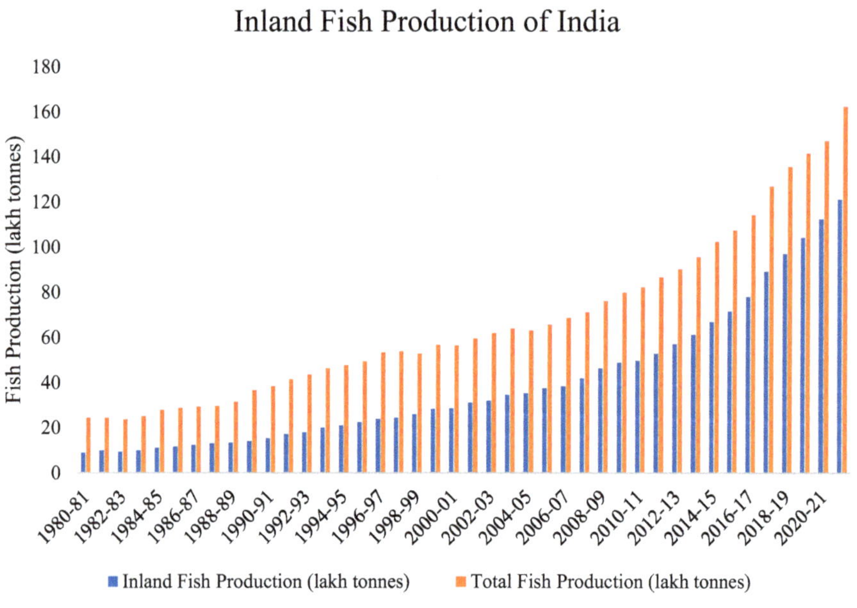

Fig. 1.1 Inland fish production of India (Handbook on Fisheries Statistics 2022)

Because of their diversity, dispersion, and social complexity, SSF are often poorly documented and poorly regulated, and many of the complex management issues remain largely unresolved. Most countries have their own definitions of small-scale fisheries. These national definitions are typically grounded in measurable criteria such as vessel length, engine capacity, gear type, or fishing zones each of which can be limiting. As a result, some genuine small-scale fishers may be left out, while larger commercial operators may mistakenly fall within the small-scale classification. This highlights the need for a more inclusive and functional definition of SSF to ensure that management, conservation, trade, and market policies do not unintentionally impede the progress of the sub-sector. The inconsistent characterization of SSF across nations has made it difficult to conduct meaningful comparisons between countries and regions. In many parts of the world, the absence of organized and consistent data on catches and related information obstructs effective fishery management and impairs our understanding of SSF contributions to sustainable development. As fish stocks face growing pressure and climate change intensifies, unresolved challenges in small-scale fisheries risk jeopardizing the livelihoods of millions who depend on them.

The catch per capita in inland fisheries is the highest in least developed countries which is 4.5 kg per person per year while that in developing countries is 1 kg per person per year and in developed countries, it is 0.4 kg per person year. Total catch values in SSF in inland fisheries particularly are likely underestimated mostly due to limited availability of information on unreported or unmonitored catch for the more remote SSF. Data collection efforts should focus on improving the understanding of the environmental impacts of SSF on aquatic environments in order to design and implement mitigation measures, while sustaining fishery yields and livelihoods. Catch in the small-scale subsector is frequently under-reported, the contributions of SSF to local and national economies are also often overlooked. Recent study on riverine fisheries in India by ICAR-CIFRI revealed that the annual fish landing of river Mahanadi was 15134.43 tonnes; total fish catch of river Tapti was 6946 tonnes per annum; total fish catch of river Krishna was 18,902 tonnes per annum; and the total fish catch of river Barak was 1593 tonnes per annum (ICAR-CIFRI 2022).

Artisanal or small-scale fisheries refer to traditional fishing practices carried out by fishing households using minimal capital, low energy inputs, and small or no vessels, often focused on meeting local consumption needs. According to FAO (2015), small-scale fisheries in developing nations are responsible for nearly 50% of the total fish caught and provide employment for approximately 90% of the global capture fishing workforce. These fisheries play a vital role in enhancing access to nutritious food at the community, national, and even global level, while also generating income for both those directly involved in fishing and those engaged in related sectors. Processing and trade activities within small-scale fisheries bring in a significant gender aspect, as women often dominate these roles, especially in post-harvest operations. Despite their key role in supporting food systems and reducing poverty by creating economic opportunities, these communities frequently remain on the margins of recognition and policy support. With growing global focus on small-scale fisheries, it has become essential to gather reliable information covering ecological, social, economic, and cultural dimensions (Pita et al. 2019a, b). Small-scale fisheries not only employ the vast majority of the world's fishers but also contribute significantly to the livelihoods of coastal populations and to global food and nutritional security.

This chapter aims to present a comprehensive overview of small-scale inland fisheries (SSF) in India, highlighting their socio-economic, ecological, and cultural significance, and critically examining their historical context, current status, and future potential. By integrating together scientific evidences, field data, and policy analysis, the chapter attempts to bridge existing knowledge gaps and provide a multi-dimensional understanding of SSF in the Indian inland context.

1.2 Small-Scale Fisheries in Relation to Ancient Indian Mythology and Folklore

Fish and fisheries have held a significant place in Indian culture, with numerous mentions in ancient texts, legends, and epics. In Indian mythology, Lord Vishnu is believed to have taken several incarnations to restore order during times of chaos, with *Matsya* (the Fish) being the first of these ten primary incarnations, who saved the first man, Manu (Fig. 1.2). The celebrated sage Veda Vyasa, credited with

Fig. 1.2 Pictorial art depicting the Matsya Avatar (the fish reincarnation), the first incarnation of Lord Vishnu

composing the *Mahabharata*, was born to a woman from a fishing community. Particularly in eastern and southern regions of India, fisheries are deeply rooted in cultural traditions and history (Hora 1951). A range of classical narratives, including poetic literature known as *Kavya*, reference open-water fishing practices. Traditional texts such as *Manasa Mangal Kabya* also highlight various fish species and describe customary cooking methods that were popular during ancient times.

1.3 Global Status of Small-Scale Fisheries

Most of the existing literature on small-scale fisheries (SSF) in India focuses primarily on marine systems. However, despite inland capture fisheries contributing significantly—representing nearly half of global capture fisheries production (Pita et al. 2019a, b), comprehensive studies on India's inland SSF remain scarce. In India, nearly all of the estimated 1.46 million tonnes of inland capture fisheries output (FAO 2020) originates from small-scale fishing practices. While SSF across the Global North and South vary by context, many common traits are shared, including socioeconomic vulnerability and adaptive livelihoods (Jentoft 2023). One such trait is occupational pluralism fishers frequently engage in supplementary occupations to cope with the seasonal and uncertain nature of fishing work (Williams 2019). In many developing regions, fisheries are deeply embedded in community life and contribute substantially to local economies and food systems, yet despite numerous interventions and reforms, small-scale fisheries continue to face persistent threats to their long-term sustainability. In Asia, much of the inland fish consumption originates from lake systems, with major riverine fisheries dominating only specific regions such as the lower Mekong basin (Amarasinghe and De Silva 2015). In the Philippines, increasing pressure from coastal populations, food insecurity, escalating demand for fish, and a high dependency on fishing for livelihoods have posed significant challenges for effective fisheries management (Barboza et al. 2024). In Sri Lanka, the fisheries sector supplies roughly 70% of the country's animal protein requirements (Amarasinghe 2013). The development of Culture-Based Fisheries (CBF) in small reservoirs has been recognized as a promising approach to enhance inland fish production. The Maldives, known for its robust tuna fishery, exports largely to markets in Sri Lanka, South Korea, and Japan. However, growth in this sector is hindered by issues such as fragmented collection systems, inadequate processing infrastructure, and political interference affecting companies managing processing operations (Giri 2018). In Bangladesh, inland open water bodies span approximately 45,000 square kilometers, supporting the livelihoods of over 12 million people (Hossain et al. 2006). Declines in capture fisheries have been linked to ineffective governance, competing sectoral interests, and flawed access rights. However, during 2014–2016, inland aquaculture experienced rapid growth driven by the adoption of scientific and technological innovations, outpacing production from wild capture fisheries. Artisanal fishing continues to be a defining feature of Bangladesh's small-scale fisheries sector (Islam 2023) (Fig. 1.3).

Fig. 1.3 Fish drying as a small-scale fisheries activity in Tanguar Haor (wetland), Bangladesh

Expert opinions were used in the USA to improve knowledge about SSF stock status (Charysafi et al. 2019). Expert opinions were used to derive stock status as a data-limited stock assessment method in the wake of large-scale stock assessments (Newman et al. 2015). In North America, governance in SSF has been given importance in research since the 1980s (Chuenpagdee et al. 2012). Small-scale indigenous and artisanal fisheries, such as the outport fishers in Canada, are an example of the practical knowledge and fish harvesting methods that enabled the sustainable outport fishery of Newfoundland and Labrador's past (Darby and Forsey 2023). SSF in Canada spans a wide range of scale of operations, such as intermittent harvesting for family-level subsistence fishing, "moderate livelihood" commercial harvesting to a few opportunities for higher incomes.

Pita et al. (2019a, b) reconstructed historical landings and analyzed socioecological changes in data poor shellfisheries at Galicia in Spain. The production of native clam species was replaced by non-native species to satisfy seafood market demand for the economic survival of the small-scale fishing communities (Fig. 1.4).

In Barbados, part of the Caribbean region, estimates of annual catch from small-scale fisheries (SSF) obtained through on-site interviews significantly exceeded official statistics, pointing toward severe overfishing pressures (Gill et al. 2019). In Colombia's Caribbean and Pacific coastal areas, fishers' adherence to both formal regulations and customary practices has been found to rely heavily on their involvement in governance processes, how well the rules align with community socio-economic conditions, the strength of local institutions, and the presence of monitoring and enforcement systems (Jimenez and Saavedra-Diaz 2019). In Spain, the establishment of a large marine protected area (MPA) in the Llevant de Mallorca-Cala Rajada

Fig. 1.4 Small-scale fishing at Duma Wetland, North 24 Parganas, West Bengal, India

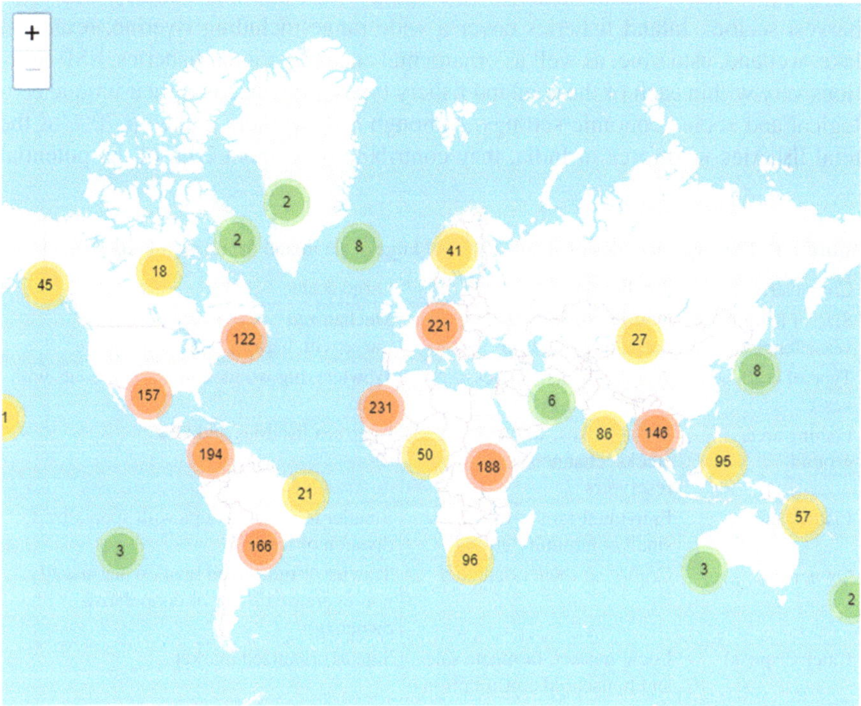

Fig. 1.5 Records of SSF case studies across the globe as per TBTI records (the numbers in the circles denote the number of case studies conducted) (https://issfcloud.toobigtoignore.net/, last accessed on 6 February 2024)

Marine Reserve brought notable shifts to traditional SSF activities. These included the disappearance of certain customary fishing methods, a partial move of fishing efforts to offshore zones, lengthening of the fishing season, broader spatial fishing activity, and greater prominence of larger boats (Mallol and Goni 2019).

The Information System on Small-scale Fisheries (ISSF), a web-based open data portal to collect and disseminate knowledge on various aspects of SSF and governance developed by Chuenpagdee et al. (2012), builds on crowd-sourced knowledge and captures dataset on global SSF (Fig. 1.5).

1.4 Characteristics of Inland Fisheries in India

India does not have a formal definition for small-scale fisheries (SSF). According to the Food and Agriculture Organization (FAO) of the United Nations, SSF are characterized by low capital and energy input, use of small vessels, reliance on household labor, and production aimed mainly at local consumption or local market sales. In the Indian context, inland open-water fisheries are considered small-scale due to their lower investment in terms of time and money (DebRoy et al. 2021; Roy and Das 2022). SSF plays a key role in ensuring access to nutritious food at local, national, and even global levels, while also offering significant employment opportunities both directly and indirectly (Jena and George 2018).

Fisheries in India are broadly classified into freshwater aquaculture, brackish water aquaculture, marine fisheries (including island and mariculture), and post-harvest sectors. Inland fisheries cover a wide range including riverine, reservoir, lake, wetland, estuarine, as well as ornamental and recreational fisheries. SSF functions vary within each of these inland fishery types, influenced by their unique ecological and socio-economic settings. Although SSF workers make up 70% of the total fisheries workforce in India, they contribute only about 20% of the potential

Table 1.1 The characteristics of small-scale and large-scale inland fisheries in India

Characteristics	Small-scale	Large-scale
Size of fishing vessel/engine	Non-motorized Small 3–7 m <12m	Mechanized Large with high horse power engine
Type of craft/vessel	Wooden boats, canoes, coracles, dingi	Trawlers, big wooden motorized boats with deck
Fishing area/ground	Rivers, lakes, wetlands, creeks, channels, small reservoirs	Deep sea fisheries and EEZ
Fishing unit	Individual/family members/small community group	Smaller to larger groups with specialized division of labor
Ownership	Craft/gear owner operated	Trawler or motorized boat owner, usually non-operators; share or cooperative ownership
Catch disposal	Local market, farmgate sale and household consumption	Sale to organized market
Utilization of catch	Fresh consumption	For consumption, commercial dry fish production, fish meal
Technology/knowledge	Traditional knowledge, scientific management	Experience, manual and mechanized gears, motorized boats, cage technology, scientific knowledge
Factors of production	Labor intensive	Labor and capital intensive
Benefits	Direct consumption Income	Income, profit and tax

harvestable yield (Jena and George 2018). Women play a vital role in SSF, particularly in processing and trade, thereby bringing a strong gender aspect to the sector (Jena and George 2018).

The inland fisheries sector in India operates under private ownership, community systems, or cooperative models. Despite its importance, India's inland SSF remains largely unorganized with minimal official data collection. There is inadequate information regarding the actual number of fishers engaged in this sector and limited insights into their employment status or income levels. Additionally, data on small-scale aquaculture practices and their respective contributions to local consumption and commercial supply remain sparse (Table 1.1).

Fish is consumed regularly by fishers for their sustenance which contributes to nutritional security on family level as well as community level (Seema 2015).

1.5 Diversity in Indian Small-Scale Fish Catch

Out of approximately 30,700 fish species identified globally, India's inland aquatic systems are home to 930 species across 326 genera (Bhattacharjya 2022). Indian Major Carps (IMCs), which include catla (*Labeo catla*), rohu (*Labeo rohita*), and mrigal (*Cirrhinus mrigala*), are indigenous to the Indus-Ganges River systems and the broader Indo-Gangetic plains, and these three species account for about 60% of total carp output. Together, these major carp species contribute around 5.96 million tonnes annually. Minor carps account for about 0.489 million tonnes, exotic varieties such as silver, grass, and common carps yield approximately 9.81 million tonnes, murrels (*Channa* spp.) produce 2.01 million tonnes, and catfishes including *Wallago attu* and *Pangasius* provide nearly 4.32 million tonnes. In freshwater aquaculture, IMCs remain dominant, comprising more than 85% of total output, followed by exotic and minor carps, catfish, and trout. Of the 266 carp species native to India, only about 34 are considered commercially important, and most of these are harvested from natural water bodies (Handbook on Fisheries Statistics 2022). Fewer than ten species are routinely farmed or captured for food and commercial purposes. Minor carps such as reba (*Cirrhinus reba*), bata (*Labeo bata*), fringe-lipped carp (*Labeo fimbriatus*), calbasu (*Labeo calbasu*), white carp (*Cirrhinus cirrhosus*), and Cauvery carp (*Labeo kontius*) also hold ecological and economic value. A wide range of inland water bodies like ponds, floodplains, tanks, wetlands, streams, rice paddies, and beels are rich in Small Indigenous Fish Species (SIFs). Of the 1035 native freshwater fish species identified in India, 450 are classified as SIFs. The northeastern region leads the country in SIF diversity, followed by the Western Ghats and Central Indian landscapes. Around 104 of these species, or 23%, are particularly valuable for nutrition, ornamental use, household sustenance, and livelihood security.

1.6 ICAR-CIFRI's Intervention in Small-Scale Fisheries

ICAR-CIFRI (Central Inland Fisheries Research Institute) has undertaken research on socio-economic and livelihood aspects of small-scale fisheries (SSF) across various inland water bodies such as rivers, estuaries, oxbow lakes, floodplain wetlands, reservoirs, backwaters, and lagoons. Findings revealed that the general living conditions of fisher households were poor (Jena and George 2018). For most of these households, fishing accounted for around 60% of the total income. Nonetheless, fishing alone was not enough to ensure a sustainable livelihood, pushing fishers to seek additional income through non-fishing occupations. Across the country, the majority of inland fishers operate at a small scale and depend on open-water fishing for both income and nutritional needs. ICAR-CIFRI has conducted assessments on the socio-economic conditions and livelihood strategies of small-scale fishers in various inland water systems including rivers, estuaries, oxbow lakes, wetlands, reservoirs, and lagoons. Recent studies have been conducted in rivers such as the Ganga, Mahanadi, Gandak, Brahmaputra, Torsa, Cauvery, Tapti, Siang, and Chaliyar; estuaries like Narmada, Hooghly, and Mahanadi; reservoirs including Kangsabati, Chandil, Patratu, Tenughat, and Ukai; and wetlands located in Assam, Manipur, Bihar, Uttar Pradesh, West Bengal, and the Chilika Lagoon in Odisha. These studies confirmed the persistently low living standards among fisher families. Education levels, land holdings, and household incomes were considerably below the threshold for a modest standard of living. While fishing was the primary livelihood source for most families, accounting for roughly 60% of their income, it remained insufficient on its own, prompting diversification into other economic activities.

In the context of reservoir fisheries, CIFRI has evaluated both ecological and fishery-related parameters across numerous reservoirs in India. Demonstrations of culture-based fisheries were carried out with active participation from local Fishermen Cooperative Societies. Numerous awareness and training initiatives were held to promote improved management practices in reservoir fisheries. The institute showcased cage culture techniques in reservoirs across Karnataka, Chhattisgarh, Jharkhand, Telangana, and Madhya Pradesh (Table 1.2). In support of tribal fishers, FRP coracles and gill nets were distributed. In Kerala's Karapuzha reservoir, a small-scale culture-based fishery achieved notable success through ICAR-CIFRI's support, with major carps making up to 60% of the catch after stocking interventions. This approach enabled tribal communities in Wayanad District to generate a stable and respectable income ranging from Rs. 2000 to 3000 per day from fishing activities. Techniques for raising stocking material through cage culture have been optimized, especially for air-breathing species like Magur, Singhi, and Koi in deeper wetland environments. In 2017–18, ICAR-CIFRI developed and introduced the CIFRI-GI Cage, a cost-effective and robust cage system with each unit measuring $5 \times 5 \times 4$ m^3. The institute has also focused on the fisheries and ecology of floodplain wetlands, which are vital fishery resources in eastern and north-eastern parts of the country. Its research has yielded substantial ecological and fisheries data for beels across varying agro-climatic regions. These studies have

Table 1.2 Types of inland fisheries in India and ICAR-CIFRI interventions

Type of inland fishery	Representative waterbodies/ states	Key ICAR-CIFRI interventions
Riverine fisheries	Ganga, Mahanadi, Gandak, Brahmaputra, Torsa, Cauvery, Tapti, Siang, Chaliyar	Socio-economic surveys, catch assessments, livelihood diversification
Estuarine fisheries	Narmada, Hooghly, Mahanadi Estuaries	Biodiversity studies, community-based management approaches
Reservoir fisheries	Kangsabati, Chandil, Patratu, Tenughat, Ukai, Karapuzha (Kerala)	Cage culture, FRP coracles/net distribution, stock enhancement, cooperative model promotion
Floodplain wetlands (beels)	Assam, Bihar, West Bengal, Manipur, UP	Pen culture (CIFRI-HDPE pens), stock enhancement, climate-resilient fishery under NICRA, doubling fishers' income
Tribal and TSP interventions	Burdwan (WB), Wayanad (Kerala), Gardanmari wetland, NE states	Livelihood improvement, yield increase from 300 kg to 3 tonnes, inclusive governance
Canal fisheries	Sundarbans (West Bengal)	Revitalization of unutilized canals, yield up to 800 kg/ha, enhanced community livelihoods
Deep wetlands	Assam, Manipur, Goalpara (Assam), Namsai (Arunachal), Loni (MP), Meghalaya	Air-breathing fish (magur, singhi, koi) rearing using cage culture, pen-based fingerling production
Lagoon fisheries	Chilika Lagoon (Odisha)	Social-ecological assessment, community-based management

shown that the average fish production in unstocked wetlands is around 221 kg/ha/year, significantly lower than the estimated potential of 1000–1500 kg/ha/year. Techniques for fish stock enhancement in wetlands have been established. Pen culture methods have been developed, refined, and tested in these water bodies. A large-scale demonstration of pen culture was implemented in 40 beels across Assam in collaboration with the state fisheries department. The CIFRI-HDPE pen, commercialized in 2018–19, offers a cost-effective and durable pen system covering an area of 1000 m^2. These pens have already been installed in Assam and Manipur, with widespread adoption observed in the northeast. In Bihar, the institute recently worked on enhancing the incomes of fishers in five wetlands, aiming to raise fish yields from the current 300 kg/ha/year to between 800 and 1000 kg/ha/year. This initiative sought to double the income of about 2000 fishers over a two-year period through production enhancement. On the governance side, ICAR-CIFRI's engagement led to more equitable distribution of benefits and strengthened cooperative management systems. Under the NICRA initiative, the institute also piloted climate-resilient fisheries interventions in the wetlands of Assam and West Bengal.

CIFRI has made several interventions for improvement of livelihood of tribal fishers under TSP. The activities have increased the productivity to 3.0 ton from 300 kg in Gardanmari wetland, Burdwan, West Bengal, and enhanced the livelihood of 120 fishes depending upon it. Apart from wetland management aspects, pen has been provided in different wetlands for in situ production of advance fingerlings as the stocking material for the wetland. ICAR-CIFRI is also working in wetlands, Loni, Madhya Pradesh; Goalpara, Assam; West Garo Hills, Meghalaya; Namsai, Arunachal Pradesh. The unutilized canals of Sunderbans, West Bengal, were also targeted for improvement of productivity through community motivation and mobilization. The fish production was achieved up to 800 kg/ha from the unutilized canals. ICAR-CIFRI intervention has provided enhanced livelihoods to more than 1200 fishers under canal fisheries development. Various revenue and welfare-based models based on socio-economic and socio-ecological conditions were depicted by ICAR-CIFRI. This showed the similarities and trade-offs in fisheries management. Also, the marketing channels in fisheries in India suffer from inefficiencies (Kumar et al. 2008). The problems in SSF marketing are noted as insufficient landing center facilities, wholesale and retail market infrastructure, and cold chains and logistic shortfalls (Jena and George 2018). Hygienic handling of fish, efficient storage, and transportation are lacking. There is a need to channelize long-term investments to improve infrastructural investments for enhancing the efficiency of fish markets.

1.7 Social-Ecological Changes and Key Drivers

Fisheries are complex social-ecological systems that involve multiple actors from various cultural groups (Berkes 2011). Inland fisheries management principles do not adopt holistic approaches that consider the complex interplay between fish, humans, and the environment (Nguyen et al. 2016). Resource use conflicts in Chilika lagoon resulted in marginalization of small-scale fishers impacting the ecology of the lagoon, and in turn contributing to national poverty and food insecurity. Unregulated development catalyzing sequential exploitation in one area after another (Huitric et al. 2002); loss of livelihoods of artisanal fishers displaced by aquaculture development (Nayak and Berkes 2011); environmental injustice (Nayak and Berkes 2010); and outmigration of fishers (Robson and Nayak 2010) are unspoken social-economic impacts having greater impact on ecologies of aquatic ecosystems.

Key drivers leading to fish stock decline are increased human demand due to population growth and rising incomes, overfishing, use of better technology by large fishing fleets, IUU fishing, poor management decisions in some fisheries, and impacts of climate change and biodiversity loss (Worm et al. 2009; Worm et al. 2006; Rockström et al. 2009; Pauly et al. 1998; Costello et al. 2016). Effects of drivers are a two-way process in which global, local, and regional drivers impact each other (Nayak and Berkes 2014). A driver is any natural or human-induced factor that directly or indirectly causes a change in an ecosystem (MA 2003). Drivers have both top-down and bottom-up effects in a system based on scales where they operate (Nayak and Berkes 2014). If other larger drivers of unsustainable fisheries need to be addressed, we need to analyze the economic impacts of fisheries conservation programs and create alternatives to diversify income sources (Barboza et al. 2024). SSF have fewer negative impacts on the ecosystem in comparison to other fishing methods. This is due to the fact that the fishing gears used are more selective and less destructive that render less by-catch.

1.8 Policy and Governance

1.8.1 Existing Schemes and Policies in India

Fisheries is a state subject in India; centrally sponsored schemes (Table 1.3) support fishers through model fishing villages, accident insurance, saving-cum-relief programs, and training initiatives. The 2018 Committee on Doubling Farmers' Income highlighted the importance of diversifying away from traditional field crops and

Table 1.3 Major policies and programs relevant to Small-Scale Inland Fisheries (SSF) in India

Policy/program	Year launched	Implementing agency	Core objective	Relevance to SSF
Blue Revolution Scheme	2015–16	Department of Fisheries, GoI	Comprehensive growth and management of the fisheries sector	Pioneered efforts in developing inland aquaculture, enhancing infrastructure, and improving fishers' livelihoods
Fisheries and Aquaculture Infrastructure Development Fund (FIDF)	2018–19	NABARD, NCDC, NFDB	Development of essential infrastructure like hatcheries, markets, and landing points	Facilitates infrastructure access for small fishers and cooperative societies
Pradhan Mantri Matsya Sampada Yojana (PMMSY)	2020–21	Department of Fisheries, GoI	All-round advancement in fisheries including production, infrastructure, and marketing	Specifically benefits inland SSF through input subsidies, training, and livelihood support
Group Accident Insurance Scheme (GAIS)	Ongoing	Department of Fisheries, GoI (CSS)	Offers insurance to active, registered fishers	Stabilizes earnings for SSF households affected by seasonal income fluctuations
Saving-cum-Relief Scheme	Ongoing	Department of Fisheries, GoI (CSS)	Provides financial assistance during non-fishing periods	
Model Fishermen Village Scheme	Ongoing	Department of Fisheries + State Governments	Provides basic amenities like housing, safe water, and sanitation in fishing villages	Infrastructure and dignity for marginalized SSF families, often prone to disaster and displacement
Training and Extension Programs	Ongoing	ICAR, CIFRI, CIFE, NFDB	Training in sustainable fishing, aquaculture practices, and post-harvest handling	Improves technical skills and offers alternative livelihood opportunities to SSF

placing greater emphasis on allied sectors, identifying fisheries as offering the highest potential returns. As part of this shift, the Blue Revolution Scheme was launched in 2015–16 with a budgetary allocation of Rs. 3000 crore to promote fisheries development. This was followed by the establishment of the Fisheries and Aquaculture Infrastructure Development Fund (FIDF) in 2018–19, which committed Rs. 7522 crore to enhance sectoral infrastructure. In 2020–21, the government introduced the Pradhan Mantri Matsya Sampada Yojana (PMMSY) with a substantial allocation of Rs. 20,050 crore aimed at the holistic growth of the fisheries sector. Key reforms and interventions under PMMSY have led to increased investment across the value chain, adoption of modern technologies, strengthened policy frameworks, improved access to finance, and better use of land and water resources for alternative livelihoods and entrepreneurship opportunities. While Article 262 of the Indian Constitution addresses the resolution of inter-state water disputes, the broader legal and policy framework continues to underrepresent the inland fisheries sector, leaving significant gaps in its governance.

1.9 Emerging Issues and Challenges

The root issues in inland aquatic systems are larger and systemic. Biological and ecological resource uncertainty, conflicting social and economic objectives, and lack of pre-appraisal of the impacts of fishery management scheme regulations are the main constraints in present day fisheries management (Caddy and Cochrane 2001; Caddy 2000). One of the significant threats to inland aquatic ecosystems in India is the increasing presence of non-native fish species, which poses serious risks to the country's endemic aquatic biodiversity (Sandilyan 2022). According to reports from the Zoological Survey of India, a total of 31 aquaculture fish species, 600 ornamental varieties, and 2 larvicidal species have been introduced into the ecosystem. Of these, 14 species are considered invasive and have reportedly established populations in ecologically sensitive areas, including biodiversity hotspots, as noted by the National Biodiversity Authority. Additional contributing factors—such as climate change, river interlinking projects, the Genetically Improved Farmed Tilapia (GIFT) initiative, and the ornamental fish trade—are expected to further increase the risk of invasive species introductions (Sandilyan 2022). The lack of access to credit facilities and minimal engagement from formal banking systems continues to limit the financial resilience of small-scale fishers. There is a growing need to ensure that these fishers are actively involved in shaping policy, participating in management, and engaging in legislative discussions (Mariette 2016, 2017). A collaborative approach, such as co-management and community-based resource governance, is essential to ensure better compliance and equitable responsibility-sharing in fisheries governance (Jena and George 2018). Challenges around tenure security and equitable access to fishing grounds are ongoing issues in inland fisheries, mirroring similar struggles in the marine sector. Furthermore, the sector faces institutional limitations, vulnerability to external pressures, and gaps in generating stable employment and income through fisheries-based livelihoods. Accurate and comprehensive data on small-scale inland fisheries remain scarce, and the limited information available often raises concerns about its reliability (Pita et al. 2019a, 2019b). This data deficiency contributes to the marginalization of SSF in national assessments and policy priorities. Recognizing this, the FAO has recently emphasized the urgent need to strengthen attention and investment in the small-scale fisheries subsector at the global level.

1.10 Future Perspective

Addressing the multifaceted challenges facing small-scale fisheries (SSF) management in India requires an integrated and inclusive strategy. Systemic issues such as biological and ecological unpredictability, competing socio-economic interests, and the absence of comprehensive impact assessments of fishery regulations underscore the need for structural reform. Moreover, the persistent lack of access to institutional credit and the limited engagement of formal financial mechanisms weaken the economic foundation of SSF communities, reinforcing the need for targeted and equitable policy interventions. Active involvement of SSF stakeholders in the design and execution of policies, legislative frameworks, and governance mechanisms is critical to advancing sustainable fisheries management. Participatory governance not only strengthens transparency and trust but also facilitates co-management and community-based models that distribute responsibility and reinforce local compliance. Despite this, inland SSF continues to face significant constraints, including insecure tenure and access rights, weak institutional capacities, exposure to external pressures, and unmet goals of livelihood enhancement and job creation. Looking ahead, a resilient future for SSF in India depends on the adoption of a multi-dimensional and inclusive governance model.

Stakeholders must be empowered to shape co-management agendas that balance equity, competence, and ecological sustainability. The post-harvest segment—particularly in processing and trade—presents untapped potential for pro-poor growth and food system resilience. To harness this, coordinated investments from both public and private sectors are essential to scale up local market systems, improve value chains, and support nutritional and livelihood security, especially in rural and marginalized areas. Strengthening this domain also contributes significantly to rural development and the economic empowerment of women. Effective implementation of these strategies calls for coordinated cross-sectoral governance. Policymakers must develop clear, enforceable legislative frameworks that define administrative jurisdictions, formalize tenure rights, and establish mechanisms for planning, enforcement, and accountability. Legal empowerment of fisher collectives and local organizations is fundamental to enabling self-determined management and sustainable resource use. Furthermore, gender equity must be institutionalized within SSF governance. Recognizing and integrating women's roles in both community dynamics and fisheries value chains is vital for building inclusive and durable co-management systems. Strengthening the participation of women enhances the legitimacy, social impact, and long-term viability of fisheries reform.

Equally important is the expansion of reliable assessment methodologies that can accurately quantify the contributions of SSF to food security, livelihoods, and socio-economic development. In sum, aligning India's SSF governance with inclusive, evidence-based, and community-centered approaches is essential to addressing the sector's evolving challenges and unlocking its full development potential.

1.11 Conclusion

Small-scale fisheries (SSF) form a vital component of India's food and nutritional security. Strengthening governance systems and regulatory approaches that cater to the specific needs of the SSF sector is crucial for ensuring its long-term viability. However, the sector is increasingly burdened by a host of challenges, including competition for water resources, sediment accumulation in water bodies, encroachment on aquatic habitats, disrupted connectivity between wetlands and rivers, climate-related impacts, loss of biodiversity, and unsustainable fishing practices. The management of SSF involves both social and legal dimensions, with policies often shaped by wider political agendas. Given the diverse and region-specific nature of SSF across India, existing governance frameworks tend to be fragmented and inconsistent. It is important to reassess, streamline, and potentially integrate the numerous laws currently regulating fisheries at various administrative levels. Furthermore, valuing the economic and ecological contributions of inland SSF can help convey their significance to policymakers and development planners, especially in the context of inland open water ecosystems.

Despite their crucial role in supporting livelihoods and reducing poverty among marginalized and vulnerable groups, SSF often receive limited attention in national food security and rural development strategies. Unlocking the full potential of inland SSF requires policy reforms aimed at improving access to credit and financial services, promoting sustainable fishing practices, enhancing income-generating opportunities, encouraging participatory governance, and enabling women to take on leadership roles in fisheries management. Such integrated and inclusive reforms will not only improve the resilience of fishing communities but also contribute to sustainable resource use, gender equality, and balanced rural development.

Points to Remember

- Small-scale inland fisheries are highly diverse, scattered, region-specific, and culture-specific in nature.
- In India, SSF constitutes 70% of the fishing workforce but harvests only 20% of the potential yield, highlighting inefficiencies and underutilization.
- Though small in scale, inland fisheries contribute to a crucial role in livelihood of millions of rural poor in India.

References

Amarasinghe U (2013) Fisheries resources in alleviation of hunger and malnutrition in Sri Lanka – accomplishment and challenges. Sri Lanka J Aquat Sci 18(23):1–15
Amarasinghe US, De Silva SS (2015) Fish and fisheries of Asian inland lacustrine waters. In: JF Craig (eds) Freshwater fisheries ecology, pp 384–403
Barboza A, Macusi ED, Borazon EQ, Santos MD, Muallil RN, Nallos IM (2024) Small-scale fisheries (SSF) management and conservation schemes and their application in the Philippines. Mar Policy 161:106018
Berkes F (2011) Restoring unity: the concept of marine social-ecological systems. World Fisheries: A Soc Ecol Anal 9–28
Bhattacharjya BK (2022) Inland open water fisheries of India: emerging challenges. J Inland Fish Soc India 53(3–4):115–118
Caddy JF (2000) Marine catchment basin effects versus impacts of fisheries on semi-enclosed seas. ICES J Mar Sci 57(3):628–640
Caddy JF, Cochrane KL (2001) A review of fisheries management past and present and some future perspectives for the third millennium. Ocean Coast Manag 44(9–10):653–682
Charysafi A, Cope J, Kuparinen A (2019) Eliciting expert knowledge to inform stock status for data-limited stock assessments. Mar Policy 101:167–176
Chuenpagdee R, Greene R, Britton E, Rogers E, Ivany I, Ramirez-Luna V, Hart K (2012) The state of small-scale fisheries research: North America [Poster]. Too big to ignore: global partnership for small-scale fisheries research
Costello C et al (2016) Global fishery prospects under contrasting management regimes. Proc Natl Acad Sci USA 113(18):5125–5129
Darby B, Forsey H (2023) Echoes: Newfoundland outport fisheries and Indigenous traditions. In: Andrews EJ, Knott C (eds) Thinking big about small-scale fisheries in Canada. TBTI Global Publication Series, St. John's
DebRoy P, Das BK, Parida PK, Roy A, Chakraborty S (2021) Economic injustice in small-scale wetland fisheries in West Bengal, India. In: Kerezi V, Chuenpagdee R (eds) Blue justice for small-scale fisheries: a global scan, vol 2. TBTI Global Publication Series, St. John's
FAO (2020) The state of world fisheries and aquaculture 2020: sustainability in action. FAO, Rome
FAO (2015) Voluntary guidelines for securing sustainable small-scale fisheries in the context of food security and poverty eradication. Food and Agriculture Organization of the United Nations, Rome
Gill D, Oxenford H, Turner R, Schumann P (2019) Making the most of the data-poor fisheries: low cost mapping of small island fisheries to inform policy. Mar Policy 101:198–207
Giri SS (2018) An overview of small-scale fisheries in South Asia. In: Giri SS (ed) Small-scale fisheries in South Asia. SAARC Agriculture Centre, Dhaka
Handbook on Fisheries Statistics (2020) Fisheries Statistics Division, Department of Fisheries, Ministry of Fisheries, Animal Husbandry & Dairying, Government of India
Handbook on Fisheries Statistics (2022) Department of Fisheries, Ministry of Fisheries, Animal Husb Dairy, GovMent India, New Delhi. pp. 198
Hora SL (1951) Knowledge of the ancient Hindus concerning fish and fisheries of India. J Asiat Soc Lett 17:145–169
Hossain MM, Islam AM, Ridgway S, Matsuishi T (2006) Management of inland open water fisheries resources of Bangladesh: issues and options. Fish Res 77(3):275–284
Huitric M, Folke C, Kautsky N (2002) Development and government policies of the shrimp farming industry in Thailand in relation to mangrove ecosystems. Ecol Econ 40:441–455
ICAR-CIFRI (2022) Annual report ICAR – Central Inland Fisheries Research Institute, Barrackpore, Kolkata
Islam MM (ed) (2023) Advancing knowledge of small-scale fisheries in Bangladesh. TBTI Global Book Series, Sylhet Agricultural University

Jena JK, George G (2018) Small-scale fisheries in India: an appraisal. In: Giri SS (ed) Small-scale fisheries in South Asia. SAARC Agriculture Centre, Dhaka

Jentoft S, (2023) The gift of community TBTI Global Book Series

Jimenez JM, Saavedra-Diaz LM (2019) Evaluating formal and informal rules as a basis for implementing coastal marine artisanal fisheries management in Colombia. Mar Policy 101:225–236. https://doi.org/10.1016/j.marpol.2018.09.019

Kumar M, Bhatt G, Duffy CJ (2008) An efficient domain decomposition framework for accurate representation of geodata in distributed hydrologic models. Int J Geogr Inf Sci 23(12):1569–1581

MA (2003) Ecosystems and human well-being: a framework for assessment millennium ecosystem assessment. World Resources Institute, Island Press, Washington

Mallol S, Goni R (2019) Unintended changes of artisanal fisheries métiers upon implementation of an MPA. Mar Policy 101:237–245

Mariette C (2016) WFF-WFFP-ICSF-CIC: national workshop on capacity-building for the implementation of the voluntary guidelines for securing sustainable small-scale fisheries (SSF guidelines). ICSF, Chennai

Mariette C (2017) Report on workshop on enhancing capacities of women fish workers in India for the implementation of the SSF guidelines. ICSF, Chennai

Nayak PK, Berkes F (2010) Whose marginalization? Politics around environmental injustices in India's Chilika Lagoon. Local Environ 15:553–567

Nayak PK, Berkes F (2011) Commonization and decommonization: understanding the processes of change in Chilika Lagoon, India. Conserv Soc 9:132–145

Nayak PK, Berkes F (2014) Linking global drivers with local and regional change: a social-ecological system approach in Chilika Lagoon, Bay of Bengal. Reg Environ Chang 14:2067–2078. https://doi.org/10.1007/s10113-012-0369-3

Newman D, Berkson J, Suatoni L (2015) Current methods for setting catch limits for data-limited fish stocks in the United States. Fish Res 164:86–93

Nguyen VM, Lunch AJ, Young N, Cowx IG, Beard TD Jr, Taylor WW, Cooke SJ (2016) To manage inland fisheries is to manage at the social-ecological watershed scale. J Environ Manag 181:312–325

Pauly D et al (1998) Fishing down marine food webs. Science 279(5352):860–863

Pita P, Fernandez-Marquez D, Macho G, Antelo M, Villasante S (2019a) Socio-ecological changes in data-poor S-fisheries: a hidden shellfisheries crisis in Galicia (NW Spain). Mar Policy 101:208–224

Pita C, Villasante S, Pascual-Fernandez JJ (2019b) Managing small-scale fisheries under data poor scenarios: lessons from around the world. Mar Policy 101:154–157

Robson JP, Nayak PK (2010) Rural out-migration and resource-dependent communities in Mexico and India. Popul Environ 32:263–284

Rockström J et al (2009) A safe operating space for humanity. Nature 461(7263):472–475

Roy A, Das BK (2022) Resource degradation and conflicts affecting small-scale wetland fishers of West Bengal, India. In: Kerezi V, Chuenpagdee R (eds) Blue justice for small-scale fisheries: a global scan, vol 3. TBTI Global Publication Series, St. John's

Sandilyan S (2022) Alien fish species in Indian inland wetlands: current status and future challenges. Wetl Ecol Manag 30:423–437

Seema S (2015) Report of the ICSF-BOBLME India (East Coast). In: Workshop on implementing the FAO voluntary guidelines for securing sustainable small-scale fisheries. ICSF, Chennai

Sharma AP, Chandra G (2015) Issues & challenges for sustainable small-scale fisheries in inland fisheries sector of India. ICSF-BOBLME Workshop, Chennai

Williams R (2019) A future for the fishery: crisis and renewal in Canada's neglected fishing industry. Nimbus, Halifax

Worm B et al (2006) Impacts of biodiversity loss on ocean ecosystem services. Science 314(5800):787–790

Worm B et al (2009) Rebuilding global fisheries. Science 325(5940):578–585

Craft and Gear in Small-Scale Fisheries in the River Gandak and Middle Stretch of Ganga

2

Pranab Gogoi, Deependra Singh, Abhilash Wodeyar K., Sanjenbam Bidyasagar, and Basanta Kumar Das

Abstract

The use of fishing craft and gear in the Gandak and Ganga Rivers reflects the traditional knowledge and adaptive practices of local fishing communities. In addition, small-scale fishers employ region-specific crafts, such as flat-bottomed boats (plank-built/galvanized iron boats) and canoes, along with selective and eco-friendly gears that are appropriate for the dynamic flow and depth variations of these rivers. The current study explored six types of non-motorized fishing craft and two types of customized motorized crafts. In addition, 5 types of major fishing gears (gill net, falling gear, lift net, bag net, and drag net), including fishing gears line fishing (5 types) and traps (12 types) are recorded under 6 broad categories from the river Gandak and the central section of the Ganga. The different fishing methods relied on the seasonal availability of fish species and their behavioral patterns. The livelihoods of the local fishermen are greatly dependent on the traditional practices. However, regulating the use of gear, safeguarding habitats, and combining conventional practices with scientific management strategies are all necessary to guarantee long-term sustainability.

Keywords

Livelihood · Small-fishes · Fishers · Crafts and gear · River

P. Gogoi (✉) · D. Singh · Abhilash Wodeyar K. · B. K. Das
ICAR—Central Inland Fisheries Research Institute (ICAR-CIFRI), Barrackpore, Kolkata, India

S. Bidyasagar
College of Fisheries, Dr. Rajendra Prasad Central Agricultural University, Pusa, Bihar, India

B. K. Das et al. (eds.), *Small-scale Inland Fisheries in India*, Sustainability Sciences in Asia and Africa, https://doi.org/10.1007/978-981-95-3849-2_2

2.1 Introduction

Small-scale inland fisheries (SSIF) in India play a crucial role in sustaining rural livelihoods, food security, and employment for millions of people. SSIF are rely mostly on traditional craft and gear systems that that differ by regions according to local water bodies, target species, and community practices (FAO 2010). Common fishing crafts include non-motorized and motorized country boats made of wood or fiber-reinforced plastic, while gears range from simple hand nets and cast nets to more complex gill nets, drag nets, and traps. The design and use of these crafts and gears are often eco-friendly and low-cost, making them accessible to marginalized fishers. Additionally, the adaptability of these gears allows efficient exploitation of diverse aquatic environments such as rivers, floodplains, lakes, tanks, and reservoirs. Despite their small size, such fisheries make an important contribution to India's inland fish supply and require policy support to upgrade equipment while retaining traditional expertise. In the Indian context, small-scale fisheries (SSF) are not formally defined. In India, inland open-water fisheries are classified as small-scale since they demand less time and money (DebRoy et al. 2021; Roy et al. 2020). In light of this, the current chapter discusses the craft and gears used in small-scale fisheries, particularly in the middle stretch of the Ganga and the Gandak River, India.

According to the Food and Agricultural Organization (FAO 2005–2011), "Artisanal, or small-scale fisheries, are traditional fisheries involving fishing households, using relatively small amount of capital and energy, relatively small fishing vessels (if any), making short fishing trips, close to shore, and mainly for local consumption. They can be for subsistence or commercial." There is no single definition that elaborately understands the diversity and complexity of what constitutes the SSF. It is factual that SSF serve as the backbone of local economies and are deeply rooted in cultural traditions. According to the report of FAO, Duke University and WordFish (2023) "Illuminating Hidden Harvest – The contributions of small-scale fisheries to sustainable development" stated that every year, SSF generates 37 million tons of aquatic foods (of which 68% marine catch and 32% inland catch), accounting for almost 40% of the global catch. Of the 92 million tonnes of global fisheries catch, 55 million tonnes shares (60%) from large scale fisheries. It was estimated that 492 million people were somewhat related to SSF for their livelihoods, with an around 60 million people working full- or part-time with SSF, 53 million are involved in subsistence fishing and around 379 million are additional household members, accounting for 90% of capture fisheries employment, globally. Of the 77 billion USD total revenue of SSF, 19 billion USD shares from inland SSF catch. Valuing women workforce, 45 million women engaged in SSF globally, of which majority are involved in post-harvest (50%), followed by subsistence fishing (45%), pre-harvest (15%) and commercial fishing (19%) (FAO, Duke University and WordFish 2023). According to the FAO (2015), inland fisheries account for around 15% of total global employment.

2.2 Indian Scenario

In India, fisheries sector contributes 1.07% in the national Gross Domestic Product (GDP) and 6.8% of agricultural Gross value-added (GAV) in 2023–2024. Inland fisheries sector of India is dominated by small-scale fisheries, which are widely diversified in terms of seasons and geography. The inland fisheries, a sunrise fisheries sector, involving about 2,31,17,820 people with women accounting for 43.7% (Fig. 2.1). Inland shares 17% and 19% of the total population of fishermen and fisherwomen, respectively (Handbook of Fisheries Statistics 2022). Inland fisheries include riverine, reservoir, lake, wetland, and estuarine fisheries, as well as ornamental and recreational fishing. Depending on the location and context, SSF in each of these sectors has unique characteristics. The inland fisheries sector in India is either private, community-based, or cooperative. The unorganized nature of fishing in the inland open waters limits the inland data source. The number of fishermen engaged in small-scale inland fishing in the country, as well as the employment and livelihood of the sector, fisherman wages, small-scale aquaculture, and its proportionate contribution to local consumption and commercial purposes, are therefore somewhat lacking (Das et al. 2022). In India's inland sector, SSF is operationally described as "an uncomplicated individual or household (family) venture, entail low levels of technology and synonymously employed as artisanal or traditional fisheries" (Das et al. 2022).

The average annual income of small-scale inland fishermen in the nation is Rs. 30,690/–, which is quite low income (Sharma et al. 2013). Cash income from inland fishing is earned by fishing households. The fishermen sell their catch between 75 and 90%, enabling the fishermen to make money for essentials like clothing, food, medical care, and education (Das et al. 2020). According to the report of ICAR-Central Inland Fisheries Research Institute, Barrackpore (Annon 2023), the literacy rate of inland fishers is ranging from 63.32–79.10% with an average 70.70%. Further, the report says that the annual income of individual reservoir fishers' is

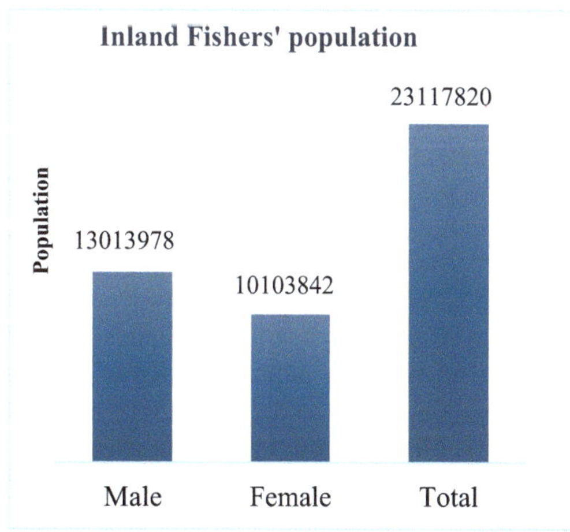

Fig. 2.1 Inland fishers' population in India. (Handbook on Fisheries Statistics 2022)

comparatively higher (approx. Rs. 34,696/–) than that of river (Rs. 31,311/–) and wetland (Rs. 15,164/–). The governance and institutional arrangement in small-scale inland open water fisheries in India are based on two models, i.e., welfare-based model and revenue-based model. For instance, revenue-based model includes tanks, ponds, wetlands and reservoirs mainly in Uttar Pradesh, Assam, and some other states. The welfare-based model includes resources such as river fisheries, wetlands and reservoirs in West Bengal and Kerala, large reservoirs in Madhya Pradesh, and medium and large reservoirs in Chhattisgarh.

2.2.1 Types of Small-Scale Fisheries in India

(i) *Artisanal fisheries or traditional fisheries:* sometimes synonym with small-scale fisheries. Individual or group fishing families collect aquatic organisms from inland waters like rivers, lakes, marshes, and reservoirs using traditional methods and with less capital involvement. Artisanal fisheries can be either subsistence or commercial, producing for local use or export. Small-scale artisanal fisheries are dominated by fishermen, primarily from the fishing community. These fishermen communities are among the poorest and least structured in India (Sharma and Chandra 2015).

(ii) *Subsistence fisheries:* In general, it refers to the adoption of low-tech "artisanal" fishing techniques by the individuals. Fishing is frequently part of a livelihood that includes small-scale agriculture and other forms of income, as well as the sale of fish. While artisanal fishers can sell some of their catch in markets, they mainly fish for subsistence. This small-scale activity is more environmentally benign and poses a lower risk of damaging fish populations.

(iii) *Recreational (sport) fisheries*: Recreational fishing, often known as sport fishing, is done for fun, leisure, and recreation. It is not related to commercial or subsistence fishing. It is now gaining popularity among amateurs/hobbyists and has the potential to generate income for the fishing community.

2.2.2 Fishing Method in Small-Scale Inland Fisheries

Inland open waters in India support a wide array of traditional and modern/customized fishing activities. The types of fishing craft and gear used in inland water bodies vary based on the local geographic conditions, target species, and socio-economic status of fishermen. Manna and Das (2024) previously identified 12 categories of fishing gear that largely targeted indigenous small fish (ISF) species in Indian inland waters, with gill nets having the highest diversity, followed by traps and bag nets. The fishing methods largely employed in inland fisheries were (i) without gear (hand picking, de-watering), (ii) grappling and wounding device, (iii) stupefying device, (iv) line fishing, (v) fishing trap, (vi) aerial fishing trap, (vii) gill netting, (viii) drive-in-gill net, (ix) falling gear, (x) lift net, (xi) bag net, and (xii) dragged gear.

Numerous reports have been available on the study of crafts and gear used for fishing in small-scale inland fisheries in India, with the majority of those reports focusing on fishing gear. Some of the reports (Samajdar and Saikia 2014; Das and Barat 2014; Purakit et al. 2018; Sandhya et al. 2019; Ganguly et al. 2022; Manna et al. 2024) were largely concerned with Gangetic plain fishing methods. Sandhya et al. (2019) mentioned 12 types of fishing gears that are frequently used in *beels* in lower Gangetic plains. Those active gears were (i) seine net (*ber jal*), (ii) cast net (*khapla jal*), (iii) lift net (*bhasal jal*), (iv) push net (*khuchni jal*), (v) hook and line (*borshi*), (vi) lantern net (*chak jal*), (vii) cover pot (*polo*) and passive gears viz., (viii) gill net (*phasa jal*), (ix) small meshed gill net (*punti jal*), (x) long line (*dhaon borshi*), (xi) box trap (*ghuni, atol*), and (xii) spindle traps (*chai*). Most of those active fishing gears (*ber jal, khapla jal, bhasal jal*) were targeted IMCs, exotics, murrels, *Wallago attu* and ISFs (*Puntius* spp., *Pethia* spp., *Trichogaster* spp., *Parambassis* spp., *Chnada nama, Salmostoma bacaila, Amblypharyngodon mola, Xenentodon cancila, Macrognathus puncalus, Anabus testudineus, Mystus cavasius, M. tengra, Channa punctata, Glossogobius giuris* etc.). Furthermore, passive gears (gill net, long line, box traps etc.) primarily target ISF's and carps. Specialized fishing techniques (behavior pattern) such as set barrier with light trap (aggregating behavior towards light), aerial trap (locally known as "*Chali tana*" at Nabadwip), setting of gill net with manual disturbance (predator avoidance and respiration), cast net with light (fish attracts towards light) are popularly used for catching *Rhinomugil corsula* in the lower estuarine stretch of Ganga (Manna et al. 2024).

In hill streams, cast nets, varying mesh size gill nets, traps, hook and line, groping, impoundments, noose fishing, and dynamiting are all prevalent fishing techniques (Gurumayum and Choudhury 2009; Das et al. 2015; Dutta et al. 2019). In hilly areas, certain traditional fishing methods are age-old practise, and very effective for catching specific types of fish. Fish are caught by locals using locally available resources, such as bamboo and small wooden logs, for their own consumption or commercial use in the fast-flowing hill streams by using their own traditional knowledge. Since it is difficult to catch fish in these large, perennial fast-flowing rivers, fishermen occasionally resort to illegal methods like poisoning and dynamiting, because modern fishing equipment is not readily available (Gurumayum and Choudhury 2009). For instance, the "*Nocte*" tribe in Tirap district, Arunachal Pradesh, practices the traditional fish aggregating method known as "*Longphongtook,*" which is mostly utilized for catching migratory hill stream fish (Dutta et al. 2019). "*Longphongtook*"—a fish catching method consists mostly of aggregating cobbles and boulders during winter months (2.5–2.75 diameter circular structure) at a depth of 0.5–0.6 m in the river-bed of *Namsang* stream (Tirap). After a pre-determined period (35–45 days), encircled the aggregate boulders by a mosquito net with the help of bamboo poles, and hiding fishes are caught by hand picking. The local population has been doing this prior to their "*Jaatcha*" festival (Dutta et al. 2019). Another exhaustive study performed by Baruah et al. (2013) reported 28 varieties of fishing traps based on the principle of capture, design, and operational methods under 5 broad categories from Brahmaputra valley, and the broad categories were (i) hiding place (artificial arrangements or bunch of brushes, twigs etc.), (ii) barricades and barriers, (iii) tubular traps, (iv) box traps, and (v) trap nets.

2.2.3 Crafts in Small Scale Fisheries

In inland small-scale fisheries, common fishing vessels include dug-out canoes, plank-built boats, thermocol rafts, coracles, *dingi*, and *dhoni*. In wetlands, lakes, and river environments, dug-out canoes are typical composed of locally accessible "*Shisham*" wood from a single tree trunk. Coracles are a common craft in Indian reservoirs for long lines and gill netting that are operated by two people. Primarily utilized in Madhya Pradesh, Tamil Nadu, and Karnataka, the coracles range in size from 1.5 to 2.0 m. Nowadays, thin sheets of aluminium, tin, and other materials are used to make these circular crafts, coracle. In addition, retrofitting of traditional boats with outboard or inboard motor engines, particularly in plank-built boats, to improve efficiency and range. Various authors have studied the fishing crafts that are employed in small-scale fisheries. Bose et al. (2019) reported eight types of fishing crafts from Madhya Pradesh, which includes (i) dug-out canoe, (ii) plank-built boat, (iii) rubber tube platform, (iv) thermocol raft, (v) bamboo raft, (vi) boat covered on wooden frame or bamboo frame, (vii) large boat (group fishing), and (viii) wooden frame iron tin/aluminium sheet boat. The authors found that, rafts (plank-built boat, large boat and wooden frame iron tin/aluminium sheet boat) are most popular among fishermen. Plank-built boat, large boat and wooden frame iron tin/aluminium sheet boat are mostly used in reservoirs, while dug-out canoes are mainly in rivers. According to Sudhessa et al. (2023), the most popular boat among fishermen along the entire stretch of the Mahanadi River is a plank-built boat, known locally as a "*donga*." These dongas are typically 25 × 3.5 feet in size and can accommodate two to three people when fishing. Additionally, inflated rubber tubes are frequently used in nearshore waters to operate gill nets in rivers.

2.3 Note on Craft and Gears Operated in River Gandak and Middle Stretch (Bihar) of Ganga

Fishermen use fishing craft to operate a variety of fishing gear, particularly in deeper waters or when catching fish. Caste nets and gill nets are among the gears that fishermen use without boats in shallow riverbeds, floodplain wetlands, and reservoirs. Plank-built boats, dug-out canoes, and specially made iron boats are examples of inland crafts that typically have flat bottoms for effortless mobility in both shallow and deep water. In addition to being motorized, some of these are non-motorized (modified with an outboard motor engine). These crafts typically measure between 3.0 and 10.0 m in length, 1.0–1.5 m in width, and 0.20–0.60 m in depth (Manna et al. 2024).

2.3.1 Traditional Non-motorized Boat

(i) *Plank-built boat:* These traditional wooden boats are ranging from 18 to 35 feet in length and 4 to 6 feet in breadth (Fig. 2.2a). These boats are extensively used for fishing and river transport; typically handcrafted from *sal, babool,* or mango wood. Operated manually using long oars or bamboo poles, while common fishing practices include the use of gill nets, cast nets, long lines, drag nets, traps like *pinjara* and *duaribritti*, and hook-and-line systems, and operations often performed in pairs or group. The boat is capable of withstanding the hydrodynamic conditions and turbulence of river.

(ii) *Dug-out canoe:* There are fewer dug-out canoes used for fishing than plank-built boats in Gandak River. Only a few areas of Gandak River have dug-out canoes, which are mostly used for gill netting and line fishing and are typically manned by one person or two.

(iii) *Wooden framed iron/tin/aluminum sheet boat:* Locally referred to as "*Nauka*" or "*Donga*," it refers to wooden-framed vessels constructed with sheets of iron, tin, or aluminum. The sizes and designs of these boats vary based on their intended use and the types of fishing to be conducted. A maximum of one or two fishers can be accommodated simultaneously on communal boats of 3–5 m in length and 1.5–2 m in width.

(iv) *Galvanized iron sheet boat:* Galvanized iron-sheeted canoes, measuring approximately 16–30 ft. in length and 3.5–5.0 ft. in breadth, are increasingly preferred by fishers due to their durability, light weight, and low maintenance (Fig. 2.2b). These crafts are operated manually using long oars or bamboo poles. Commonly used for diverse fishing methods such as gill netting, cast netting, drag netting, and hook-and-line fishing, they are employed both in

Fig. 2.2 Typical plank-built boat (**a**), a galvanized iron boat (**b**), a local made tin boat (**c**) and a haul of gill net catch in tin boat (**d**)

Fig. 2.3 Gill net operation using an inflated rubber tube

shallow and deeper river sections, with operations often carried out in pairs or groups, adapting to the fast current and disturbed hydrodynamic conditions of river.

(v) *Tin boat:* Locally referred to as "*Nao,*" this fishing boat was originally prevalent in the reservoir but has gradually diminished in significance. The dimensions of the boat are around 3.3 m × 1.2 m × 0.42 m and it is operated by 1 to 2 individuals. This vessel previously captured small fish and prawns utilizing gill nets in marginal zones of the rivers. The cost of the boat ranges from Rs. 3000/– to 4000/– with a lifespan of 3 to 5 years (Fig. 2.2c–d).

2.3.2 Others

(i) *Inflated rubber Tube:* Inflated rubber tubes, mostly repurposed from truck and car tyres, are primarily used for gill net operations in the inland waters of Bihar, especially in the Gandak and Ganga River, where individual fishers sit on the tubes and manually manoeuvre them using their hands or legs and performed fishing activities, i.e., gill netting, line fishing, cast netting, and trap setting in calm to mild flowing waters (Fig. 2.3)

2.3.3 Motorized Boat (Customized)

1. *Plank built boat with customized out-board motor engine.*

The motorized plank-built boats are similar to traditional plank-built boats. Only variation of these boats is that these boats are customized with an out-board motor engine (diesel) at the "bow" of the boat (Fig. 2.4) and headed towards its opposite direction. That apparently makes a "*power boat*" and help for fast movement in the river, especially during the high flow seasons and flooding periods. Common fishing practices include the use of gill nets, cast nets, long lines, drag nets, and

Fig. 2.4 A typical plank-built boat fitted with an outboard motor engine employed in the Gandak River

hook-and-line systems with operations often conducted in pairs or groups, adapting to the diverse hydrodynamic conditions and seasonal fish movements in these dynamic river systems.

2. *Galvanized iron boat with customized motor engine.*

The specially designed iron boats are equipped with 10–15 HP out-board diesel engines are employed for quick travel and fishing and particularly favorable during high-flow seasons. These motorized galvanized iron boats, like their non-motorized counterparts, are employed for a range of fishing methods and goods transportation. These iron boats, which are more stable and durable than plank-built boats, are often operated by two individuals or in groups, depending on the circumstance (Fig. 2.5).

2.4 Different Fishing Gears Commonly Used in Riverine Fisheries

2.4.1 Line Fishing

Hook and Line Fishing Hook and line fishing is the most common technique used in inland waters. Different kinds of hooks and lines that have been used to catch fish in open waters. Smaller hooks (size greater than No.16) are utilized to target the small fishes. Various hook and line techniques exist, including hand line without a pole, multiple set lines, pole and line and set line (drift line) made from small twigs (referred to as *"Nolborshi"* in Assam) (Bhattacharjya et al. 2004; Manna et al. 2020). There have been reports of a wide variety of baits from the Ganga that are suited to various fish groups according to their feeding preferences (Manna et al. 2020).

Fig. 2.5 A galvanized iron boat fitted with a motor engine (left corner), a motor engine (middle) and a small sized iron boat employed for crossing passengers (right corner)

Fig. 2.6 Hook and line using a bamboo pole (**a**), hook and line with bait (**b**) and long line with hook (**c**)

(i) *Pole and Line:* This kind of fishing gear consists of a long bamboo pole, typically composed of bamboo twigs, to which a line with hooks attached is attached (Fig. 2.6a). Lines are occasionally run through a tiny pulley that is fastened to the twig's terminal end. When a fish swallows the bait, the pulley enables the line to move quickly. As an indicator, a small piece of float is attached to the line.

(ii) *Multiple hook systems:* The idea behind using multiple hook techniques is to entangle more fish in one operation. Such hooks (*Joar/Barshi*) are widely used from "*nauka*," or locally designed plank-built boats. Two or three people go fishing in the river together at the same time using hooks and lines. Even small tin boats, locally known as *"Tiner donga,"* with a single individual capacity are employed for hook laying. Three types of such multiple hooking methods are mentioned below.

(a) *Long line:* Hooks ranging from 500 to 2000 in number are baited usually with species-specific bait and attached to nylon threads. These threads are suspended at fixed intervals from a thicker, longer nylon rope in a perpendicular manner. The entire structure remains stationary during operation, as heavy sinkers are used to anchor it. Due to easy availability, half bricks weighing approximately 1 to 1.5 kg are commonly used as sinkers in most longline hooking systems (Fig. 2.6b–c). These sinkers are typically attached after every 10–12 hook-bearing threads. As the number of hooks increases, the spacing between sinkers is reduced.

A plastic bottle or thermocol sheet is usually fixed at one end of the setup to serve as a surface marker, while a full brick is attached at the other end as a terminal sinker. The distance between two consecutive hooks is maintained at approximately 4.5 to 5 feet, which is sufficient to prevent tangling of adjacent threads. This method requires significantly less effort compared to pole-and-line fishing. Due to river currents, the hook-bearing threads remain inclined in an upright position, keeping the bait suspended in the water column. Although the intensity of hook deployment increases during the monsoon months (July to September), this method is practiced year-round in the studied stretch of the river.

(b) *Aggregation of two and five hooks:* Two hooks on a single line have been observed in the Gandak River. There is a main chord in the design, from which two lines with hooks diverts (Fig. 2.7b). With a similar goal, a third branch that serves as a sinker is attached to the main chord using a long (6 inch) nail or several iron nuts.

(c) *Aggregation of seven hooks:* Aggregation of hooks was found to be a significant and distinctive fishing technique in the Buxar-Balia stretch of the Ganga River. Hooks in seven lines (No. 5/0, 3/0 and 2/0) are attached to a main chord resembling a tree branch (Fig. 2.7c). As noted, specially made bait is either individually engraved onto each hook or occasionally clubbed together. To settle the hooks in the riverbed, sinkers made of heavily loaded adhesive mud are gathered around the main single chord.

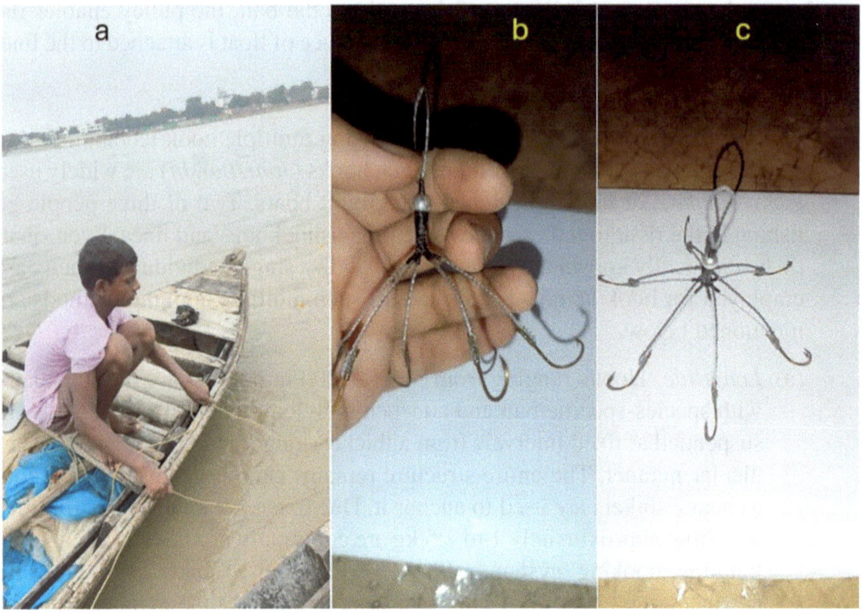

Fig. 2.7 (**b–c**) typical hook (five and seven hooks) and hand line fishing (**a**)

(iii) *Hand line:* Here, hook and line are used to catch fish instead of a sturdy supporting rod, which is somewhat different from the conventional pole and line (Fig. 2.7a). In this case, the lines with baited hooks are typically thrown far from the coast. Sinkers are made of iron nuts, which weigh about 50 g.

2.5 Fishing Gears

(a) *Gill nets*

The diversity of gill nets is largely influenced by mesh size, which is selected based on the target species. Smaller mesh nets (< 45 mm) are typically used when small indigenous freshwater species are the target. A wide range of mesh sizes is observed in gill nets used in multi-species fisheries in rivers. However, a clear seasonality is often apparent in the use of species-specific gill nets.

Various gill nets with different mesh sizes have been documented in inland open waters, such as rivers and wetlands (Table 2.1). For instance, a 12 mm gill net is used to capture *Amblypharyngodon mola*, a 20 mm mesh gill net is effective for small snakeheads (*Channa* spp.), and a 40 mm mesh gill net is typically used for minor carps (Bhattacharjya et al. 2004). There is also variation in their deployment within the water column; some gill nets are surface set, while others are bottom set, depending on the behavior of the target species.

Table 2.1 Different gill net mesh sizes are employed in inland water fisheries

Sl. No.	Local name	Mesh size (mm)	Species caught
1	*Bhasha jaal*	126	IMC, *Notopterus notopterus, Bagarius bagarius*
2	*Dora jaal*	160	IMC, *Bagarius bagarius, Sperata* spp.
3	*Sutri jaal*	14–25	*Ailia coila*
4	*Bekra jaal*	25–30	*Clupisoma garua, Eutropichthys vacha*
5	*Vachaili jaal*	26	*Eutropiichthys vacha*
6	*Gochail jaal*	140	IMC, *Chitala chitala, Notopterus notopterus*
7	*Current jaal*	55	Rohu, *Rita rita, Sperata* spp., *Labeo calbasu*
8	*Chalhati jaal*	22	*Gudusia chapra, Gonialosa manmina, Osteobrama cotio, Puntius sophore, Setipinna phasa*
9	*Chelwa jaal*	30–35	*Crossochelius latius, O. cotio, Gudusia chapra*
10	*Aungthasi jaal*	32	*Gudusia chapra, Macrognathus pancalus*

Fig. 2.8 (a–f) Different types of gill nets with varying mesh sizes employed in river Gandak and Bihar stretch of Ganga for catching small fishes and medium sized carps

These nets are the most commonly used fishing gear in rivers due to their low cost. They are typically made of polyamide (PA) multifilament, with this material currently more prevalent. These passive gears (with the exception of floating gill nets) consist of a single wall of netting held vertically in the water column by floats and sinkers. The mesh sizes range from 25 mm to 130 mm, matching the body depth of the target species. The nets can be deployed at the surface, mid-water, or bottom, and are often anchored or weighed down to keep them in place (Fig. 2.8a–f).

Fig. 2.9 Cast net throwing in mild flowing sites at Sonpur (river Gandak)

(b) *Falling gears*

Cast Net Locally known as "*fekaili jaal*" (throw net), the cast net is used in waters with a depth of 2–3 m. It is an umbrella-shaped, circular net constructed of nylon fibers (Fig. 2.9). Cast net operation is an active fish-catching technique and a falling gear. In order to haul the net during operation, a strong rope of 5–6 m in length is fastened to the apex of the net, and iron or lead sinkers are fixed along the margin.

Based on the size of the fish to be captured, the mesh size ranges from 20 to 45 mm, and the perimeter extends from 10 to 18 m. Pre-monsoon months revealed a substantially higher use of cast nets along the river, whereas monsoon months indicated a lower use of cast nets. The gear with varied mesh and pocket sizes was found to perform particularly well in shallow river surroundings, either from the bank of the river or from a boat. These nets can be operated by one person, with the fisherman throwing the net over the water either from a boat or from the riverbanks in such a way that it spreads on the water's surface completely extended at its perimeter and rope being held in hand at its apex. Due to the weight of the connected sinkers, the net sinks to the water body's bottom with a closed perimeter.

(c) *Lift net*

Chitwa ***Jall/Gyanra Jaal***

The *chitwa jaal*, or fixed lift net, is a traditional fishing apparatus prevalent in the Ganga River of Bihar. It comprises a large rectangular or square net affixed to a robust bamboo framework anchored in the riverbed. The dimensions of the frame typically range from 12 to 18 feet in width and 15 to 25 feet in length, with a net depth of approximately 6 to 10 feet. The supporting bamboo structure can extend up to 20 to 25 feet above the water surface, facilitating the vertical lifting of the net. Operated manually, the net is submerged and then abruptly raised to ensnare fish swimming above it. This passive gear is primarily utilized in shallow or marginal

Fig. 2.10 Lift net locally called as *Chitwa jaal* at Buxur (river Ganga)

river sections and is adept at capturing small indigenous species, juvenile carps, and prawns. While it is economically viable and locally sustainable, its non-selective nature poses ecological concerns due to the incidental capture of juvenile and non-target species (Fig. 2.10).

(d) *Bag net*

Bag nets are non-selective fishing apparatus that function by straining water to ensnare fish and other aquatic animals. These nets can be either anchored in the water column to permit the river's current to flow through them or controlled manually by fishers, usually through dragging or pushing. An example is the triangle push-type bag net, referred to locally as *Chipauva jaal*, frequently utilized in the Gandak River. This apparatus is predominantly utilized for the capture of prawns, crabs, and diminutive native fish species. In this technique, fishers deploy a wooden log in the water to serve as a FAD. After a period, when fish have congregated around the log, the fisher returns, removes the log, and captures the amassed fish using the *Chipauva jaal* (Fig. 2.11). This activity often involves one or two fishers, rendering it a straightforward yet efficient fishing method. A similar kind of bag net, called a "*Kunche jaal*," is used in the Hooghly River's tidal freshwater zone in Triveni. Fish species *Apocryptes bato, Glossogobius giuris, Channa punctata, Odontamblyopus rubicundus* etc. are harvested (Bose et al. 2017).

(e) *Drag nets*

Drag nets are frequently employed in riverine fisheries and are typically non-selective, frequently ensnaring a diverse array of fish types and sizes. Drag net is designed to capture medium to large-sized fish and is appropriate for usage in deeper and broader river sections. In Bihar, two conventional varieties of drag nets are extensively utilized in the river Gandak and the middle stretch of the river Ganga. These

Fig. 2.11 A typical bag net *Chipauva jaal*/*Kunche jaal* operation at Sonpur, Gandak River

are referred to locally as (i) *Paudi Jaal* and (ii) *Chatti Jaal.* The *Paudi Jaal* is a drag net characterized by a big mesh, with sizes varying from 25 to 35 mm. It often measures approximately 350 feet in length and 20 feet in width (Fig. 2.12a). Operating a *Paudi Jaal* necessitates a crew of 8 to 10 individuals owing to its substantial dimensions and the human labor required to manoeuvre the net through the water.

Conversely, the *Chatti Jaal*, also known as a mosquito net due to its exceedingly tiny mesh (almost negligible mesh size), measures approximately 150 feet in length and 15–20 feet in width (Fig. 2.12b). Drag net effectively captures juveniles and small fish due to its non-selective characteristics and ability to harvest immature specimens pose ecological issues. It necessitates the presence of 8 to 10 individuals for optimal operation. The extensive utilization of non-selective drag nets, especially the fine-meshed *Chatti Jaal,* presents difficulties for sustainable fisheries management, since it may result in the depletion of fish stocks and obstruct natural recruitment processes.

2.6 Traps

Trap fishing, a traditional and passive method, involves the use of devices into which fish voluntarily enter and become trapped, often based on their natural behaviors such as migration, feeding, or breeding. These traps may consist of one or more chambers that close after fish enter, with small traps typically being fully enclosed like cages except at the entrance while larger traps often remain open above the water surface. Traditionally constructed from bamboo splits and bound with dry grasses, cane, or bamboo strips, these traps have increasingly incorporated synthetic nets in recent times to enhance durability and efficiency (Manna and Bhattacharjya

Fig. 2.12 Operation of drag net with group of fishermen (**a**) *Paudi jaal* (mesh size 25–35 mm) (**b**) and *Chatti jaal* (mosquito net-zero mesh size)

2009). India's diverse inland fishery resources rivers, floodplain wetlands, lakes, reservoirs, ponds, and seasonal canals support such indigenous fishing practices, especially during the monsoon season when waters become extensively submerged. Trap fishing remains significant in these open waters, not only contributing to the nutritional security of rural communities but also showcasing regional artisanal craftsmanship. Despite advances in aquaculture, this age-old practice continues to play an important role in the socio-economic fabric of traditional fishing societies.

2.6.1 Bamboo-Framed Net-Lined Traps

Bamboo frame net-lined traps are a common type of passive fishing gear used in the middle stretch of the River Ganga and the Gandak River. Due to the increasing cost and limited availability of bamboo and cane materials, fishers have gradually shifted toward using netting material as sidewalls, with bamboo primarily forming the

Fig. 2.13 *Pinjara (Birti)* uncovered by net (**a**) and covered with zero meshed net lined (**c**—photo courtesy NMCG report phase 1), and *Pinjara (Jhanki)* (**b, d**) and *Khancha* (**e**)

structural framework. These traps, often circular or rectangular, are employed extensively to target a variety of fish species. Several regional variants of these traps are used across different sites. In areas such as Balliya (Uttar Pradesh) and Gopalganj, Bhagalpur and Buxar (Bihar), traps like *Pinjara, Aanta, Jhanki* are employed by local fishers. These traps are typically set in shallow or marginal waters and are constructed using bamboo or Metal rings or frames over which netting is stretched (Fig. 2.13a–e).

(a) *Chak jaal:* The *Chak jaal* is a stationary trap designed specifically to capture bottom-dwelling catfish in the River Ganga system. It comprises a robust bamboo or iron framework enveloped in fine mesh netting and is generally utilized in muddy or semi-stagnant waters, where these species are prevalent. The trap is engineered to maintain stability on the substrate, facilitating the entry of fish and their subsequent enclosure. The design and location demonstrate a comprehension of the behavior and habitat inclinations of benthic organisms (Fig. 2.14a).
(b) *Woka:* The *Woka* is a unique fixed fishing trap frequently utilized in the Inland Stationary Fishing systems of rivers such as the Ganga and Gandak in Bihar (Fig. 2.14b). This trap is acknowledged for its efficacy in extended fishing activities and is usually positioned in shallow channels or adjacent to embank-

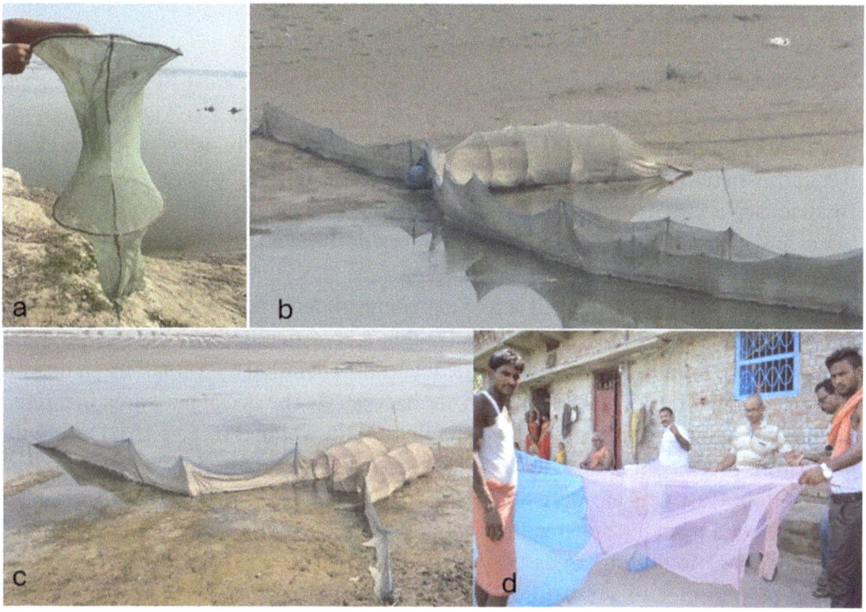

Fig. 2.14 *Chak jaal* (**a**), *Woka* (**b**), *Aarsi* (**c**) and *Jhinga jaal* (**d**). Photo "d". (Courtesy: NMCG report—Phase 1)

ments to capitalize on the natural flow of fish with the current. It possesses a conical or cylindrical form, fabricated with netting reinforced by circular hoops composed of bamboo or iron. Fish are directed through a broad entrance into the trap and progressively channelled towards a constricted exit, rendering escape challenging due to the inward taper and intricate mesh construction.

(c) *Aarsi:* The *Aarsi* is a conventional stationary fishing trap employed in the Inland Stationary Fishing system of the River Ganga in Bihar. Like the *Woka*, it comprises several circular frames typically constructed from bamboo or iron organized sequentially to create a narrowing, tunnel-like configuration (Fig. 2.14c). This design preserves form and alignment in flowing water. Located in regions with mild currents, the trap passively ensnares fish migrating downstream via a wide opening, directing them toward a constricted end that limits their escape. The *Aarsi*, prevalent in the central Ganga basin, is an effective and economical apparatus utilized in small-scale and subsistence fisheries.

(d) *Jhinga jaal:* The *Jhinga jaal* is a simple, frame-less trap made entirely of fine mesh netting and is commonly used in the Buxar region of the River Ganga. It is especially effective in shallow or receding waters, where it targets small prawns and juvenile fishes (Fig. 2.14d). Its lightweight and low-cost design make it accessible to local fishers. When used responsibly with appropriate mesh sizes, it supports sustainable fishing practices in inland waters.

2.6.2 Brush Park

Brush parks, which are mostly found in wetlands in Assam, West Bengal, and some parts of Bihar, are frequently used as fish aggregating devices (FADs). An age-old practice of capturing indigenous fish species, particularly in Assam and West Bengal, where the fish are successfully used as hiding places (Sugunan and Bhattacharjya 2000; Manna and Aftabuddin 2009). Typically, brush parks consist of twigs and submerged tree brunches within the designated area; occasionally, water hyacinth (*Eichornia crassipes*) or other surface macrophytes (*Salvenia molesta,* etc.) are used and later dispersed across a predetermined area (Manna and Aftabuddin 2009; Sharma et al. 2015). Bamboo poles that are vertically fixed form a circle that is later (1 or 2 months, depending on the location) encircled by a small mesh net that is tied to the bamboo poles before fish are harvested. The brush park area in a riverine environment is smaller than that of a floodplain wetland (Fig. 2.15). During the harvesting process, fish are captured by narrowing the encircling net with inward preposition. Cast nets are occasionally used after the tree twigs and brunches have been removed from the net area. In many locations, a small circular trench is formed in channels or low-lying areas during the lean season. Fish use these trenches as hiding places after they are filled by monsoon rains or floods. Additionally, tree brunches were positioned inside the trench and encircled by vertical bamboo poles. These brush parks offer a sustainable method of collecting fish from the wetlands and also increasing fish yield.

2.6.3 Fishing Methods for Spawn Collection

(a) *Kheral jaal:* This is a barrier trap positioned perpendicularly to the riverside, sometimes referred to as *Kheral jaal*. It comprises bamboo poles that uphold vertical mesh nets, directing fish into constricted trap chambers or baskets posi-

Fig. 2.15 A typical brush park in Ganga River (Buxar district)

tioned at regular intervals. These traps are particularly efficacious during fish migration or when water levels diminish, as fish typically follow the current and become channelled into the trap. The gear is completely passive and is generally employed in shallow marginal waterways, canals, or the peripheries of floodplains. A notable characteristic of this equipment is the incorporation of non-return valves or trap baskets, which inhibit the release of captured fish. In Bihar, especially along the Ganga and Gandak rivers, *Kheral jaal* is utilized in sandy and turbid regions to harvest fish spawn, rendering it a crucial implement for traditional seed collecting methods (Fig. 2.16a).

(b) *Ghana jaal*: The *Ghana jaal* is a finely meshed screen net, usually constructed from mosquito netting, utilized to create a barrier or partial enclosure on shallow riverbanks, floodplain regions, or marginal waters. It is engineered to limit the movement of small fish and fish spawn, facilitating effortless collecting. In contrast to large-scale nets, the *Ghana jaal* is utilized by one or two fishers and is frequently deployed temporarily during periods of strong flow. It is non-invasive, as it is employed in isolated shallow regions without disrupting the river ecosystem. These nets are extensively utilized in the Gandak River, particularly during the monsoon season, for the procurement of fish spawn and juvenile fish, significantly contributing to traditional fisheries and seed collecting while minimizing ecological impact (Fig. 2.16b).

Fig. 2.16 Operation of *Kheral jaal* (**a**), *Ghana jaal* (**b**) and *Musairi jaal/ghera jaal* (**c**) for spawn collection at Ganga and Gandak River

(c) *Musairi jaal*: This fishing structure consists of a semi-enclosed bamboo framework equipped with fine-meshed nets, creating pen-like chambers in the river. These cages are generally fixed and positioned in tranquil or stagnant waters adjacent to the riverbank. They are typically utilized for the retention of fish, fattening, or, in certain instances, trap-based capture. Fish may enter naturally via the current or may be introduced for temporary cultivation prior to harvesting. In Bihar, specifically in the Ganga River adjacent to Bhagalpur, these net pen enclosures are primarily utilized for the collection of fish spawn during the monsoon and post-monsoon periods. Their shape facilitates efficient spawn collection and supports small-scale aquaculture and seed rearing, becoming them a vital element of traditional fisheries in the region (Fig. 2.16c).

2.7 Conclusion

The prudent and flexible use of both conventional and modern fishing craft and gear is essential to SSIF's sustainability. Local fishermen resilience and indigenous knowledge are reflected in their continued use of selective and environmentally friendly fishing gear along with craft designs that are appropriate for shallow and dynamic water bodies but sustainability necessitates more than just conventional methods; it calls for efficient management, fish stock monitoring, habitat preservation, regulated fishing, and community participation in decision-making. The long-term health of inland aquatic ecosystems and the survival of small-scale fisheries for future generations can be guaranteed by promoting sustainable fishing methods, reducing destructive practices (especially use of the zero mesh size nets, dynamiting etc.) and combining traditional and modern scientific methods.

References

Anon (2023) Small-scale inland fisheries of India: status and strategies for sustainable development. ICAR-Central Inland Fisheries Research Institute, Barrackpore. https://www.icsf.net/wp-content/uploads/2022/11/INF_TOT_Keynote-Speech_BK_DAS.pdf

Baruah D, Dutta A, Pravin P (2013) Traditional fish trapping devices and methods in the Brahmaputra valley of Assam. Indian J Tradit Knowl 12(1):123–129

Bhattacharjya BK, Manna RK, Choudhury M (2004) Fishing crafts and gear of North-Eastern India. Central Inland Fisheries Research Institute (ICAR), Barrackpore., Bulletin 42:67

Bose AK, Bose R, Gupta S (2017) Kunche Jal fishing: a low-cost indigenous fishery in Ganga River Basin. Int J Fish Aquat Stud 5(4):87–90

Bose R, Gupta S, Das AK, Suresh VR, Bose AK (2019) Traditional fishing crafts and gears of Madhya Pradesh. Int Res J Biol Sci 8(3):29–36

Das RK, Barat S (2014) Fishing gears operated in lentic and lotic water bodies of Cooch Behar district, West Bengal, India. Indian J Tradit Knowl 13(3):619–625

Das BK, Singh NR, Boruah P, Kar D (2015) Fishing devices of the river Siang in Arunachal Pradesh, India. J Fish 3(2):251–258

Das BK, Sarkar UK, Manna SK, Roy A (2020) Proceedings of international webinar on small-scale fisheries in inland open waters: status and opportunities. Inland Fisheries Society of India & ICAR-Central Inland Fisheries Research Institute, Barrackpore. ISSN: 0970_616

Das BK, Parida PK, DebRoy P, Roy A (2022) A situational analysis of small-scale inland open water fisheries in India: from vulnerability to viability. V2V Working Paper 2022-8. V2V Global Partnership,. University of Waterloo, Waterloo

DebRoy P, Das BK, Parida PK, Roy A, Chakraborty S (2021) Economic injustice in small-scale wetland fisheries in West Bengal, India. In: Kerezi V, Chuenpagdee R (eds) Blue justice for small-scale fisheries: a global scan, volume 2, TBTI Global Publication Series, St John's

Dutta R, Bhagabati SK, Mili K, Bhuyan PC, Kumar A, Songtheng P, Borah S (2019) Longphongtook—an indigenous tribal fishing technique in fast flowing rivers of Arunachal Pradesh, India. Fish Technol 56:110–114

FAO (2005–2011) Fisheries and aquaculture topics. Small-scale and artisanal fisheries. Topics fact sheets. Text by Jan Johnson. In: FAO fisheries and aquaculture department [online], Rome. Updated 27 May 2005. [Cited 19 November 2011]. http://www.fao.org/fishery/topic/14753/en

FAO (2010) Inland fisheries resource enhancement and conservation in Asia. FAO Fisheries and Aquaculture Technical Paper

FAO (2015) Voluntary guidelines for securing sustainable small-scale fisheries in the context of food security and poverty eradication. Food and Agriculture Organization of the United Nations, Rome, pp 1–34

FAO, Duke University, WorldFish (2023) Illuminating hidden harvests—the contributions of small-scale fisheries to sustainable development. FAO, Rome. https://doi.org/10.4060/cc4576en

Ganguly A, Konar U, Kundu A, Chaterjee S, Nandi S, Guin RK, Duari M, Mandal A, Das Mohapatra PK (2022) Traditional fishing gears of Bankura District, WB, India: some uniqueness in fish catching. Not Sci Biol 14(1):11132

Gurumayum SD, Choudhury M (2009) Fishing methods in the rivers of Northeast India. Indian J Tradit Knowl 8(2):237–241

Handbook on Fisheries Statistics (2022) Department of Fisheries, Ministry of Fisheries, Animal Husbandry & Dairying, Government of India, New Delhi, p 218

Manna RK, Aftabuddin M (2009) Comparative limno-chemistry study of brush park (katal) and macrophyte free area in a wetland of Assam. Environ Ecol 27(4A):1796–1800

Manna RK, Bhattacharjya BK (2009) Incorporation of new construction material into indigenous technological knowledge—a case study of V-shaped fish trap of eastern India. Indian J Tradit Knowl 8(4):548–550

Manna RK, Das BK (2024) Indigenous technical knowledge (ITK) associated with fisheries of indigenous small fishes (ISFs) in Indian inland waters: present status, issues, and way forward. In: Sinha A, Roy A, Gogoi P (eds) Perspectives and applications of indigenous small fish in India. Springer, Singapore. https://doi.org/10.1007/978-981-97-1586-2_21

Manna RK, Ray A, Samanta P, Singh A, Bhowmik TS, Baitha R, Paul SK, Behera SK, Das BK (2020) Bait specificity in hook and line fishery of River Ganga and associated conservation issues. Indian J Tradit Knowl 19(1):61–72

Manna RK, Bhakta D, Nair SM, Samanta S, Das BK (2024) Unique behaviour-driven fishing strategy of freshwater mullet, Rhinomugil corsula (Hamilton, 1822) in lower and estuarine stretch of River Ganga. Fish Technol 61:144–152

Purakit S, Karmakar S, Mondal A, Pradhan SK (2018) Indigenous technical knowledge in fisheries of south 24 Parganas District of West Bengal, India. Int J Curr Microbiol Appl Sci 7(2):2793–2797

Roy A, Aftabuddin M, Bayen S, Ghosh A, Das BK, Parida PK (2020) Changing dimensions of 'fish for all' from 'food and livelihood' to 'nutritional security'. Int J Fish Aquat Stud 8(3):320–324

Samajdar I, Saikia SK (2014) Traditional fishing gears of Birbhum District, West Bengal, India. Indian J Tradit Knowl 13(1):187–194

Sandhya KM, Roy A, Hassan MA, Kumari S, Mishal P, Lianthamluaia L, Kumar V, Aftabuddin M, Bhattacharjya BK, Meena DK, Ali Y, Naskar B (2019) Traditional fishing gears, fish catch and species composition of selected floodplain wetlands of lower Gangetic plains, West Bengal, India. Fish Technol 56:101–109

Sharma AP, Chandra G (2015) Issues and challenges for sustainable small-scale fisheries in inland fisheries sector of India. In: ICSF-BOBLME India (East Coast) workshop: implementing the

FAO voluntary guidelines for securing sustainable small-scale fisheries in the context of food security and poverty eradication, Chennai. https://doi.org/10.13140/2.1.27303200
Sharma AP, Katiha PK, Sahu AK, Chandra G (2013) State of inland fisheries and inland fisher community in India. Proc Natl Strategic Workshop on Small-Scale Fisheries. BOBP-IGO, Chennai
Sharma P, Sarma J, Sarma D, Ahmed S, Phukan B, Baishya S, Kashyap D, Dutta M, Deka P, Hussain IK (2015) An indigenous fish aggregating method practiced along the Kolong River in Nagaon District of Assam. Indian J Tradit Knowl 1(1):112–117
Sudhessa D, Sajina AM, Samanta S, Paul SK, Nag S, Baitha R, Bhowmick S (2023) Fishing crafts and gears used along selected stretch of River Mahanadi. Fish Technol 60:173–180
Sugunan VV, Bhattacharjya BK (2000) Ecology and fisheries of beels in Assam. Central Inland Fisheries Research Institute, Barrackpore. Bulletin No. 104. ISSN: 0970-616X, p 66

Small-Scale Inland Fisheries in Nutritional Security of India

Piyashi DebRoy, Basanta Kumar Das, Prateep Kumar Nayak, and Dibakar Bhakta

3

Abstract

Nutritional security characterizes appropriate health and wellbeing of individuals through intake of balanced and adequate nutrition from foods consumed. SDG 2: Zero Hunger, SDG 3: Good Health and Well-being and SDG 12: Responsible Consumption and Production point toward achieving nutritional security in the world. In inland open water fisheries, the diversity of fish is high, many of whose proximal composition have been analyzed to reveal that they are micro-nutrient dense fish (NutriFishIN database developed by ICAR—CIFRI with nutritional information of food fishes of India). Hidden harvest is data on small-scale fisheries about fish catch, fishers' livelihood, dietary importance of fishes, and other facets that are not included in available statistics hindering national, regional, and global policy making. If we want to achieve SDG 2 (Zero Hunger) by 2030, we have to ensure policy level up-gradations in inclusion of fish in dietary requirements in the nutritional programs.

Keywords

Nutritional security · Hidden harvest · Fish consumption · Food policy · Inland fisheries

Healthy nutrition and equitable access to food are basic human rights towards the development of a prosperous, educated and peaceful nation

P. DebRoy (✉) · B. K. Das · D. Bhakta
ICAR–Central Inland Fisheries Research Institute (ICAR-CIFRI), Barrackpore, Kolkata, India

P. K. Nayak
School of Environment, Enterprise and Development, Faculty of Environment, University of Waterloo, Waterloo, ON, Canada

B. K. Das et al. (eds.), *Small-scale Inland Fisheries in India*, Sustainability Sciences in Asia and Africa, https://doi.org/10.1007/978-981-95-3849-2_3

3.1 Introduction

Small-scale fisheries provide nutritious aquatic food to the planet which has ensured food availability to humanity since ages (FAO and INFOFISH 2022). Food availability, food access, food utilization, and stability are the four pillars of food security (Clapp et al. 2022) which should be monitored in every country. The Food and Nutritional Security (FNS) Framework has been discussed in Fig. 3.1. Small-scale fisheries contribute to 17% of global animal protein requirement. Small fish in freshwater resources are especially nutritious and maintain nutritional security in many developing and developed nations (Fluet-Chouinard et al. 2018). Small-scale fisheries in India provide livelihood to fishers and subsistence for household requirements, and ensure that the fish-eating populace of the country and the fishers are nutritionally secure with animal-based protein sources in the diets they consume. Nutritional security characterizes appropriate health and wellbeing of individuals in the society through intake of balanced and adequate nutrition from foods consumed. The rising cost of food has been affecting lives and nutritional security in the world. Fish is an important and integral part of diets consumed by half of the Indian population. Inland open water fisheries resources in India comprise rivers, reservoirs, lakes, wetlands, lagoons, and estuaries. Fish from inland open water resources are more organic in nature than from cultured resources (Pandit et al. 2021). The fish from the small-scale inland fisheries in India are considered to be of nutritional significance in view of essential micronutrients and the potential to combat malnutrition and hunger (Mohanty et al. 2019). The future of aquatic food systems lies in harnessing the diversity of the roles played by the participants in small scale fisheries (Short et al. 2021). Small scale fisheries in the inland sector play a very promising role in minimizing the various risk factors associated with diseases resulting from malnutrition in impoverished communities. Therefore, nutrition-sensitive

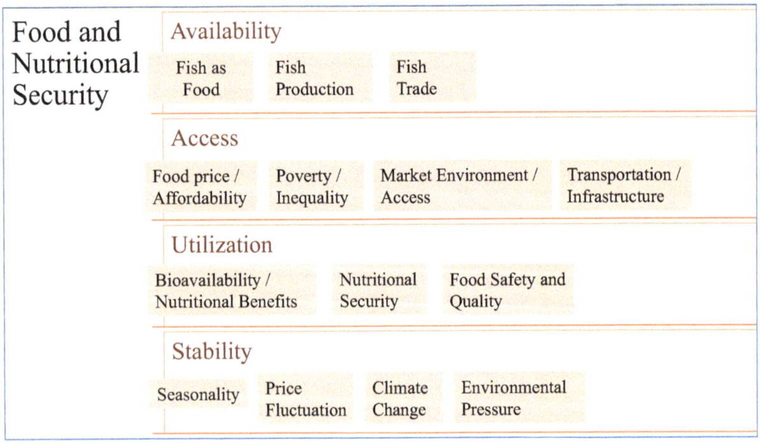

Fig. 3.1 Framework and Dimensions of Food and Nutritional Security. (Source: FAO 2006; Hasselberg et al. 2020; Wustefeld 2013)

fisheries policies are essential for community-wellbeing in the society (Thilsted et al. 2016). Recent nutritional studies hint at food policies requiring to concentrate beyond productivity gains of cultured species in commercial fisheries and aquaculture (Thilsted et al. 2016).

The challenge of absence of desired quality of fish in local markets from capture fisheries (Béné et al. 2010) has to be addressed through proper nutritional policies in the country. Therefore, it is high time that the role of small-scale fisheries in promoting nutritional security in food systems need to be acknowledged. The population of India is estimated to be 1.52 billion by 2036 growing with an increase of 25% in twenty-five years (National Commission on Population 2019). To feed such a huge population, it has been estimated by FAO and INFOFISH that the global annual per capita fish consumption will increase to 25.5 kg by 2050 from 20 kg at present. In India, fish is a source of nourishment and nutrition for 70% of the population, and an additional 14.6 million metric tonnes of fish would be required to meet fish demand by 2047–48 (Padiyar et al. 2024). Achievement of SDG 2 Zero Hunger necessitates right to food, global food security, and consumer education. A thorough understanding of household socio-economics in small-scale fisheries, and its contributions to household food security and nutrition, in turn illuminating governance and the well-being of fishers is necessary in this context.

3.2 Fish as Nutrition Rich Food

The protein from fish plays an important role in combating malnutrition or hidden hunger. A common example in this case is the omega-3 fatty acids which are available only from fish and very rarely from any other foods. Omega-3 fatty acids are important for cognitive development in infants and provide coronary protection in adults. The omega-3 fatty acids in fish caught in small-scale fisheries provide half of the recommended nutrient intake (RNI) for 773 million women in Asia (FAO, Duke University and WorldFish 2022). Each tonne of wild caught inland fish provides more protein requirement than an equivalent amount of marine or fish obtained through aquaculture to a greater number of individuals (Funge-Smith and Bennett 2019). The production of mola (SIF) *Amblypharyngodon mola* in pond polyculture systems in Bangladesh is a successful model which has been both cost-effective and reduced malnutrition (Fiedler et al. 2016). Small fish and pelagic species including cyprinids are the highly nutritious which are also easily affordable for the rural population. A positive externality arising from inland open water fisheries is that the diversity of fish is high, many of whose proximal composition has been analyzed to reveal that they are micro-nutrient dense fish (NutriFishIN database developed by ICAR—CIFRI with nutritional information of food fishes of India). Fish consumption reduces 36% mortality risk from heart disease (Rimm and Mozaffarian 2006). It has been also reported that daily consumption of 60 g fish results in 12% reduction in mortality (Zhao et al. 2015). Developed countries have already incorporated recommended levels of fish consumption in their dietary guidelines (Thilsted et al. 2016). Fish is also rich in omega-3 fatty acids whose crucial role in pregnancy reduces the risk of early pre-term delivery (Imhoff-Kunsch et al. 2012). In this context, the reference to small indigenous fishes (SIFs) is important as they are rich in micro-nutrients (Bogard et al. 2015). Dried fish, especially dried small fish, are consumed through India among the fish-eating population. Important market centers exist in the country to cater to the demand of dried fish. The North-Eastern region of the country is an important hub of dried and processed fish marketing and consumption, many of which are sourced from inland waters. Dried fish provides high quality proteins, healthy fats, long-chain omega-3 fatty acids, and essential nutrients such as iodine, zinc, copper, selenium, and calcium (Siddhnath et al. 2022). It was reported in the same study that dried fish sector contributes to 20% of total fish production in India. It was also analyzed that 100 g of dried fish yields 80% protein (Siddhnath et al. 2022). However, a major threat to achieving global nutritional standards is that the poor people around the world continue to face global inequities in fish consumption (Thilsted et al. 2016).

Consistent research and development efforts of scientific organizations in the country including the Indian Council of Agricultural Research has resulted in generation of awareness of the importance of small indigenous fish (SIF) as a health food. Historically, the fish-eating population in West Bengal, Odisha, and the North-Eastern Region of India along with the tribal populace of inland fisheries states of the country have always known the importance of consumption of SIFs. That is why SIFs were always a part of their regular diets. Recently, SIFs in these states are priced in premium. This is a double-edged sword in the way that traders harness high profits from the premium prices of SIFs in retail markets. However, this again puts the poor consumers in a disadvantaged position owing the high prices. This is also the picture in the global scenario (Belton et al. 2014; Toufique and Belton 2014).

3.3 Inland Fish Production and Consumption Trends in India

The inland fish production has increased 12-folds from 1980–81 to 2019–20 while the total fish production of the country has registered 6-folds increase during the same time. The average world per capita consumption at 20.7 kg per capita per year (FAO 2024). In India, the states with high fish consumption per capita are Tripura (25.53 kg per capita per year), Manipur (18.25 kg per capita per year), Kerala (17.93 kg per capita per year), Odisha (16.34 kg per capita per year), and Assam (11.89 kg per capita per year) (Table 3.1). Union Territories such as Lakshadweep (125 kg per capita per year), Goa (78 kg per capita per year), Andaman and Nicobar Islands (77.84 kg per capita per year), and Puducherry (18.88 kg per capita per year) have exceptionally high fish-eating per capita consumption. Among the top five fish consuming states, Tripura, Manipur, and Assam are three North-eastern states whose fisheries sector are absolutely inland and small-scale in nature. This undoubtedly shows the crucial role small scale fisheries plays in sustaining the nutritional security of these states. On the other hand, the average annual growth rate of the fisheries sector in India is 4.35% (2019–20) (Department of Fisheries, Government of India 2020) where the average annual growth rate of the inland sector is 7.37% at 2019–20.

Table 3.1 Fish production and consumption in various states of India (2021–22)

State	Total fish production (lakh tonnes)	Inland fish production (lakh tonnes)	Consumption (per capita/kg)
Andhra Pradesh	48.13	42.19	8.91
Arunachal Pradesh	0.05	0.05	3.65
Assam	4.17	4.17	11.89
Bihar	7.62	7.62	9.60
Chhattisgarh	5.91	5.91	19.70
Goa	1.16	0.05	78
Gujarat	8.74	1.86	7.44
Haryana	2.08	2.08	0.3
Himachal Pradesh	0.16	0.16	2.22
Jharkhand	2.57	2.57	10.61
Karnataka	10.74	4.85	10.55
Kerala	8.26	2.25	17.93
Madhya Pradesh	2.93	2.93	4.87
Maharashtra	5.9	1.57	4.72
Manipur	0.33	0.33	18.25
Meghalaya	0.18	0.18	9
Mizoram	0.05	0.05	3.38
Nagaland	0.09	0.09	6.06
Odisha	7.89	7.89	16.34
Punjab	1.90	1.90	0.4
Rajasthan	0.66	0.66	0.01
Sikkim (data not available)	–	–	1.16
Tamil Nadu	8.07	2.12	9.3
Telangana	3.90	3.90	8.7
Tripura	0.82	0.82	25.53
Uttar Pradesh	8.09	8.09	11.09
Uttarakhand	0.06	0.06	0.81
West Bengal	18.43	16.52	Not Reported
Andaman and Nicobar Islands (data not available)	0.44	–	77.84
Chandigarh (data not available)	–	–	Not Reported
Dadra and Nagar Haveli, and Daman and Diu (data not available)	0.30	–	1.2
Delhi	0.01	0.01	0.25
Jammu and Kashmir	0.25	0.25	6
Ladakh	0.01	0.01	0.05
Lakshadweep	0.12	0.00	125
Puducherry	0.47	0.08	18.88
India	162.48	121.21	6.31

Source: Department of Fisheries, Government of India (2022)

3.4 Gender Roles in Nutritional Security

Studies were attempted in identifying the pathways that exist between the livelihoods of fishers and their nutritional security (Kawarazuka and Béné 2010). Three pathways were identified by which small scale fisheries contribute to nutritional security namely through nutrition uptake from fish consumption; financial security through sale of fish enabling fisher households greater purchasing power to access other necessary nutritional foods; and enhancement of women's economic security where women are involved in fisheries activities ultimately empowering women in better decision-making for household nutritional security. Women constitute 30% of the small-scale fishers (20.9 million) in the world. Women's contribution in sustaining nutritional security through fish protein intake in fishers' household is very important because women are the ones who carryout household processing for everyday cooking and the decision-makers with regard to quantity of daily fish consumption and the variety of dishes that are cooked. In 2018, women accounted for 10% of the workforce engaged in capture fishing itself in Asia. Women are the majority of the fish workers who are involved in fish processing and in retail and vending business. The United Nations Decade of Family Farming (UNDFF) 2019–2028 supports healthy and productive communities around the world. Therefore, the intersection of IYAFA 2022 and UNDFF strengthen the emphasis for nutritional security in households. The ninth Global Conference on Health Promotion of the World Health Organization held in Shanghai in 2016 ushered in political commitments across countries for ensuring healthy lives and promoting well-being for all people at all ages to be achieved within 2030. Under the theme of Promoting Health for Promoting Sustainable Development, SDG 14 (Life Below Water) aims in supporting the restoration of fish stocks to improve healthy diets. Women are being not represented adequately in small scale fisheries governance with a few exceptions. This kind of hidden representation in the scenario of nutritional security secured from fisheries results in gender-biased policies and unrecognition of their roles, rights, and access in small-scale fisheries.

3.5 Hidden Harvest

FAO defines hidden harvest as data about small scale fisheries regarding fish catch, fishers' livelihood, nutritional value of fish and other facets that is accounted for during policy formulation. It was assumed that fish harvest from inland fisheries resources are underreported because of unmonitored subsistence fisheries (De Graaf et al. 2015) which led to obscured vision about the real contribution of such resources in global food and nutritional security (Fluet-Chouinard et al. 2018). The base line report about hidden harvest (World Bank 2012) brought to the forefront the nutritional significance of fish in the transformation of food systems. Hidden harvests particularly assume importance when fisheries statistics from governmental agencies do not include fish catch data about recreational fishery, subsistence fishery and the catch that fishers consume in their families from their own fishing efforts

(Kjellevold et al. 2022). Small scale inland fisheries are many a times concentrated in developing and under-developed countries where the essential contribution of such subsistence fisheries in the food and nutritional security in local economies is very much underappreciated (McIntyre et al. 2016; World Bank 2012). Subsistence fisheries which are not traded, and often not accounted, and brought home by fishers for family consumption, provide a livelihood safety net for them to overcome malnutrition (Virdin et al. 2023). The hidden harvests have resulted in under-appreciation of the contribution of inland fisheries to food security (Golden et al. 2016). This lack of data and information creates a confusion between small-scale fishers' employment data and subsistence fishing activity thereby undermining the fish catch which is not traded in markets but forms an integral part of consumption and value chain. The *Illuminating Hidden Harvests* study by FAO, Duke University and WorldFish has brought to light that small scale fisheries contribute over 40% of the global capture fish catch (FAO and INFOFISH 2022). Freshwater capture fisheries are reported to contribute only about 7% of reported global fish harvests (FAO 2016).

The first and foremost concern in calculating and accounting for hidden harvest is lack of recorded data from households, farms, co-operative societies, and communities regarding fish harvest and consumption. In general, surveys in fish landing centers and fish markets provide the basis of socio-economic data in fisheries surveys (Welcomme 2011). However, such catch estimation methods exclude the fish catch which is consumed without entering supply chains (McIntyre et al. 2016; Bartley et al. 2015). The data reporting system in some countries happen to be not much robust because of which the fisheries statistics there are generated with proxy variables which do not represent the accurate picture of the fish catch and its disposition (Welcomme 2011; Coates 2002). This necessitates improvement in data collection methods and approaches in inland fisheries (Bartley et al. 2015; De Graaf et al. 2015). Therefore, a usual approach adopted to overcome such missing data is to do back calculation to obtain fish harvest data from household consumption surveys to estimate such fish catch from freshwater, estuarine waters, inland saline waters, and other inland fisheries resources (Fluet-Chouinard et al. 2018). For example, in a study conducted by Fluet-Chouinard et al. 2018 for 42 low- and middle-income countries in the world between 1997 and 2014, a net underreporting of freshwater fish harvest to the extent of 64.8% was observed. These hidden harvests were equivalent to the total animal protein consumption of 39.6 million people. Such a kind of approach is known as Household Consumption and Expenditure Surveys (HCESs) (Funge-Smith 2016; Hortle 2007; Mills 2011). HCESs record fish, crustaceans, and molluscs which are caught or bartered or purchased by the household over a 1-day to 2-week recall period before the interview (follow Fluet-Chouinard et al. 2018 for detailed explanation). This implies that the household respondents where surveys are carried out are asked questions about their fish consumption at the most in the previous 2 weeks which is said to be a proper recall period. It is noted that surveying large number of households is a more correct approach than random sampling in landing sites and markets (De Graaf et al. 2015; Funge-Smith 2016; Mills 2011). While estimating the original harvest levels to account for the hidden harvest, fish from inland fisheries resources must be disaggregated from marine, aquaculture, and imports if applicable, and also adjusted to reflect the biomass discarded during post-harvest operations (gutting, filleting, dehydration during preservation in methods like smoking, salting, freezing etc.) (Funge-Smith 2016). HCES has been applied to account for hidden harvests in Lower Mekong Basin (Hortle 2007; IIFRD 2012) and in middle and upper Amazon basin (Bayley 1981).

Fish consumption rates are also calculated by the units of grams per person per day from traditional units (for example heaps, tin etc.) and from the inedible portions which are retrieved to convert to biomass equivalent. Euryhaline and diadromous species such as hilsa and mullets are categorized as inland fishes as per FAO guidelines.

According to the National HCES-Based Catch Calculations, the total inland fish consumption (R_i) is calculated in fresh weight equivalents for each country (i) as the sum of annual consumption (C_{ij}) of each fish product (j) weighted by its fraction of inland provenance (F_{ij}) and live-weight conversion factor (L_j), through the following formula

$$R_i = \sum_{1}^{j} F_{ij} \times C_{ij} \times L_j$$

Since the total fish consumption includes sources of fish other than inland openwater fisheries, the above formula needs more refinement. Thus, HCES-based catch (S_i) of each country by adding exports (E) of inland fish to R_i and then subtracting imports of inland fish (I) and aquaculture production.

$$S_i = R_i + E_i - I_i - A_i$$

It is to be noted that this approach sidelines the consumption of fish through non-food purposes, if any.

For person equivalent calculation, the additional number of people per country whose animal protein consumption could be met with the hidden harvests in HCES nations was calculated following published methods (McIntyre et al. 2016). The dependence D_i of each country (i) is calculated as the protein from inland fish (I_j) over the sum of protein from all fish (F_i) and all other animal protein consumption excluding fish (O_i) and then multiplied by the country's population (N_i).

$$D_i = I_i / \left(F_i + O_i\right) \times N_i$$

The dependence is calculated twice for each country using I and F from reported statistics, and then from HCES. The difference in D_i is the additional dependence.

3.6 Food Policies in Fisheries

The unexplored dimension of fisheries in different policy research has future implications for human health and sustainability (Love et al. 2017). In the lines of "nutrition-sensitive agriculture" (FAO and WHO 2014), fisheries policies and food policies should also focus on nutrition-sensitive fisheries. The current fisheries policies, programmes, schemes, and investments in the country maximize on productivity and income-generating goals. To achieve measurable impacts on nutrition, integration of small-scale inland fisheries in food policies is necessary. For example, the United States fisheries policy contributes to seafood sustainability. Nutritional quality needs to be improved by development of value chains that consider increasing the bioavailability of nutrients from fish. The disconnect that exists between health and fisheries policies needs to be bridged in to improve food security. Investment in national fish market chains can boost equitable supply of fish throughout the country. Culture-based fisheries and enclosure culture such as cage and pen culture should be upscaled throughout the country so that capture fisheries in inland open water resources are not over-harvested. Fisheries policies must be placed within a systemic approach in food policies where culturally appropriate. In this, cultural preferences, food habits, nutritional needs of women and young children, and the regional traditions should be considered. The Voluntary Guidelines for Securing Sustainable Small-Scale Fisheries in the Context of Food Security and Poverty Eradication (SSF Guidelines) are the first set of international guidelines for regulating the small-scale fisheries sector in the world. Healthy small-scale fisheries are necessary for food security, employment, pro-poor fisheries policies, and rural livelihoods (World Bank 2012). Such dynamics need to be interplayed into social, political, biophysical, and economic domains. In India, factors such as fish prices, poverty, inequality, poor market infrastructure, inadequate transportation, seasonality, price fluctuations, climate change, and ineffective government policies on fish quality and safety effect access, utilization, and consumption of fish (Vijayan et al. 2025). There is limited information about integration of fisheries in food policies in India at present which provides a larger platform for fisheries in food policy research.

3.7 Conclusions

Multi-sectoral policy solutions are necessary to realize the goal of attaining SGDs for quality foods and reduction of malnutrition. Research and case studies in nutritional security indicated that the role of inland fisheries in fulfilling nutritional security in various strata of societies is more important than what we commonly assume. Being rural, highly dispersed and informal in nature, the inland fisheries resources should be given proper attention and priorities to win over the combating resource uses that put tremendous pressures on these ecosystems. A lot of inland fish harvests in India are unaccounted in fisheries statistics contributing to hidden harvest. If these hidden harvests are illuminated on a pan-India level, the multiple dimensions of nutritional security contributed by inland fisheries will be highlighted. Fisheries management to promote optimal nutrition and efficient fish production would require new outcomes to be monitored and promote health policies to include fisheries as a component.

Things to Remember

1. Food and Nutritional Security Framework involves availability, access, utilization, and stability.
2. Small scale fisheries play an important role in combating nutritional deficiencies.
3. The hidden harvests have resulted in under-appreciation of the contribution of inland fisheries to food security.

References

Bartley DM, De Graaf GJ, Valbo-Jørgensen J, Marmulla G (2015) Inland capture fisheries: status and data issues. Fish Manag Ecol 22:71–77

Bayley PB (1981) Fish Yield from the Amazon in Brazil: comparison with African River yields and management possibilities. Trans Am Fish Soc 110:351–359

Belton B, van Asseldonk IJM, Thilsted SH (2014) Faltering fisheries and ascendant aquaculture: implications for food and nutrition security in Bangladesh. Food Policy 44:77–87

Béné C, Hersoug B, Allison EH (2010) Not by rent alone: analysing the pro-poor functions of small-scale fisheries in developing countries. Dev Policy Rev 28:325–358

Bogard JR, Hother A-L, Saha M, Bose S, Kabir H, Marks GC, Thilsted SH (2015) Inclusion of small indigenous fish improves nutritional quality during the first 1000 days. Food Nutr Bull 36(3):276–289

Clapp J, Moseley WG, Burlingame B, Termine P (2022) Viewpoint: the case for a six-dimensional food security framework. Food Policy 106:102164

Coates D (2002) Inland capture fishery statistics of Southeast Asia: current status and information needs. RAP Publications No. 2002/11, FAO Regional Office, Asia Pacific, Bangkok

De Graaf G, Bartley D, Jorgensen J, Marmulla G (2015) The scale of inland fisheries, can we do better? Alternative approaches for assessment. Fish Manag Ecol 22:64–70

Department of Fisheries, Ministry of Fisheries, Animal Husbandry and Dairying, Government of India (2020) Handbook on fisheries statistics, New Delhi

Department of Fisheries, Ministry of Fisheries, Animal Husbandry and Dairying, Government of India (2022) Handbook on fisheries statistics, New Delhi

FAO (2006) Food security policy brief Issue 2, June

FAO (2016) The state of world fisheries and aquaculture. FAO, Rome

FAO (2024) The state of world fisheries and aquaculture 2024—blue transformation in action. FAO, Rome

FAO, Duke University, WorldFish (2022) Small-scale fisheries and sustainable development: key findings from the Illuminating Hidden Harvests report. FAO, Rome

FAO, INFOFISH (2022) The small-scale fisheries and aquaculture sector in Asia—small in scale, big in value. FAO, Bangkok

FAO, WHO (2014) Rome declaration on nutrition: outcome document Second International Conference on Nutrition. FAO, Rome

Fiedler JL, Lividini K, Drummond E, Thilsted SH (2016) Strengthening the contribution of aquaculture to food and nutrition security: the potential of a vitamin A-rich, small fish in Bangladesh. Aquaculture 452:291–303

Fluet-Chouinard E, Funge-Smith S, McIntyre PB (2018) Global hidden harvest of freshwater fish revealed by household surveys. Proc Natl Acad Sci USA 115(29):7623–7628

Funge-Smith S (2016) How national household consumption and expenditure surveys can improve understanding of fish consumption patterns within a country. In: Taylor WW, Bartley DM,

Goddard CI, Leonard NJ, Welcomme R (eds) Fish and the future. FAO, Michigan State University, American Fisheries Society, pp 121–130
Funge-Smith S, Bennett A (2019) A fresh look at inland fisheries and their role in food security and livelihoods. Fish Fish 20:1176–1195
Golden CD et al (2016) Nutrition: fall in fish catch threatens human health. Nature 534:317–320
Hasselberg AE, Aakre I, Scholtens J, Overa R, Kolding J, Bank MS, Kjellevold M (2020) Fish for food and nutrition in Ghana: challenges and opportunities. Glob Food Secur 26:100380
Hortle KG (2007) Consumption and the yield of fish and other aquatic animals from the Lower Mekong Basin., MRC Technical Paper No. 16. Mekong River Commission, Vientiane
Imhoff-Kunsch B, Briggs V, Goldenberg T, Ramakrishnan U (2012) Effect of n-3 long-chain polyunsaturated fatty acid intake during pregnancy on maternal, infant, and child health outcomes: a systematic review. Paediatr Perinat Epidemiol 26:91–107
Institute for Inland Fisheries Research and Development (IIFRD) (2012) Food and nutrition security vulnerability to mainstream hydropower dam development in Cambodia. Inland Fisheries Research and Development Institute, Phnom Penh
Kawarazuka N, Béné C (2010) Linking small-scale fisheries and aquaculture to household nutritional security: an overview. Food Secur 2:343–357
Kjellevold M, Kuhnle GA, Iversen SA, Markhus MW, Mancha-Cisneros MDM, Gorelli G, Nedreaas K (2022) Small-scale fisheries contribution to food and nutrition security—a case study from Norway. Npj Ocean Sustain 1:5
Love DC, da Silva PP, Olson J, Fry JP, Clay PM (2017) Fisheries, food and health in the USA: the importance of aligning fisheries and health policies. Agric Food Secur 6:16
McIntyre PB, Reidy Liermann CA, Revenga C (2016) Linking freshwater fishery management to global food security and biodiversity conservation. Proc Natl Acad Sci USA 113:12880–12885
Mills DJ, (2011) Under-reported and undervalued: small-scale fisheries in the developing world In: Pomeroy RS, Andrew NL (Eds.) Small-scale fisheries management: frameworks and approaches for the developing world CAB International, Wallingford: 1–15
Mohanty BP, Mahanty A, Ganguly S, Mitra T, Karunakaran D, Anandan R (2019) Nutritional composition of food fishes and their importance in providing food and nutritional security. Food Chem 293:561–570
National Commission on Population (2019) Population projections for India and states, 2011–2036. Ministry of Health and Family Welfare, New Delhi
Padiyar PA, Dubey SK, Bayan B, Mohan CV, Belton B, Jena J, Susheela M, Murthy LN, Karthikeyan M, Murthy CK (2024) Fish consumption in India: patterns and trends. WorldFish, New Delhi
Pandit A, Das BK, DebRoy P (2021) Marketing of Indian inland open water fishes as 'green fish': strategies and policy options for efficient value chain. J Inland Fish Soc India 53(3–4):123–131
Rimm EB, Mozaffarian D (2006) Fish intake, contaminants, and human health: evaluating the risks and the benefits. JAMA 296:1885–1899
Short RE et al (2021) Harnessing the diversity of small-scale actors is key to the future of aquatic food systems. Nat Food 2:733–741
Siddhnath RA, Mohanty BP, Saklani P, Dora KC, Chowdhury S (2022) Dry fish and its contribution towards food and nutritional security. Food Rev Int 38(4):508–536
Thilsted SK, Thorne-Lyman A, Webb P, Bogard JR, Subasinghe R, Phillips MJ, Allison EH (2016) Sustaining healthy diets: the role of capture fisheries and aquaculture for improving nutrition in the post-2015 era. Food Policy 61:126–131
Toufique KA, Belton B (2014) Is aquaculture pro-poor? Empirical evidence of impacts on fish consumption in Bangladesh. World Dev 64:609–620
Vijayan V, Suresh A, Sabu SS, Padi HP (2025) Fish for food and nutrition security in India: a comprehensive framework analysis. Food Secur 17(9):749–765
Virdin J et al (2023) Fishing for subsistence constitutes a livelihood safety net for populations dependent on aquatic foods around the world. Nat. Food 4:874–885
Welcomme R (2011) Review of the state of the world fishery resources: inland fisheries. FAO Fish and Aquaculture. Circular. No. 942

World Bank, (2012) Hidden harvest: the global contribution of capture fisheries. Economic and Sector Work No. 66469

Wustefeld M (2013) Food and nutrition security: UNSCN meeting of the minds—nutrition impacts of food systems. Geneva

Zhao LG, Sun JW, Yang Y, Ma X, Wang YY, Xiang YB (2015) Fish consumption and all-cause mortality: a meta-analysis of cohort studies. Eur J Clin Nutr 69(2):144–152

4 Governance Patterns and Institutional Frameworks: How They Influence Small-Scale Inland Fisheries

Ganesh Chandra

Abstract

Inland fisheries are predominantly small-scale in nature due to the absence of commercial and intensive fishing practices. These water bodies are considered common property resources and are governed under varied institutional and management regimes. Multiple organizations exert ownership and control, resulting in overlapping responsibilities and limited impact from fisheries development initiatives, primarily due to the shared and multi-use character of these resources. Despite these challenges, there is significant potential for improved governance to enhance resource management and support poverty alleviation. The sector involves numerous stakeholders with different mechanisms for accessing and utilizing fisheries resources, making governance frameworks crucial. Regulations and policies governing small-scale fisheries in states like Assam, Bihar, Uttar Pradesh, Madhya Pradesh, Chhattisgarh, Gujarat, and West Bengal have been analyzed, along with recent reforms. A key recommendation is to restructure the lease tenure system to better incentivize fishers, encouraging them to adopt scientific and sustainable fisheries management practices.

Keywords

Governance · Fisheries · Institutions · Stakeholders · Access · Leasing · Policy

Equitable access to fisheries resources strives sustainable livelihood enhancement for the community

G. Chandra (✉)
ICAR–Central Inland Fisheries Research Institute (ICAR-CIFRI), Barrackpore, Kolkata, India

B. K. Das et al. (eds.), *Small-scale Inland Fisheries in India*, Sustainability Sciences in Asia and Africa, https://doi.org/10.1007/978-981-95-3849-2_4

4.1 Introduction

Fishing is one of the oldest forms of utilizing natural resources for sustaining human life. Even today, it continues to be a crucial livelihood activity for millions of rural poor, offering food and nutritional security along with various income and employment opportunities. Inland fisheries in India benefit from a wide range of water bodies including rivers, floodplain wetlands (beels), reservoirs, swamps, canals, and tanks. Fishing activities in these water bodies remain primarily small-scale, as commercial and intensive fishing practices are largely absent (Sharma and Chandra 2015).

Governance is comprised of three interrelated systems: economic, political, and administrative. It extends beyond formal government actions such as resource allocation, regulation, and management, and encompasses the ways power is exercised and distributed among different societal actors concerning public issues. According to Swiderska et al. (2008) and Bene and Neiland (2006), governance reflects the broader mechanisms through which authority operates. Delmas and Young (2009) describe governance as a societal function aimed at guiding communities away from collectively harmful outcomes—such as the tragedy of the commons—towards more socially desirable results like environmental sustainability. Similarly, Oakerson (1992) views environmental governance as the process by which societies make decisions that influence or directly affect the environment. The institutional framework is central to natural resource management, as it shapes the entire decision-making process. Two main schools of thought dominate institutional analysis. The first, New Institutionalism, emphasizes the role of "rules-in-use" (Ostrom 1990) and focuses on economic and political dimensions of natural resource governance. The second is the Collective Action approach, which highlights the importance of shared norms and institutional rules designed to manage common-pool resources sustainably (Agrawal 2002; Berkes 1989; Ostrom 1990; Ostrom et al. 1994).

Fisheries governance encompasses the legal, social, economic, and political systems used to regulate fisheries. It operates at international, national, and local levels, incorporating both legally binding regulations and customary practices (FAO 2001). Achieving sustainable fisheries development requires balancing social, environmental, and economic objectives, which in turn demands integrated, participatory approaches involving various stakeholders. Institutions play a crucial role in facilitating this coordination (Mitchell 1990; Bandaragoda 2006; Rydin and Falleth 2006). There is a limited body of international literature on the governance and institutional structures of inland open water fisheries, focusing mostly on the political and management aspects of fisheries in different global river basins (Baird and Flaherty 2005; Werners et al. 2009; Sneddon and Fox 2012) and wetlands in Bangladesh (Ahmed et al. 2008; Kabir et al. 2011). In India, while extensive studies exist on aquaculture governance (Marothia 2002), relatively few examine river basins and reservoirs (Singh 1994; Sinha and Katiha 2002; Katiha et al. 2007; Sugunan and Sinha 2001; Chandra 2009; Vass et al. 2011; Katiha et al. 2013;

Chandra and Bhattacharyya 2016; Chandra and Ekka 2015). A significant number of studies have been conducted on floodplain wetlands, though they primarily focus on fisheries biology (Jhingran 1991; Vass and Bhattacharjya 2008; Vass et al. 2011) and enhancement techniques (Sugunan and Bhattacharjya 2000; Vinci et al. 2003; Chandra 2007, 2010, 2014; Chandra et al. 2013; Sarkar et al. 2021). Under the constitutional framework of India, fisheries fall under the jurisdiction of state governments. The ownership of inland open water bodies such as rivers, reservoirs, and wetlands lies with the respective state authorities. These resources are managed as common property and fall under diverse governance regimes and policy structures (Chandra 2011; Chandra and Das 2019). According to the public trust doctrine, the government is entrusted with the power to lease, allocate, or transfer public resources as long as it does not compromise public interest. Various state agencies are empowered to manage fisheries activities in reservoirs and wetlands. In many states, lease durations granted to fisher communities range from 1 to 7 years, though frequently leases are only valid for 1–3 years. Lease amounts differ from state to state and are determined by factors such as fish production, water spread area, and effective fishing zones. Additional considerations include the size of the water body, traditional usage rights, physical geography, accessibility, and hydrological connections, all of which influence property rights and lease valuations. Inland water bodies serve multiple uses—both consumptive and non-consumptive and exhibit complex interactions between their social and ecological systems.

This chapter is based on both primary and secondary data collected between 2011 and 2018 from various inland fisheries across Assam, Bihar, Uttar Pradesh, Madhya Pradesh, Chhattisgarh, and Gujarat. The data was gathered using participatory rural appraisal methods (Chandra 2010), alongside structured and informal engagements with fishers, lessees, local elders, community leaders, and other key stakeholders. Interactions also took place with government officials, and focused group discussions were conducted to understand governance mechanisms. Relevant laws and policies, especially those concerning resource access and tenure rights, were thoroughly reviewed.

4.2 Stakeholder in Inland Open Water Bodies

A wide range of stakeholders are involved in the use and management of wetlands and reservoirs, not only from the fisheries sector but also from related areas such as agriculture, forestry, animal husbandry, and flood control (Chandra and Bhattacharyya 2016) (Table 4.1).

Stakeholders function at various levels of management, resulting in multiple management domains, each with distinct elements and outcomes (Table 4.2). Among these, fishers are the most crucial stakeholders, as their livelihoods and daily sustenance are directly tied to the sustainable use of these resources.

Table 4.1 Stakeholders in inland fisheries

Sl. No.	Stakeholders	Roles
1.	Fishers, Lesees	Direct user of the resources
2.	Villagers, residents of adjoining areas of water resources	Indirect interest in the management of the resources
3.	Share fishers/laborers	Remote user of the resources
4.	Lessee, middle man and fish traders	Commercial direct user of the resources
5.	Government agencies (DoF, panchayat, revenue, water resources etc.)	Government departments with responsibility for management of some aspect of fisheries resources
6.	Research institutions, development organization	Providing assistance to users and authorities
7.	Consumers	End users of fish products

Table 4.2 Management domains in reservoirs and wetlands. (Modified from Chandra and Das 2019)

Sl. No.	Domains of analysis	Components	Outcomes
1.	Natural process	Nutrient assemblage Auto stocking Connectivity to river Biodiversity Flooding	Productivity potential Technological and management options Production enhancement possibilities In-situ conservation Automatic recruitment Sustaining assemblage and ecosystem Ecosystem integrity
2.	Human intervention	Management regime Fishing practices Fishing efforts Weed control Enclosure culture	Sustainable production Productivity enhancement Resource use Sustainability In-situ seed production
3.	Agencies of intervention	Individuals Formal and informal groups Community Govt. departments Scientific organization Traders and commission agents Financial institutions	Efficient resource use Community participation Employment opportunity Livelihood and nutritional security Fishing ban and holidays Credit availability Technological improvement Improved resource monitoring
4.	Institutions	Social interaction and process Property relationship Social institutions Market agencies	Rules and norms Negotiated understanding Conflict resolution Increased participation Access to resources Community based management Information sharing and exchange Capacity utilization

(continued)

Table 4.2 (continued)

Sl. No.	Domains of analysis	Components	Outcomes
5.	Policies	Leasing policy Infrastructure development Control of effort Resource conservation Institution building Security mechanism Welfare measures	Resource allocation, user access right, Sustainability Long term welfare Input support Improved decision making Conservation of endangered species Participative and precautionary measures Improved enforcements measures Awareness among fishers Biodiversity conservation

4.3 Fisheries Management Regimes of Inland Open Water Bodies

Inland open water fisheries are governed under various management regimes, broadly classified into four categories based on property rights and access: Private Management (by individuals or groups), Fishers' Cooperative Management, Community-Based Management, and Open Access with no formal management.

4.3.1 Private Management

In this model, water bodies such as wetlands are managed by private lessees who exercise de facto control. Lease durations typically range from 3 to 7 years, depending on the policies of different states. The lessee is responsible for managing the fishery, which includes activities such as fish stock enhancement, aquatic weed control, and harvesting. Other fishers may gain access to these water bodies either by paying a fee to the lessee or through a profit-sharing arrangement where a portion of the catch is given as compensation. In many wetlands (beels) of Assam, private management systems are commonly practiced under individual fisher leases (Chandra 2010; Chandra et al. 2013).

4.3.2 Cooperative Management

In many states, priority for leasing fisheries is given to primary fishermen cooperative societies. Under cooperative management, fishery operations are organized and executed by society members in accordance with rules and guidelines established by the cooperative itself. Access to the resource is regulated through society membership. Two main revenue-sharing models exist: in some cooperatives, the total fish catch is collected, sold collectively, and a portion of the income is distributed among the members. In others, individual fishers contribute around half of their catch to the society as a form of revenue sharing. A prominent example of this model is found in the wetlands of West Bengal, where primary fishermen cooperative societies actively manage fisheries through direct participation of their members (Roy et al. 2025).

4.3.3 Community-Based Fisheries Management

In several states, especially in floodplain wetland areas, there has been a growing interest in community-based and co-management approaches, evolving from earlier systems of traditional and lease-based management. These models aim to involve local communities directly in both fisheries management and conservation efforts. Government departments, particularly the Department of Fisheries, have initiated pilot projects to implement community-based fisheries management in selected water bodies. In such initiatives, local fishermen and women from nearby villages form Waterbody Development Committees, which are responsible for overseeing fisheries activities. These committees elect executive members to handle day-to-day operations, while the Department of Fisheries plays a supportive, facilitative role in ensuring effective management.

4.3.4 Open Access

Water bodies like rivers, rivulets, and streams are generally categorized as open-access systems, implying no formal restrictions on entry or usage. However, in practice, access is often informally governed by local norms, such as requiring that fishers belong to nearby villages or the same community. These water bodies support only capture fisheries, which depend on natural fish recruitment during flood seasons. In Assam, a large number of unregistered wetlands function under this open-access model, where fishing practices occur without structured oversight or formal lease agreements.

4.4 Governance and Policies at State Levels

4.4.1 Assam

In Assam, floodplain wetlands are classified based on ownership into two categories: government-owned and privately managed fisheries. The rights to manage registered wetlands (beels) are granted through a lease to the highest bidder, which can be either an individual or a fishermen cooperative society (Fig. 4.2). Only individuals identified as fishermen—defined as those engaged in fishing, fish trading, or members of fishermen cooperative societies—are eligible to lease these water bodies. This same policy applies to the leasing of reservoirs.

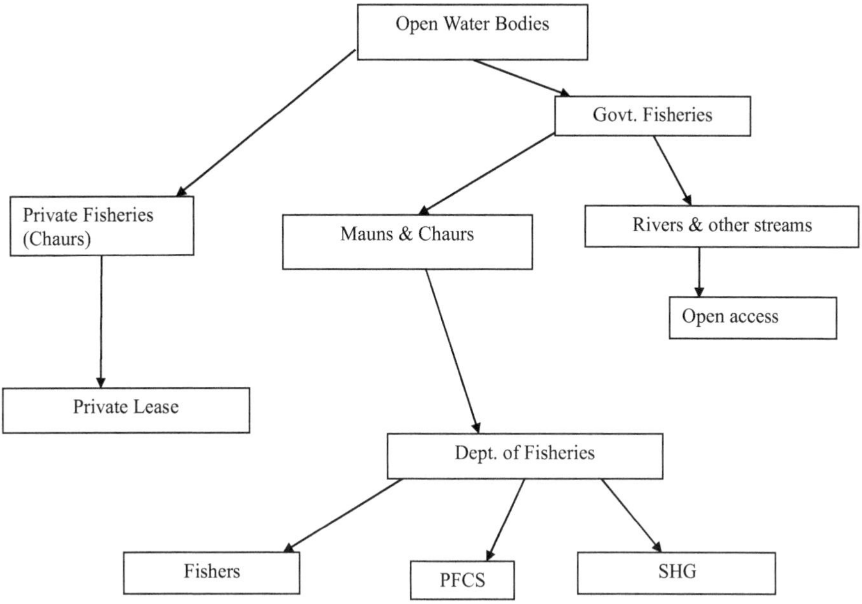

Fig. 4.1 Formation and transfer of fishing right in Bihar. (Chandra and Ekka 2015)

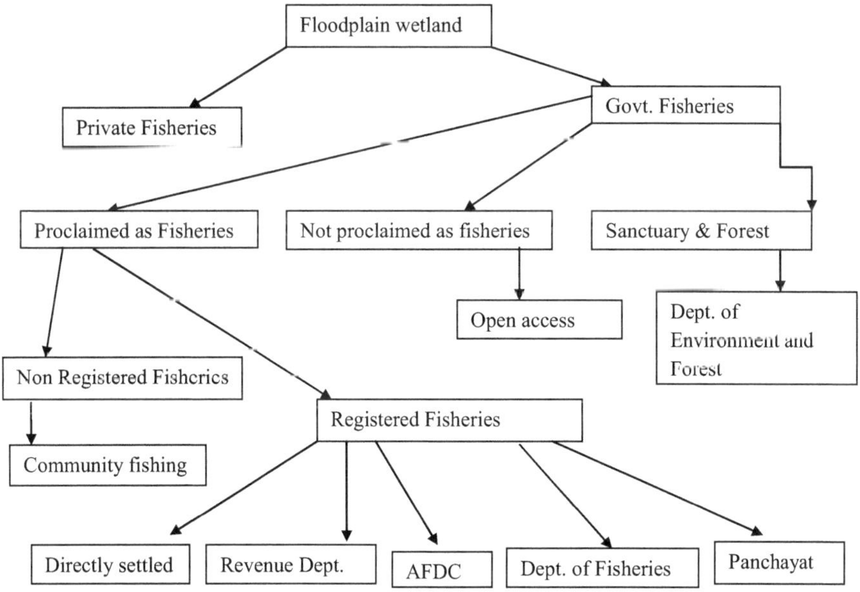

Fig. 4.2 Formation of ownership and transfer of fishing rights in Assam. (Chandra 2011)

4.4.2 Bihar

In Bihar, the Department of Fisheries holds authority over open water bodies. Management rights for government-owned wetlands (mauns and chaurs) are typically leased to primary fishermen cooperative societies for a 5-year period. Rivers are accessible to traditional fishers and fishermen cooperative societies (see Fig. 4.1), with the latter receiving preferential consideration in the leasing process for fishing rights.

4.4.3 Uttar Pradesh

Water body ownership in Uttar Pradesh is distributed across several state departments. The Irrigation Department controls 45 reservoirs, while the Fisheries Department oversees 40. These reservoirs are categorized into four classes for fisheries leasing purposes. The first two classes have higher base prices for lease, and the management rights are awarded to the highest bidder through an auction. Leases for these categories are limited to 3 years and are open to both cooperative societies and private contractors. In Classes 3 and 4, the lease period extends to 5 years, and local fishermen cooperative societies are given preference. If the minimum bid criteria are not met, both cooperatives and contractors are allowed to participate. Compared to other states, Uttar Pradesh offers relatively less favorable access rights to fisher communities, especially in higher-tier water bodies where contractors often dominate.

4.4.4 West Bengal

Fisheries in West Bengal are regulated under the West Bengal Inland Fisheries Act, 1984. Fishing access in wetlands is reserved for traditional fishers and cooperative societies. A cooperative-based management system is prevalent across most water bodies, and primary fishermen cooperative societies in the state manage operations with a high level of professionalism and organization.

4.4.5 Madhya Pradesh

Water body ownership in Madhya Pradesh is distributed among various government departments. Leasing rights are tiered based on the size of the water body:

- 0–10 ha: Granted by the Gram Panchayat
- 10–100 ha: Allocated by the Janpad Panchayat
- 100–1000 ha: Managed by the Zila Panchayat
- 1000–2000 ha: Controlled by the State Fisheries Department
- Above 2000 ha: Managed by the Madhya Pradesh Fisheries Federation.

Priority in leasing is given to different groups based on the size of the water body. Individual traditional fishers have access to water bodies under 1 ha, while fisherman cooperative societies and self-help groups (recognized by the Fisheries Department) can lease areas between 1 and 5 ha. Registered cooperatives or self-help groups (with registration required within a year) can lease water bodies up to 1000 ha.

4.4.6 Chhattisgarh

In Chhattisgarh, various government departments own the water bodies, and management rights are granted through a tiered system.

- Up to 200 ha: Rights are allocated by the three-tier Panchayati Raj system.
- 200–1000 ha: Managed by the Department of Fisheries
- 1000–5000 ha: Allocated to the Chhattisgarh Fisheries Federation
- Above 5000 ha: Handled again by the Department of Fisheries.

Primary fishermen cooperative societies are the designated entities for managing fisheries in these water bodies.

4.4.7 Gujarat

In Northern Gujarat and Saurashtra, fishing rights for water bodies up to 200 ha are granted exclusively to Fish Farmers Development Agency (FFDA) beneficiaries. For reservoirs above 200 ha, rights are leased through open tender, awarded to the highest bidder.

In Southern Gujarat, particularly in tribal regions, fishing rights are structured as follows:

- Up to 20 ha: Allocated to individual tribal fishers.
- 20–1000 ha: Given to tribal fishermen cooperative societies
- Above 1000 ha: Managed by federations of tribal cooperative societies.

Preference in the leasing process is given to local cooperative societies and beneficiaries under the Swarnjayanti Gram Swarozgar Yojana (SGSY). The Institutional arrangement and governance in different states are given in the (Table 4.3) below.

4.4.8 Preference to Fishers/Fishermen Cooperative Societies

To promote the overall well-being of the fishing communities, many state governments have implemented welfare-oriented policies that grant preferential access to fishers in the management of inland water bodies. In several states such as Bihar,

Table 4.3 Institutional arrangement and governance in different states

State	Ownership of waterbodies	System of transfer and license	Lease tem	Governance model	Management system	Fishing right and equity
Assam	Department of revenue, fisheries, AFDC, panchayat, environment	Open tender system, license to highest bidder,	7	Revenue model	Individual and cooperative	Fishers oriented
Bihar	Department of fisheries	Licence to PFCS	5	Welfare model	Cooperative	Fishers oriented
Uttar Pradesh	Irrigation, fisheries	Open tender system, license to higher bidder, fishers and contractors	3–5 years	Revenue model	Private and cooperative	Favouring private parties over fishers
West Bengal	Revenue, fisheries	Fisherman cooperating society	5–10 years	Welfare model	Cooperative	fishers
Madhya Pradesh	Irrigation Fisheries Panchayat	Fisherman, PFCS	5 years	Welfare model	Individual and cooperative	Fishers oriented
Chhattisgarh	Water resources, fisheries, panchayat	Fisherman PFCS	5 years	Welfare model	Cooperative	Fishers oriented
Gujarat	Water resources, fisheries, panchayat	Open tender system, fishers, PFCS and contractors in North Gujarat Tribal fishers and cooperatives in South Gujarat	5 years	Revenue model	Private and cooperative	Revenue oriented

Madhya Pradesh, Gujarat, West Bengal, Uttar Pradesh (except for Class 1 and 2 water bodies), Kerala, Chhattisgarh, and Odisha, exclusive fishing rights are granted to primary fishermen cooperative societies or individual fishers.

Some states also offer financial concessions to encourage the participation of cooperative societies. For example, a concession of 7.5% in Assam, 10% in Haryana, and 5% in Rajasthan is extended to primary fisheries cooperatives if they place bids close to the highest bid amount during the leasing process.

In most parts of India, rivers are generally treated as open-access resources, allowing fishers to operate freely. However, in states such as Andhra Pradesh, parts of Assam, Uttar Pradesh, and other southern states, leasing of river stretches for fisheries is also practiced.

This preferential approach is intended to enhance equity and empower the fisher community by involving them directly in the sustainable management of fisheries resources.

4.5 Policy Implications of Institutional Arrangements for Small-Scale Inland Fisheries

India's inland fisheries governance generally follows two models: the revenue-based model and the welfare-based model. Most states favor a revenue-driven approach, leasing fishing rights to the highest bidder. In contrast, states like Chhattisgarh and Bihar adopt a more welfare-oriented governance framework, where fishing rights are allocated preferentially to primary fishermen cooperative societies to support traditional fisher communities.

In Uttar Pradesh, the involvement of the fishing community in managing Class 1 and 2 water bodies is minimal due to the dominance of private contractors, who often prioritize profit over sustainability. Even in Class 3 and 4 categories, contractors frequently compete with cooperatives for access, limiting equitable opportunities for local fishers. As a result, the state's system leans heavily towards a revenue-centric model, offering limited inclusion to the fishing community.

Assam grants exclusive fishing rights to fishers and their cooperatives, ensuring their central role in fisheries governance.

In Bihar, traditional fisher groups have legally protected access to open water resources like rivers, and cooperative societies are given exclusive rights to manage wetlands. However, many of these cooperatives are dominated by their secretaries, who often exert disproportionate control, undermining collective ownership.

In Chhattisgarh and Madhya Pradesh, exclusive fishing rights are granted to fishers and cooperatives, demonstrating a clear preference for inclusive governance.

In Gujarat, particularly in tribal regions, large reservoirs are managed under a cooperative framework. Fishing rights are granted to tribal and non-tribal fisher cooperatives, and a welfare-oriented, revenue-sharing model is followed. Lessees pay royalties, making it a blend of welfare and income-based governance.

4.5.1 Reforms

Institutional arrangements across various states indicate that decisions regarding the management and utilization of inland water bodies are shaped within a framework that blends both welfare-oriented and revenue-driven approaches. Over the years, several state governments have initiated significant reforms in the governance of small-scale inland fisheries. These reforms reflect a shift away from a purely revenue-based model toward a more integrated welfare- and development-centric governance system.

It is now imperative to acknowledge and strengthen the role of local communities and indigenous groups in conserving, restoring, and co-managing aquatic ecosystems. Their active involvement is essential for ensuring the sustainable use of fisheries resources. This requires continuous and meaningful consultation with local stakeholders to guarantee their free, informed, and effective participation in all stages of decision-making that impact their livelihoods. Furthermore, there is an urgent need to establish robust and credible mechanisms for resolving conflicts to support equitable and sustainable fisheries governance.

4.6 Conclusion

Inland fisheries are dynamic, multi-use resources that support the livelihoods of a wide range of actors. Understanding the roles and involvement of different stakeholders is crucial for integrating them into effective governance frameworks. State-level fisheries laws generally uphold the customary rights of traditional fishers to access open water bodies. Across India, small-scale fisheries are managed under both revenue-based and welfare-oriented models. In many states, priority in the allocation of fishing licenses is granted to primary fishermen cooperative societies or individual fishers.

Despite the existence of regulatory frameworks, fisher communities continue to face significant challenges particularly regarding equitable access and resource allocation. There is a pressing need for governance reforms that shift toward a development-focused, welfare-driven model, ensuring that fishers are central to fisheries management. Policies must prioritize sustainability over revenue generation, especially in cases such as leasing open-access water bodies like rivers, where ecological conservation is more critical than catch maximization.

Another important reform area is the standardization and extension of lease durations across states. Currently, lease periods vary from 1 to 7 years depending on state policy. Extending lease terms to at least 5 years or more would encourage fishers to adopt sustainable and responsible management practices, while also promoting equity across regions.

States are gradually moving away from purely revenue-driven fisheries governance, and transitioning towards a development-oriented model that emphasizes sustainable production, particularly from wetlands and reservoirs. Reforming lease terms and incentivizing long-term scientific management practices will play a key role in empowering fisher communities and ensuring long-term sustainability of inland fisheries resources.

Points to Remember

- Achieving sustainable fisheries development requires a balanced approach that integrates social, environmental, and economic considerations, alongside the active involvement of diverse stakeholders.
- Although still limited in scope, governance reforms in small-scale fisheries are gradually shifting from revenue-driven models to development-oriented frameworks that emphasize community participation and long-term sustainability.
- Leasing and access policies for water bodies vary significantly across states. There is an urgent need for a more unified, inclusive, and coherent policy framework to ensure equitable access and effective management.
- Establishing a transparent, reliable, and efficient conflict resolution mechanism is essential to address disputes and support fair governance in inland fisheries.

References

Agrawal A (2002) Common resources and institutional sustainability. In: Ostrom E, Dietz T, Dolsak N, Stern PC, Stonich S, Weber EU (eds) The drama of the commons. National Academy Press, Washington DC, pp 293–321

Ahmed A, Deaton J, Sarker R, Virani T (2008) Wetland ownership and management in a common property resource setting: a case study of Hakaluki Haor in Bangladesh. Ecol Econ 68(1–2):429–436

Baird I, Flaherty M (2005) Mekong River fish conservation zones in southern Laos: assessing effectiveness using local ecological knowledge. Environ Manag 46(3):439–454

Bandaragoda DJ (2006) Status of institutional reforms for integrated water resources management in Asia: indications from policy reviews in five countries. Working Paper 108. International Water Management Institute (IWMI), Colombo

Bene C, Neiland A (2006) From participation to governance: a critical review of governance, co-management and participation in natural resources management. In: The challenge program on water and food & WorldFish center, Policy economics and social science discussion paper series. WorldFish Center, Penang

Berkes F (1989) Common property resources: ecology and community-based sustainable development. Belhaven Press, London

Chandra G (2007) Entrepreneurship in beel fisheries of Assam: a case study of Haribhanga beel. In: Kohli MPS (ed) Ecology and fisheries in wetlands in India. Indian Society of Fisheries Professionals, Mumbai, pp 173–179

Chandra G (2009) Fisher management regimes and fisheries governance in floodplain wetlands of Assam. Available at SSRN: https://ssrn.com/abstract=2086483

Chandra G (2010) Impact of adoption of pen culture technology on well-being of fishers of Haribhanga wetland in Assam. Indian Res J Ext Educ 10(2):61–65

Chandra G (2011) Management regimes and institutional arrangement in floodplain wetlands fisheries of Assam: an evaluation. Indian J Ext Educ 47(1–2):27–33

Chandra G (2014) Management regime and its impact on the wetland fisheries management in Assam. J Inland Fish Soc India 46(1):62–68
Chandra G, Bhattacharyya U (2016) Institutions and governance in fisheries of Indian Brahmaputra River Basin. J AgriSearch 3(1):51–56
Chandra G, Das BK (2019) Governance and institutional arrangement in floodplain wetlands fisheries of India: comparative study of Assam and Bihar. J Inland Fish Soc India 51(1):82–90
Chandra G, Ekka A (2015) Tenurial rights and institutional arrangement in floodplain wetland fisheries of Bihar. Conference: ISEE Golden Jubilee National Seminar. At: BHU, Varanasi. https://doi.org/10.13140/RG.2.1.1897.9283
Chandra G, Sharma AP, Sahu SK (2013) Impact of pen-culture technology on fish productivity of floodplain wetlands in Asom. Indian J Anim Sci 83(2):209–215
Delmas M, Young O (2009) Governance for the environment: new perspectives. Cambridge University Press, New York
FAO (2001) FIGIS topics and issues fact sheet: fisheries governance. Fishery policy and planning division. FAO, Rome, p 5
Jhingran VG (1991) Fish and fisheries of India, 3rd edn. Hindustan Publishing Corporation, New Delhi, p 727
Kabir GM, Yew TS, Noh KM, Hook LS (2011) Assessing fishers' empowerment in inland openwater fisheries in Bangladesh. Ecol Econ 70(11):2114–2123
Katiha PK, Vass KK, Shrivastava NP, Sharma VK, Sarkar S (2007) Inventory of reservoirs in Indian Indo-Gangetic Basin for fishery enhancements. Presented at 12th World Lake Conference, 28 October–3 November 2007, p 226
Katiha PK, Sharma AP, Chandra G (2013) Institutional arrangements in fisheries of Ganges River system. Aquat Ecosyst Health Manag 16(4):465–472
Marothia DK (2002) Alternative governance in common pool resource. In: Marothia DK (ed) Institutionalizing common pool resources. Concept Publishing Company, New Delhi
Mitchell B (1990) Integrated water management: international experiences and perspectives. Belhaven Press, London
Oakerson RJ (1992) Analyzing the commons: a framework. In: Bromley D et al (eds) Making the commons work: theory, practice and policy. ICS Press, Oakland
Ostrom E (1990) Governing the commons: the evolution of institutions for collective action. Cambridge University Press, Cambridge
Ostrom E, Gardner R, Walker J (1994) Rules, games, and common-pool resources. University of Michigan Press
Roy A, Das BK, Debnath S, Parida PK, Karnatak G, Borah S, Pandit A, Das AK, Bhattacharya BK, Bhattacharya S, Chandra G, Mondal K, Chakraborty S, Chandra P (2025) A conceptual framework for sustainable governance of self-recruiting small indigenous fishes in the lower Gangetic floodplain wetlands of Eastern India. Sustainability 17(5):2226
Rydin Y, Falleth E (2006) Networks and institutions in natural resource management. Edward Elgar Publishing, Cheltenham
Sarkar UK, Mishal P, Borah S, Karnatak G, Chandra G, Kumari S, Meena DK, Debnath D, Yengkokpam S, Das P, DebRoy P, Yadav AK, Aftabuddin M, Gogoi P, Pandit A, Bhattacharjya BK, Tayung T, Lianthuamluaia L, Das BK (2021) Status, potential, prospects, and issues of floodplain wetland fisheries in India: synthesis and review for sustainable management. Rev Fish Sci Aquac 29(1):1–32. https://doi.org/10.1080/23308249.2020.1779650
Sharma AP, Chandra G (2015) Issues and challenges for sustainable small-scale fisheries in inland fisheries sector of India, Chennai. https://doi.org/10.13140/2.1.2730.3200
Singh K (1994) Managing common pool resources: principles and case studies. Oxford University Press, New Delhi
Sinha M, Katiha PK (2002) Management of inland fisheries resources under different property regimes. In: Marothia DK (ed) Institutionalizing common Pool resources. Concept Publishing Company, New Delhi
Sneddon C, Fox C (2012) Inland capture fisheries and large river systems: a political economy of Mekong fisheries. J Agrar Change 12:279–299

Sugunan VV, Bhattacharjya BK (2000) Ecology and fisheries of beels in Assam. Bulletin 97. Central Inland Fisheries Research Institute, Barrackpore

Sugunan VV, Sinha M (2001) Sustainable capture and culture-based fisheries in fresh waters of India. In: Pandian TJ (ed) Sustainable Indian fisheries. National Academy of Agricultural Sciences, New Delhi, pp 43–70

Swiderska K, Row D, Siegele L, Grieg-Gran M (2008) The governance of nature and the nature of governance: policy that works for biodiversity and livelihoods. IIED, London

Vass K, Bhattacharjya BK (2008) Consultancy assignment on establishment of population of exotics common, grass and silver carps in the natural water bodies of Assam. Final Technical Report. Central Inland Fisheries Research Institute, Barrackpore

Vass KK, Das MK, Tyagi RK, Katiha PK, Samanta S, Srivastava NP, Bhattacharjya BK, Suresh VR, Pathak V, Chandra G, Debnath D, Gopal B (2011) Strategies for sustainable fisheries in the Indian part of the Ganga-Brahmaputra River Basins. Int J Ecol Environ Sci 37(4):157–218

Vinci GK, Jha BC, Bhaumik U, Mitra K (2003) Fisheries management of floodplain wetlands in India. Bulletin 125. Central Inland Fisheries Research Institute, Barrackpore

Werners SE, Flachner Z, Matczak P, Falaleeva M, Leemans R (2009) Exploring earth system governance: a case study of floodplain management along the Tisza River in Hungary. Glob Environ Change 19(4):503–511

Sectoral Conflicts in Small-Scale Fisheries and Inland Fisheries

5

Basanta Kumar Das, Pranaya Kumar Parida, Liton Paul, and Piyashi DebRoy

Abstract

India's small-scale inland fisheries sector is a vital source of livelihood, nutrition, and cultural identity for millions. However, it faces various sectoral conflicts that threaten its sustainability and the well-being of dependent communities. These conflicts arise from socio-economic disparities, environmental pressures, and governance challenges. Socio-economic conflicts stem from competition over limited resources, leading to disputes among fishers, aquaculture operators, and other stakeholders. The lack of equitable access to fishing grounds exacerbates tensions, particularly when industrial and semi-industrial operations encroach upon areas traditionally used by small-scale fishers. Environmental conflicts are fueled by overfishing, pollution, and habitat degradation, often resulting from unregulated activities and inadequate enforcement of conservation measures. Governance-related conflicts emerge due to weak regulatory frameworks, poor compliance, and policies that favor large-scale operations over small-scale fisheries. The absence of comprehensive legislation for inland water bodies further complicates management efforts. Addressing these conflicts requires a holistic approach that encompasses improved resource management, participatory governance, and conflict resolution mechanisms. The Food and Agriculture Organization's Voluntary Guidelines for Securing Sustainable Small-Scale Fisheries provide a framework for empowering small-scale fishers, promoting sustainable practices, and ensuring equitable access to resources. Implementing

In the tangled nets of inland fisheries and aquaculture, livelihoods clash over dwindling resources, entangling tradition in a tug-of-war for the future, where sectoral conflicts brew beneath the tranquil surface.

B. K. Das (✉) · P. K. Parida · L. Paul · P. DebRoy
ICAR–Central Inland Fisheries Research Institute (ICAR-CIFRI), Barrackpore, Kolkata, India

B. K. Das et al. (eds.), *Small-scale Inland Fisheries in India*, Sustainability Sciences in Asia and Africa, https://doi.org/10.1007/978-981-95-3849-2_5

these guidelines necessitates collaboration among stakeholders, investment in infrastructure, and recognition of the socio-cultural significance of small-scale fisheries. By embracing inclusive policies and fostering community engagement, India can mitigate sectoral conflicts and pave the way for a resilient and sustainable small-scale inland fisheries sector.

Keywords

Access · Equity · Resilience · Co-management · Livelihood resilience

5.1 Introduction

Sectoral conflicts within India's small-scale fisheries sector have emerged as a significant challenge, reflecting clashes of interests among diverse stakeholders engaged in fishing activities. These conflicts often stem from intense competition for finite resources, disputes over access to fishing grounds, and conflicting priorities within the sector. The intricate web of conflicts involves various subgroups within the small-scale fishing community, each vying for a share of limited resources (Murshed-e-Jahan et al. 2009; Spijkers et al. 2021).

The small-scale fisheries landscape in India is marked by numerous complexities, including disputes over access to specific fishing areas, disagreements on fishery management approaches, and varied goals related to conservation and economic development. These conflicts manifest in different forms, from verbal disputes to protests, physical altercations, and even violence. The repercussions extend beyond the immediate, negatively impacting the livelihoods of fishers, the sustainability of fish stocks, and the overall well-being of fishers' communities (Mills 2023). In the realm of fisheries governance in India, challenges arise from weak regulations and poor compliance, with a notable exception being coastal seasonal fishing bans or closures. Unfortunately, many other policies, such as fuel subsidies and incentives for deep-sea fishing, tend to prioritize production over conservation, disproportionately benefiting industrial and semi-industrial operations at the expense of the small-scale fisheries sector. Notably, inland water bodies in India face a legislative gap, further exacerbating the complexities surrounding fisheries management.

Understanding and addressing sectoral conflicts in small-scale fisheries management is pivotal for the sustainable development of India's fishing industry. The small-scale fisheries sector often finds itself marginalized, lacking the resources and influence to compete effectively with the industrial sector. These power dynamics result in conflicts over fishing resources and access to critical fishing grounds. The existing weak regulatory framework and compliance issues further intensify these conflicts. Here is a problem tree that outlines the issues surrounding sectoral conflicts in small-scale fisheries management in India (Fig. 5.1).

Assessing and comprehending these conflicts becomes imperative to promote sustainable resource management and safeguard the livelihoods of small-scale fishers. The recently introduced Voluntary Guidelines for Securing Sustainable Small-Scale Fisheries by the Food and Agriculture Organization (FAO) serve as a normative foundation for strengthening the small-scale fisheries sector and protecting its

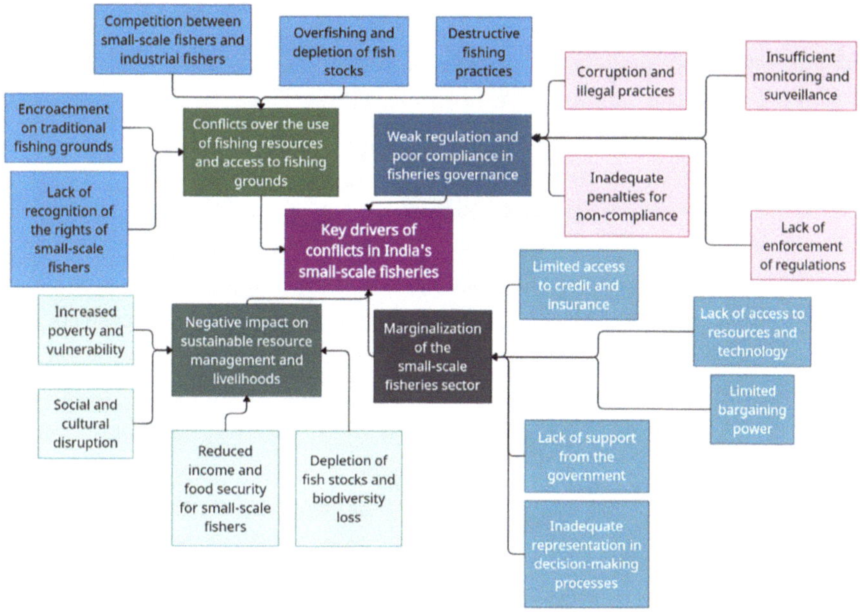

Fig. 5.1 Problem tree of key drivers of conflicts in inland small-scale fisheries

resources (Food and Agriculture Organization of the United Nations (FAO) 2024). Additionally, recognizing the diversity and values embedded in small-scale fisheries is crucial, necessitating a departure from overly simplified economic abstractions in policy formulations. To explore into the specific areas where conflicts emerge, a comprehensive understanding of the issues is essential. Utilizing a problem tree approach, the root causes and interconnections surrounding sectoral conflicts in small-scale fisheries management in India can be identified (Fig. 5.2).

Addressing these issues requires a multifaceted approach. From improving resource management practices and enhancing participatory governance to establishing effective conflict resolution mechanisms and supporting alternative livelihoods, the path to sustainable small-scale fisheries in India is intricate but essential. The subsequent sections will direct into each aspect, offering insights into strategies and interventions aimed at mitigating sectoral conflicts and fostering a balanced and sustainable future for small-scale fisheries in India (Spijkers et al. 2021).

5.2 Socio-economic Conflicts: Competition for Resources in Inland Fisheries

Socio-economic conflicts plaguing India's small-scale fisheries reveal competition for resources as a pivotal driver. This conflict unfolds as a complex interplay between traditional small-scale fishers, utilizing age-old methods and smaller boats, and

Fig. 5.2 Major components of sectoral conflicts

industrial fishers equipped with larger vessels and advanced technology (Seto et al. 2023). Still, small-scale fisheries inland are in a traditional format and need automation and grading to meet upcoming challenges.

5.2.1 Factors Leading to Social Conflict

(a) *Unequal Landscape:*

The disparities in resource capacities between traditional and industrial fishing become stark, amplifying the challenges faced by small-scale fishers. Industrial fishing vessels, with their higher capacity, swiftly deplete fish stocks, leaving meager catches for small-scale fishers who heavily rely on these resources for their subsistence and income. This stark contrast in technological prowess creates an asymmetry that not only affects the economic landscape but also perpetuates a cycle of disadvantage for small-scale fishers who struggle to compete in the face of industrial dominance.

(b) *Access to Fishing Grounds:*

Traditional fishing grounds, integral to the livelihoods of small-scale fishers, face encroachment by industrial operations. The competition over access to these preferred areas further constrains small-scale fishers, diminishing their catch and income potential. This encroachment not only limits the spatial range of small-scale fishers but also disrupts the delicate balance of ecosystems that have evolved over generations. It is a clash of methods, with traditional practices giving way to industrial intrusion, impacting both the environment and the cultural heritage of small-scale fishing communities.

(c) *Destructive Fishing Practices:*

The adoption of destructive industrial fishing practices, such as bottom trawling (Bag net), electrofishing, the use of explosives, and poisoning, not only degrades riverine ecosystems but also undermines the long-term sustainability of fish stocks. These methods disrupt aquatic habitats, reduce biodiversity, and compromise the ecological balance essential for healthy fisheries. Additionally, the growing issue of plastic pollution, particularly through abandoned or lost fishing gear (ghost gear), further exacerbates environmental degradation. The cumulative impact of these practices transcends mere resource competition, threatening the very foundation of inland fisheries and posing long-term risks to both small-scale and industrial fishers who depend on these shared ecosystems for their livelihoods.

(d) *Livelihoods at Stake:*

The consequences of this competition extend beyond mere economic losses. Small-scale fishers face significant income reduction, pushing them into poverty and food insecurity. The intergenerational impact is notable, as younger generations find it increasingly challenging to sustain a living from fishing due to resource depletion and heightened competition.

This economic vulnerability not only disrupts the immediate livelihoods of fishers but also jeopardizes the transmission of traditional knowledge and skills to future generations, risking the cultural continuity.

(e) *Vulnerability to Exploitation:*

The limited bargaining power of small-scale fishers makes them more susceptible to exploitation by middlemen or larger fishing companies. This vulnerability stems from their immediate dependence on fishing income, creating a precarious economic situation. This power dynamic exposes small-scale fishers to exploitation, perpetuating a cycle where external actors capitalize on their economic vulnerability. The lack of bargaining power not only affects the immediate economic transactions but also contributes to a broader pattern of systemic disadvantage.

(f) *Food Security*:

The competition for fish resources directly threatens the food security of fishing communities, where fish serves as a primary source of protein and sustenance. Depleted stocks exacerbate this challenge, putting vulnerable populations at risk.

Food security becomes a critical concern, especially for marginalized fishing villages in river banks heavily reliant on fish as a primary dietary component. The competition-induced scarcity amplifies nutritional vulnerabilities, impacting the overall well-being of these communities.

(g) *Social Tensions:*

Conflicts arising from competition between small-scale and industrial fishers can escalate into social tensions within communities. As competition intensifies, there is a potential for conflicts to escalate into physical altercations or even violence. The social fabric of fishing communities and culture based fishers are strained as competition unfolds, leading to strained relationships, mistrust, and in extreme cases, violence. The very communities that should ideally collaborate face internal divisions, hampering the collective well-being of the entire community.

5.2.2 Addressing the Social Conflict

To navigate the multifaceted challenges posed by competition, a strategic and holistic approach is imperative.

(a) *Sustainable Resource Management:*

Implementing comprehensive fisheries management plans, encompassing quotas, protected areas, and responsible fishing practices, is vital. Such measures can foster sustainable fish stocks that benefit all stakeholders. It goes beyond immediate conflict resolution, it lays the foundation for long-term ecological health, ensuring that fish stocks can replenish and support the livelihoods of fishers for generations to come.

(b) *Equitable Access Rights:*

Clearly defining and enforcing fishing rights based on traditional practices and ecological considerations is crucial. This step aims to level the playing field, ensuring that small-scale fishers retain access to their traditional fishing grounds. Equitable access rights not only address immediate conflicts over space but also acknowledge the historical and cultural significance of certain fishing grounds to specific communities. It's a step towards recognizing and respecting the traditional knowledge embedded in these spaces.

(c) *Participatory Governance:*

Including small-scale fishers in decision-making processes related to fisheries management is a key component of conflict resolution. This inclusivity empowers fishers, ensures their voices are heard, and addresses their specific needs. Participatory governance fosters a sense of ownership and responsibility among small-scale fishers. It acknowledges their expertise and insights, creating a collaborative approach to fisheries management that considers the diverse needs of all stakeholders.

(d) *Livelihood Diversification:*

Supporting alternative income sources, such as aquaculture, ecotourism, or craftwork, is integral to reducing dependence on overexploited fisheries. This diversification improves economic security for small-scale fishers. Livelihood diversification not only provides economic resilience for fishers but also contributes to the overall socio-economic stability of fishing communities. It enables them to weather the uncertainties of fluctuating fish stocks and market conditions.

(e) *Finding Common Ground:*

While competition for resources is inherent in shared spaces, acknowledging the complex socio-economic dimensions of this conflict paves the way for sustainable solutions. Striving for a future where both small-scale and industrial fisheries coexist alongside healthy inland ecosystems and secure livelihoods is the ultimate goal. Finding common ground requires a collective commitment to shared sustainability. It involves recognizing the interconnectedness of all stakeholders and fostering collaboration that transcends immediate conflicts, envisioning a future where the prosperity of small-scale and industrial fisheries aligns with the health of inland ecosystems.

5.3 Environmental Conflicts: A Shadow Over India's Small-Scale Inland Fisheries

In the serene embrace of India's inland water bodies, a disquieting reality unfolds. Environmental conflicts, intricately entwined with the socio-economic challenges previously discussed, loom large over the lives of small-scale fishers. These conflicts cast a somber shadow, posing a threat to the delicate equilibrium of these ecosystems. Let's delve into the three pivotal environmental conflicts that add complexity to the struggles faced by small-scale fisheries (Charles 1992a, b).

5.3.1 Contributors of Environmental Conflicts

(a) *Overfishing and Depletion:*

In the pursuit of survival and success, unregulated and unsustainable fishing practices contribute to overfishing, draining water bodies of vital fish species. The repercussions extend beyond immediate economic concerns, disrupting the intricate food web and jeopardizing the long-term sustainability of fisheries. This depletion not only impacts the livelihoods of current fishers but poses a threat to the well-being of future generations.

(b) *Ecosystem Destruction:*

The collateral damage inflicted by destructive fishing practices, such as bottom trawling and the use of harmful chemicals, extends far beyond the targeted fish. These practices wreak havoc on the entire ecosystem, damaging sensitive habitats and disrupting the balance of aquatic life. The destruction of mangroves, seagrass meadows, and other coastal ecosystems compounds the issue, depriving fish of essential nursery grounds and shelter, exacerbating the broader ecological degradation.

(c) *Pollution and Contamination:*

The tranquil waters of inland fisheries face contamination from various sources, including industrial waste, agricultural runoff, and untreated sewage. This pollution introduces toxins and pathogens that compromise the health of fish and disrupt the delicate food chains within these ecosystems. The pervasive plastic plague, with its microplastic components, further threatens the balance, as these particles accumulate in the food chain, posing health risks to both fish and humans.

5.3.2 Addressing Environmental Conflicts

A comprehensive and strategic approach is imperative to tackle the environmental conflicts that threaten the sustainability of India's inland water bodies and small-scale fisheries (Sievanen 2014).

(a) *Sustainable Fishing Practices:*

The implementation of quotas, closed seasons, and community-based management plans stands as a crucial step in regulating fishing activities. By promoting responsible resource use, these practices not only safeguard fish populations but contribute to the overall health of the aquatic ecosystems.

(b) *Public Awareness:*

Educating communities about the importance of sustainable fishing practices and responsible waste disposal fosters a sense of ownership. By instilling awareness, communities become active participants in the preservation of their ecosystems, contributing to the collective efforts needed for long-term sustainability. Acknowledging the severity of these environmental conflicts, the collective commitment to finding solutions becomes crucial. By working towards a future where India's inland water bodies once again teem with life, supporting vibrant fisheries, healthy ecosystems, and thriving communities, a step towards sustainable coexistence is taken. This is an ongoing and complex issue, and the nuances of its dynamics demand continuous research and collaboration with communities, government agencies, and NGOs.

5.4 Governance Conflicts: Unraveling the Threads

In the intricate tapestry of India's inland fisheries, sectoral conflicts manifest through a web of intertwined challenges, perpetuating discord within small-scale fisher communities. These conflicts are rooted in various factors that disrupt traditional practices, livelihoods, and ecological equilibrium (Kelkar 2023).

5.4.1 Key Reasons Behind These Sectoral Conflicts

(a) *Dam Construction and Downstream Impacts:*
- *Sardar Sarovar Dam (Narmada River):* The colossal construction of the Sardar Sarovar Dam led to the displacement of upstream fisher folk. Downstream, altered water flow patterns disrupted fish populations and traditional fishing practices, sparking conflicts among displaced communities, the government, and environmental groups (Sahoo et al. 2014).
- *Tehri Dam (Bhagirathi River):* Similar downstream impacts and issues of displacement arose with the Tehri Dam. Conflicts ensued over compensation, rehabilitation, and the sustainable management of the river ecosystem, involving fishers, environmental activists, and dam authorities.

(b) *Inter-State Water Sharing Disputes:*
- *Cauvery River Dispute (Karnataka and Tamil Nadu):* Long-standing conflicts stem from the allocation of Cauvery River water for irrigation, affecting water levels in reservoirs and rivers. This dispute impacts fish populations and the livelihoods of fishing communities, leading to tensions between Karnataka and Tamil Nadu (Vanham et al. 2011).
- *Krishna River Dispute (Andhra Pradesh and Telangana):* Tensions and legal battles emerged over water sharing from the Krishna River, affecting downstream ecosystems and fisheries. The conflict revolves around competing claims, impacting fisher livelihoods in Andhra Pradesh and Telangana.

(c) *Friction between Irrigation and Fisheries Departments:*
   - *Competition for Water Resources:* Prioritizing irrigation demands over fisheries needs results in reduced water availability for fish breeding and habitat maintenance. This competition fuels conflicts between the irrigation and fisheries departments.
   - *Lack of Coordination and Shared Management:* Siloed approaches and limited communication between departments lead to conflicting policies and unsustainable resource use, exacerbating tensions between irrigation and fisheries authorities.

(d) *Unsustainable Fishing Practices and Enforcement:*
   - *Government Policies* vs. *Traditional Practices:* Conflicts arise when government regulations clash with traditional fishing practices handed down through generations. The divergence between policy objectives and traditional knowledge creates confusion and resistance among fishers, triggering conflicts with enforcement agencies.
   - *Inadequate Monitoring and Corruption:* Weak enforcement capabilities and corruption within government agencies allow unsustainable fishing practices, such as overfishing and harmful gear usage, to persist. This lax enforcement contributes to tensions between authorities and responsible fishers.

These inherent conflicts highlight the need for comprehensive solutions that address the root causes, foster collaboration, and ensure the sustainable coexistence of diverse interests within India's inland fisheries sector.

5.5 Managing Sectoral Conflicts in Indian Inland Small-Scale Fisheries: A Beacon of Hope

5.5.1 Frameworks to Overcome the Issues

Conflicts over the use and management of fisheries resources are widespread. However, their causes, impacts, and potential management strategies remain inadequately understood. A holistic understanding of such conflicts is crucial. Charles (1992a, b) introduced three paradigms, conservation, rationalization, and social paradigms, which together form a useful framework for analyzing fisheries-related conflicts (Fig. 5.3). These paradigms align with the policy goals pursued by various user groups of fisheries resources. The three corners of the triangle represent the extreme manifestations of these paradigms, with each advocating a distinct policy focus. For instance, the conservation paradigm emphasizes the protection and sustainable use of resources as the primary objective.

Bellanger et al. (2020) identified two broad categories of frameworks for resolving cross-sectoral conflicts: the Social-Ecological System (SES) framework developed by Ostrom (2007, 2009), and transaction cost economics as introduced by Coase (1960). The allocation of property rights and the role of bargaining among user groups are central to resolving resource-use conflicts, as emphasized by Coase.

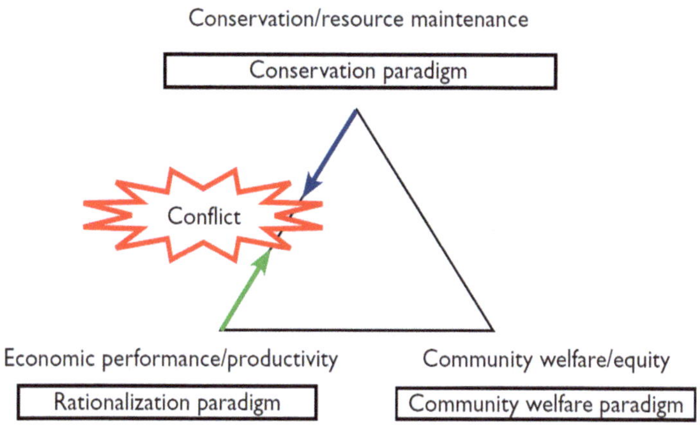

Fig. 5.3 Framework for understanding conflict

In contrast, the SES framework employs Ostrom's eight design principles for collective action to support robust institutions in complex resource systems (Ostrom 1990). This approach begins with community-based conceptualization of their socio-ecological reality, recognition of power dynamics, and an appreciation for system interdependencies. Key features include clearly defined user groups, resource boundaries, collective-choice arrangements, and equitable sharing of responsibilities and benefits.

India's inland small-scale fisheries sector characterized by overlapping jurisdictions and competing resource claims, demands innovative and participatory conflict resolution strategies (Thomson 2009; Spijkers et al. 2021). Three strategic approaches emerge as promising pathways, offering a vision for sustainable and inclusive governance:

(a) Collaborative Management Approaches
- Co-management: Empowering fishing communities to collaborate with government agencies fosters shared ownership and promotes sustainable resource use. Joint decision-making and mutual respect form the core of this model.
- Community-based Management: Granting community's autonomy to manage their fisheries within a regulatory framework reinforces traditional knowledge and incentivizes responsible stewardship of local ecosystems.

(b) Establishing Clear Rules and Regulations
- Resource Allocation: Defining access rights and quotas based on ecological and traditional considerations prevents overlap and ensures fairness.
- Prohibited Practices: Enforcing restrictions on destructive activities (e.g., overfishing, use of toxic chemicals, habitat degradation) are essential to maintain long-term ecosystem integrity.
- Transparent Monitoring and Enforcement: Regular monitoring and strong enforcement mechanisms ensure compliance and help sustain community trust in the system.

(c) Promoting Dialogue and Negotiation
- Stakeholder Forums: Platforms that facilitate open communication among small-scale fishers, industrial operators, government bodies, and civil society help bridge gaps and identify shared goals.
- Conflict Resolution Mechanisms: Transparent and easily accessible mechanisms for dispute resolution reduce tensions and promote peaceful coexistence.
- Capacity Building: Training stakeholders in conflict resolution, negotiation, and sustainable resource use builds long-term resilience and participatory governance.

Implementing these strategies requires tailored, transparent, and inclusive approaches, with a strong emphasis on local context. Long-term commitment and genuine stakeholder engagement are critical to success.

According to Stephenson et al. (2019), nine essential components underpin effective integrated fisheries management:

1. Recognition of the need for integrated management.
2. A shared vision among stakeholders and decision-makers.
3. Legal and institutional frameworks for coordination.
4. Inclusive stakeholder engagement processes.
5. Clearly articulated and comprehensive operational objectives.
6. Consideration of trade-offs and cumulative effects.
7. Flexibility to adapt to evolving conditions.
8. Mechanisms for regular review and refinement.
9. Adequate resources, leadership, and tools for implementation.

These elements provide a benchmark to assess how well existing or proposed frameworks address the dual goals of conflict resolution and sustainable governance.

Several frameworks have been proposed to address such complex issues, including ecosystem-based management (Pikitch et al. 2004), adaptive co-management (Armitage et al. 2008), integrated management (Stephenson et al. 2019), and collaborative governance (Emerson and Nabatchi 2015). The suitability of any specific approach depends on contextual factors, such as the available resources, stakeholder values, and their interrelationships. Often, a hybrid strategy, combining authoritative regulation, cooperative governance, and negotiated solutions, is required to address the deeply rooted and multidimensional nature of fisheries conflicts.

5.6 Conclusion

In the area of India's inland water bodies, tradition and livelihoods interlace with the challenges of socio-economic and environmental conflicts. The unequal playing field between small-scale and industrial fishers, exacerbated by technological disparities, threatens both economic stability and cultural heritage. However, amidst these challenges, collaborative and holistic approaches emerge as beacons of hope, emphasizing sustainable practices, participatory governance, and livelihood diversification. The vulnerability of small-scale fishers extends beyond economics, posing a threat to cultural continuity and environmental well-being. Environmental conflicts, including overfishing and pollution, demand a holistic approach involving sustainable practices, habitat protection, and public awareness. Transitioning to sectoral conflict management, nuanced solutions such as collaborative governance, clear regulations, and inclusive decision-making are highlighted. Case studies illustrate successful conflict resolution, emphasizing context-specific solutions, collaboration, and diversification. The concluding chapter envisions a future where empowered fishers, through strengthened rights and infrastructure investment, collaborate in managing specific zones. The synthesis of holistic strategies, including quotas, pollution control, and inclusive governance, forms a blueprint for restoring balance and shaping empowered communities and thriving ecosystems. In reflection, the challenges faced by India's inland small-scale fisheries necessitate a nuanced, integrated approach that transcends economic concerns to embrace ecological health and cultural continuity. The outlined solutions offer a roadmap for sustainable resource management, envisioning a future where India's inland fisheries thrive ecologically, economically, and culturally. The symphony of solutions, resonating through collaborative efforts, holds the potential to navigate the complex web of conflicts and usher in a harmonious era for India's inland small-scale fisheries.

Things to Remember

- Small-scale fisheries in India face sectoral conflicts: socio-economic, environmental, and governance.
- Conflicts stem from resource competition, habitat destruction, pollution, displacement, and livelihood loss.
- Integrated and participatory frameworks to resolve complex sectoral conflicts in India's inland small-scale fisheries,
- Proposed solutions: sustainable management, participatory governance, conflict resolution, alternative livelihoods.

References

Armitage D, Marschke M, Plummer R (2008) Adaptive co-management and the paradox of learning. Glob Environ Change 18:86–98. https://doi.org/10.1016/j.gloenvcha.2007.07.002

Bellanger M, Speir C, Blanchard F, Brooks K, Butler JR, Crosson S, Fonner R, Gourguet S, Holland DS, Kuikka S, Le Gallic B (2020) Addressing marine and coastal governance conflicts at the interface of multiple sectors and jurisdictions. Front Mar Sci 7:544440. https://doi.org/10.3389/fmars.2020.544440

Charles AT (1992a) Fishery conflicts. Mar Policy 16(5):379–393. https://doi.org/10.1016/0308-597X(92)90006-B

Charles AT (1992b) Fishery conflicts: a unified framework. Mar Policy 16(5):379–393

Coase RH (1960) The problem of social cost. In: J Law Econ, vol 3, pp 1–44. https://doi.org/10.1002/9780470752135.ch1

Food and Agriculture Organization of the United Nations (FAO) (2024) Voluntary guidelines for securing sustainable small-scale fisheries in the context of food security and poverty eradication. Rome: FAO. https://www.fao.org/3/a-i4356e.pdf

Emerson K, Nabatchi T (2015) Collaborative governance regimes. Georgetown University Press, Washington, DC

Kelkar N (2023) Inland capture fisheries, dam reservoirs, and protected areas for wildlife conservation in India: conflicts and ways forward. Rev Fish Biol Fish. https://doi.org/10.1007/s11160-023-09758-7

Murshed-e-Jahan K, Salayo ND, Kanagaratnam U (2009) Managing fisheries conflicts through communication planning: experience from inland fisheries of Bangladesh. Fish Res 99(2):112–122

Mills EN (2023) The politics of transnational fishers' movements. J Peasant Stud 50:665–690. https://doi.org/10.1080/03066150.2021.1975271

Ostrom E (2007) A diagnostic approach for going beyond panaceas. Proc Natl Acad Sci USA 104:15181–15187. https://doi.org/10.1073/pnas.0702288104

Ostrom E (2009) A general framework for analyzing sustainability of social-ecological systems. Science 325:419–422. https://doi.org/10.1126/science.1172133

Ostrom E (1990) Governing the commons: the evolution of institutions for collective action. Cambridge: Cambridge University Press. https://doi.org/10.1017/CBO9780511807763

Pikitch E, Santora C, Babcock EA, Bakun A, Bonfil R, Conover DO (2004) Ecosystem-based fishery management. Science 305:346–347. https://doi.org/10.1126/science.1098222

Sahoo T, Prakash U, Sahoo MM (2014) Sardar Sarovar dam controversy: a case study. Glob J Finance Manage 6(9):887–892

Seto K, Easterday K, Aheto D, Asiedu G, Sumaila UR, Gaynor K (2023) Evidence of spatial competition, over resource scarcity, as a primary driver of conflicts between small-scale and industrial fishers. Ecol Soc 28(1):art6. https://doi.org/10.5751/ES-13650-280106

Sievanen L (2014) How do small-scale fishers adapt to environmental variability? Lessons from Baja California, Sur, Mexico. Marit Stud 13(1):9. https://doi.org/10.1186/s40152-014-0009-2

Spijkers J, Merrie A, Wabnitz CCC, Osborne M, Mobjörk M, Bodin Ö, Selig ER, Le Billon P, Hendrix CS, Singh GG, Keys PW, Morrison TH (2021) Exploring the future of fishery conflict through narrative scenarios. One Earth 4(3):386–396. https://doi.org/10.1016/j.oneear.2021.02.004

Stephenson RL, Hobday AJ, Cvitanovic C, Alexander KA, Begg GA, Bustamante RH (2019) A practical framework for implementing and evaluating integrated management of marine activities. Ocean Coast Manage 177:127–138. https://doi.org/10.1016/j.ocecoaman.2019.04.008

Thomson K (2009) Development policies, state interventions and struggles for livelihood rights in coastal communities in Kerala, India: a case study of the Cochin clam fishery. Ocean Coast Manage 52(11):586–592. https://doi.org/10.1016/j.ocecoaman.2009.07.004

Vanham D, Weingartner R, Rauch W (2011) The Cauvery river basin in southern India: major challenges and possible solutions in the 21st century. Water Sci Technol 64(1):122–131

Inland Fisheries Resources of North-East India

6

Simanku Borah, Sullip Kumar Majhi,
and Basanta Kumar Das

Abstract

Inland fisheries are not only a vital source of livelihood and nutrition for millions of people, particularly among rural and indigenous communities, but also play a decisive role in the socio-economic fabric of the country in general and the North-eastern region in particular. This chapter provides a comprehensive overview of the inland fisheries resources of North-east India, highlighting their richness in terms of ichthyofaunal diversity, ornamental fisheries, present production, and potential.

Keywords

Fish diversity · Fisheries resource · Ornamental fisheries · Fish production

6.1 Introduction

India possesses a diverse range of fisheries resources, extending from deep seas to lakes, rivers, reservoirs, wetlands, derelict water bodies, ponds, tanks, canals, and more. Fisheries sector, holds paramount significance, plays a crucial role in contributing substantially to the national GDP, exports, food, livelihood, and nutritional security, as well as generating avenues for employment within the Indian economy. It remains a pivotal and rapidly expanding agricultural allied sector, maintaining its

Scientific intervention, judicious utilization, need-based ecological restoration, and conservation of resources should go hand in hand for a sustainable future

S. Borah (✉) · S. K. Majhi · B. K. Das
ICAR–Central Inland Fisheries Research Institute (ICAR-CIFRI), Barrackpore, Kolkata, India

B. K. Das et al. (eds.), *Small-scale Inland Fisheries in India*, Sustainability Sciences in Asia and Africa, https://doi.org/10.1007/978-981-95-3849-2_6

status as one of the most vital contributors to the country's economic growth. The North-eastern region of India, located between latitude 21°57′—29°30′ N and longitude 89°46′—97°30′ E, covering an area of 262,179 km^2, constitutes roughly 8% of India's geographical area (Das et al. 2023). Encompassing the eight sister states of Arunachal Pradesh, Assam, Manipur, Meghalaya, Mizoram, Nagaland, Sikkim, and Tripura, the North-eastern region of the country is globally recognized for its abundant biodiversity (WCMC 1998), and is a geographical and ecological mosaic of immense significance. Agricultural activities, including fisheries, play crucial role in the overall development of the North-eastern region. It is noteworthy that the fisheries sector is slowly thriving, making a substantial contribution to the region's GDP and is contributing towards generation of employment opportunities. The eco-climatic zones of freshwater fisheries in the region range from the warm waters in the plains and valleys of Assam, Manipur and Tripura to the upland cold waters of Arunachal Pradesh and Sikkim (Das et al. 2023).

Rich in biodiversity and cultural heritage, this region is home to a vast network of rivers, lakes, wetlands, floodplains, and hill streams that support a vibrant and diverse inland fisheries sector. Brahmaputra and Barak rivers, along with their numerous tributaries, criss-cross the region and are the principal rivers of North-east India. These rivers, along with associated wetlands, form the backbone of natural fisheries in the region (Borah et al. 2014). The unique topographical features, high rainfall, and extensive water resources contribute to the abundance and diversity of fish species, many of which are endemic to the region.

6.2 Fish Diversity

With 3188 indigenous fish species (ICAR-NBFGR 2022), India is one of the world's richest fish diversity repositories, and most of the country's ichthyofauna is concentrated in Western Ghats and North-east India, the two major biodiversity hotspots. Alongside various other diverse life forms, North-east India is graced with a wide array of fish species, flourishing in its intricate, distinctive, untouched, and pristine habitats. Research on the fish diversity of North-east India has been ongoing since the early part of the nineteenth century, with increased efforts in recent decades. Sinha (1994) compiled a list of 230 species of fish from the North-eastern part of India, reflecting the region's diverse aquatic life. Sen (2003) reported that the region harbours 291 fish species, while Goswami et al. (2012) listed 422 species of fish from North-eastern India, categorized into 133 genera and 38 families, highlighting the rich fish diversity of the region. They reported 325 fish species from Manipur, 311 from Assam, 271 species from Arunachal Pradesh, 231 from Meghalaya, 210 fish species from Nagaland, 202 from Mizoram, 199 species from Tripura, and 194 from Sikkim (Fig. 6.1), which highlights the rich fish diversity each state of the region possess. The rich and unique fish diversity of the region is by virtue of being

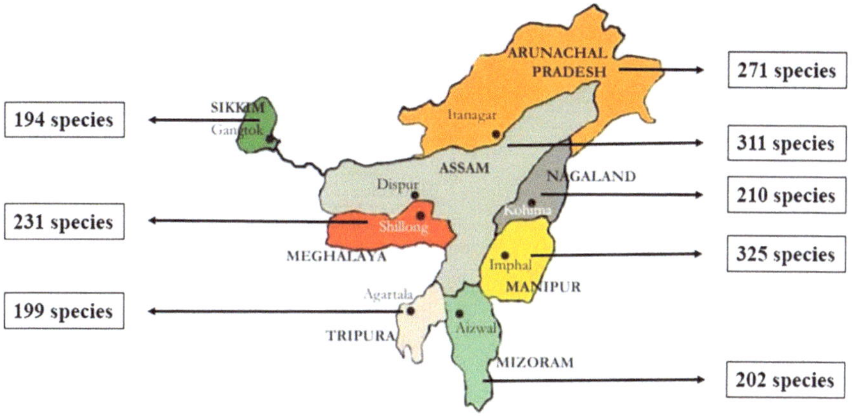

Fig. 6.1 Fish species reported from states of North-east India. (Source: Goswami et al. 2012)

the meeting point of two significant global hotspots for fish biodiversity, namely the Eastern Himalayas and the Indo-Burma (Sarma et al. 2018). Native fish diversity of the region plays a critical role in food and nutritional security, livelihood of fishers and contributes to the micro-economy.

North-east India is home to many endemic fish species which further magnifies its importance as a hotspot for biodiversity. Sen (2003) recorded 35 endemic species from the region. Recent study made by Sarma et al. (2018) revealed the presence of 138 endemic fish species from the region which is roughly 32.7% of all fish species (422 species) reported from North-east India. With 51 species or 37% of the endemic species, the family Cyprinidae has the highest endemicity followed by Nemacheilidae with 26 species (19%), and Sisoridae with 21 species (15%). Details of endemic species under different families reported from the region are given in Table 6.1.

Among states, Manipur has the highest concentration of endemic fish species (91 species), followed by Assam with 29 species, Arunachal Pradesh (26 species), Meghalaya (23 species), Nagaland (14 species), Mizoram (8 species), Tripura (7 species), and Sikkim (5 species) (Table 6.2) (Sarma et al. 2018). Further Sarma et al. (2018) also reported that 7.97% of the endemic fish species reported from the region are listed as Endangered (EN) as per IUCN, 24.64% are under Vulnerable (VU) category, 10.14% are Near Threatened (NT), 0.72% are Critically Endangered (CN), 34.06% are categorized as Least Concern (LC), 18.84% are Data Deficient (DD), and 3.62% are Not Evaluated (NE) (Fig. 6.2). Higher percentage of endemic species reported from North-east India is the result of longer periods of geographical isolation along with their intricate evolutionary history (Darlington Jr 1957). Fish diversity of the region is intricately woven into the cultural and traditional fabric of the indigenous communities.

Table 6.1 Family-wise count of endemic ichthyofauna reported from North-east India

Family	Number of endemic species
Cyprinidae	51
Nemacheilidae	26
Sisoridae	21
Bagridae	6
Cobitidae	5
Badidae	5
Amblycipitidae	4
Channidae	3
Siluridae	3
Balitoridae	3
Psilorhynchidae	2
Akysidae	2
Choudhuriidae	2
Mastacembelidae	1
Clupeidae	1
Synbranchidae	1
Osphronemidae	1
Botiidae	1

Source: (Sarma et al. 2018)

Table 6.2 State-wise count of endemic ichthyofauna in North-east India

State	Number of endemic species
Manipur (Irrawaddy-Chindwin basin)	91
Assam (Brahmaputra-Barak basin)	29
Arunachal Pradesh	26
Meghalaya	23
Nagaland	14
Mizoram	8
Tripura	7
Sikkim	5

Source: (Sarma et al. 2018)

6.3 Ornamental Fish Diversity of North-East India

Ornamental fishes exhibit a distinct colour or band pattern either during breeding season, early life stages, or throughout their entire life and has a unique or appealing appearance (Sharma et al. 2022). As documented by Goswami et al. (2012), 292 fish species reported from NE India are having ornamental value which is roughly 69.19% of the total ichthyofaunal diversity of the region. Raja et al. (2019) reported that 58 native and prized ornamental fish species are exported from North-east India

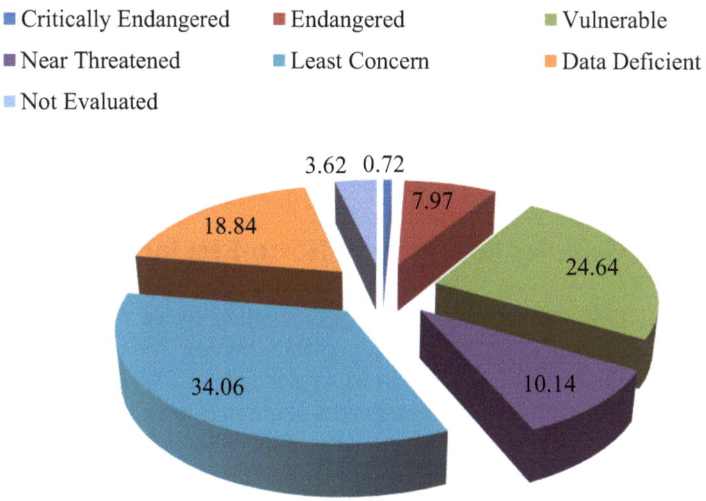

Fig. 6.2 Percentage contribution of endemic species reported from North-east India to different IUCN categories. (Source: Sarma et al. 2018)

and commands a high price in international trade. The importance and potential of this region in ornamental fish trade can be gauged from the fact that approximately 85% of India's ornamental fish species are found here. Majority of fish species with ornamental value in North-eastern region are members of the families Cyprinidae, Sisoridae, Cobitidae, Balitoridae, Bagridae, Ambassidae, Channidae, and Mastacembellidae (Goswami et al. 2012; Biswas et al. 2015). Snakeheads, or members of the Channidae family, are an important part of the global aquarium fish trade and the region is home to many species of snakeheads. There are roughly 17 Channidae species in the region, some of which are endemic, uncommon and challenging to locate. Perusal of literature shows that the state of Manipur has the most diverse variety of ornamental fishes (183 species) among North-eastern states, closely followed by Assam with 180 ornamental fish species, Arunachal Pradesh (159 species), Meghalaya (136 species), Nagaland (116 species), Mizoram (105 species), Sikkim (99 species), and Tripura (98 species) (Goswami et al. 2012). The data presented reveals the high potential of ornamental fisheries and trade in the region. However growing enthusiasm and awareness on ornamental fishes and their value has been causing rampant exploitation of these valued species, putting them at risk of getting endangered. Captive breeding and culture may make it feasible to produce these high-value fishes sustainably.

These ornamental fishes are collected and exported to South-east Asian and European countries via Kolkata. Ornamental fishes command a wide range of price depending on their appearance, color, availability, endemism etc. Price can range from few dollars to several thousand as in the case of *Channa barca*. Ornamental fisheries and small-scale fishers of NE India are intertwined. Ornamental fishes from the region are frequently harvested in live condition from the wild either for food or for aquarium trade by small-scale fishers using traditional fishing gears, primarily bamboo traps. Despite being highly priced and boosting a billion-dollar market, the irony lies in the fact that the primary collectors of these ornamental fishes gain very little remuneration for their efforts, while lion's share goes to the middleman.

6.4 Inland Fisheries Resources of North-East India

North-east India possess rich water resources including rivers, reservoirs, tanks, ponds, floodplain wetlands, and other derelict water bodies. The region has a vast network of rivers and canals, which extend to a length of 33,824.76 km (Anon. 2022). However, distribution of these resources is not uniform among states of the region. Assam, despite constituting nearly 30% of the geographical area of North-east India, accounts for a lion's share of region's water resources. With regard to capture fisheries, floodplain wetlands or beels are significant resources of Assam, covering around 100,815 ha (Sarkar et al. 2021). Manipur boasts the second-highest capture fisheries resources among the states in the region, with 29,161 ha of beels (Anon. 2022). Additionally, substantial areas under derelict water bodies with untapped potential are present in Assam, Tripura, and Arunachal Pradesh, which are currently underutilized (Das et al. 2023).

There is considerable potential for aquaculture development in ponds and tanks, especially in relatively low-altitude areas of the region, excluding Sikkim and Arunachal Pradesh. The fisheries resources in ponds and tanks are primarily concentrated in the North-eastern states of Assam, Tripura, and Manipur. The region also shows immense potential for creating impoundments and expanding water areas under reservoir fisheries. Total area under reservoirs in the region is 19,115.07 ha. Area under paddy-cum-fish culture practices is another typical resource of the region and is well-established in Arunachal Pradesh and Nagaland, with traditional paddy-cum-fish culture also notable in certain areas of Assam and Manipur (Das et al. 2023).

6.4.1 Rivers

North-eastern region of India comes under two major river basins, Brahmaputra-Barak and Chindwin-Irrawaddy. The region is rich in riverine resources and is intersected a number of major rivers and tributaries, in addition to numerous rivulets, hill streams, and canals having a combined length of 33,824.76 km (Anon. 2022) (Table 6.3). Brahmaputra is the largest as well as longest river in North-east India and is the lifeline of fisheries in the region (Borah et al. 2014). The major rivers of Arunachal Pradesh include Kameng, Dikrong, Ranganadi, Subansiri, Siang, Dibang, Lohit, Na-dihing, and Tirap, which are characterized by steep and turbulent flows, often navigating through deep gorges (Das et al. 2023). These rivers, along with

Table 6.3 Riverine resources of North-east India

States	Rivers and Canals (km) (Anon. 2022)	Potential yield (kg/km/year) (Das et al. 2017)
Arunachal Pradesh	11,157	200
Assam	10607.29	900 for River Brahmaputra and 450 for others
Manipur	1647	500
Meghalaya	4383.47	200
Mizoram	1750	200
Nagaland	1600	200
Sikkim	1600	150
Tripura	1080	400
Total	33824.76	

their tributaries, and numerous other canals, streams, and rivulets have a combined length of 11,157 km in the state (Anon. 2022). Brahmaputra is the major river draining through the middle of the entire state right from Sadiya to Dhubri, effectively dividing Assam into two parts, one towards the northern bank and another in its southern bank. Length of the river in Assam spans 640 km, receiving drainage from 42 significant tributaries both from the northern bank (27 tributaries), Subansiri, Ranganadi, Jia-bhoroli, Pagladiya, Beki, Manas, Saalbhanga, Sonkosh, and from the southern bank (15 tributaries), Lohit, Buri-dihing, Disang, Dikhow, Dhansiri, Kopili, Kulsi, Jinjiram (Bhattacharjya et al. 2017). The river is a rich source of fish diversity and harbours 141 fish species (Bhattacharjya et al. 2017). River Barak and its tributaries drain the southern part of Assam, comprising the districts of Cachar, Karimganj, Hailakandi, and the hill district of Dima Hasao. Combined length of rivers, streams, canals, in Assam is 10,607.29 km. Manipur boasts two major rivers, Barak and Imphal, with the Barak river and its tributaries dissecting the hills, while the Imphal river, also known as Manipur river in lower stretches, drains the valley districts (Das et al. 2023). Estimated total length of rivers and canals in Manipur is 1647 km (Anon. 2022). Meghalaya's extensive network of 4383.47 km of rivers and canals (Anon. 2022) mainly comprises north-flowing rivers such as Umiam, Digaru, Dudhnoi, Krishnai, and Jinjiram (all southern tributaries of River Brahmaputra), along with south-flowing rivers like Balat, Kynchiang, and Simsang (Das et al. 2023). A total of 21 rivers flows through Mizoram, including major north-flowing ones like Tlawng, Turirial, Tut, Tian, Tuichawng, Tuivawl, Teirei, Tuirini, and Serlui, which ultimately drain into Barak river and south-flowing rivers like Karnaphuli and Kolodoyne, draining into the Bay of Bengal through Bangladesh or Myanmar (Das et al. 2023). These rivers, along with tributaries, streams, and rivulets, collectively run for 1750 km, serving as a crucial fisheries resource for the hilly state (Anon. 2022). Jhanji, Dikhow, Diphu, Doyang, Dhansiri (all tributaries of River Brahmaputra), Barak, and Tizu are the main rivers of the state of Nagaland, with a combined length of 1600 km (Anon. 2022; Das et al. 2023) and act as critical habitats for wide variety of fish species. A recent study made by Ongh et al. (2025)

Fig. 6.3 Reservoirs of North-east India

revealed 69 fish species from the Nagaland stretch of Dhansiri River. Sikkim features two major rivers, Teesta and Rangeet, originating from glaciers located in north and western part of Sikkim (Das et al. 2023). Total length of rivers and canals in the state is around 1600 km (Anon. 2022). These rivers, along with their tributaries, support a diverse and rich fish fauna. Important rivers of Tripura include Longai, Juri, Deo, Manu, Dhalai (towards north), Khowai, Howrah (towards west), Gumti, Mahuri, and Feni (towards south), and along with other small streams, canals and rivulets, contribute to a combined length of 1080 km in the state (Anon. 2022; Das et al. 2023).

6.4.2 Reservoirs

India is blessed with a vast expanse of reservoirs of about 4.03 million ha numbering over 14,000. The reservoirs are predominantly located in the peninsular and central Indian states of Tamil Nadu, Karnataka, Andhra Pradesh, Kerala, Odisha, Maharashtra, Uttar Pradesh, and Madhya Pradesh, with a few small and medium reservoirs located in North-east India (Borah et al. 2023). Reservoirs of the region (Fig. 6.3) are developed primarily for hydropower generation, and only in the recent

Table 6.4 Reservoir resources of North-east India

State	Reservoirs (ha) (Anon. 2022)	Potential yield (kg/ha) (Das et al. 2017)
Arunachal Pradesh	226	500
Assam	1863	200
Manipur	2142	200 for medium and 500 for small
Meghalaya	716.73	500
Mizoram	8010	200
Nagaland	2258	200
Sikkim	850	500
Tripura	3049.34	200
Total	19115.07	

past fisheries garnering some attention. The state of Arunachal Pradesh has 4 small reservoirs; Assam has 2 medium reservoirs; Manipur has 5 small and 1 medium reservoir; while Meghalaya has 7 small reservoirs. The state of Mizoram has 3 small and 2 medium reservoirs; while Doyang and Dumbur (also known as Gumti) are the sole medium reservoirs located in Nagaland and Tripura respectively. The state of Sikkim has 3 small reservoirs (Anon. 2022). The reservoirs of the region have a very low fish yield as seen in case of Mizoram (30 kg/ha/year) and Manipur (50 kg/ha/year), which is far below their potential yield and needs immediate attention (Das et al. 2017). Only in the case of Doyang reservoir, Nagaland, fish yield of 158.5 kg/ha/year is slightly higher (Bhattacharjya et al. 2018). This has been possible due to implementation of fish stock enhancement protocols of ICAR-CIFRI by Department of Fisheries and Aquatic Resources, Government of Nagaland. The hilly terrain of the region provides ample scope for creation of reservoirs, which needs to be exploited. Development of reservoir fisheries can contribute immensely to fish production and towards livelihood and nutritional security of fishers and fish farmers of the region. Details of reservoir resources of the region are given in Table 6.4.

6.4.3 Floodplain Wetlands

The country is gifted with huge floodplain wetland resources to the tune of 0.5 million ha area, concentrated in its eastern and north eastern region (Sarkar et al. 2021). These resources are characterized by variable physical attributes, ecology, biota, and support rich diversity of fishes (Sarkar et al. 2021; Das et al. 2025). In the recent past, enhancement technologies and management measures are increasingly implemented in inland open waters with varying success (Yadav et al. 2021; Borah et al., 2022). Floodplain wetlands in the form of oxbow lakes/beels, tectonic lakes, swamps, marshes, derelict waters, river pool habitats, located along the floodplains of rivers are one of potential fisheries resources of India offering substantial scope and opportunities for fish production, employment and income to fishers (Sarkar et al. 2021). Among the North-eastern states, Assam has the highest area under floodplain wetlands (100,815 ha), followed by Manipur (29,161 ha). Floodplain

Table 6.5 Floodplain wetland resources of North-east India

State	Floodplain wetlands/lakes (ha) (Anon. 2022)	Potential yield (kg/ha) (Das et al. 2017)
Arunachal Pradesh	3277	1000
Assam	100,815	1500
Manipur	29,161	1000
Meghalaya	286.18	1500 for floodplain wetlands and 500 for lakes
Mizoram	10	
Nagaland	1110	
Sikkim		
Tripura	100	
Total	134759.18	

wetlands in Assam are primarily formed due to river meandering activity of Brahmaputra, Barak, and its tributaries and are locally called beels. The state of Arunachal Pradesh has 3277 ha of floodplain wetlands, located in the foothills. The state of Meghalaya has 286.18 ha of floodplain wetlands of which 225.23 ha are beels and 60.95 ha are lakes. In addition to these, the state has 66.94 ha of swamps and lowlying areas. Nagaland has 1110 ha of upland lakes, while area of upland lakes in Mizoram is 10 ha (Anon. 2022). Tripura has 100 ha of floodplain wetlands (Table 6.5). Besides the state also has the Rudrasagar Lake covering an area of 240 ha. Floodplain wetland resources has the potential to contribute immensely to food-basket of the region. Fish yield potential of these resources range from 1000–1500 kg/ha/year. However, majority of these resources are under-utilized and under-exploited till date, except for Assam and to some extent in Manipur. In case of Assam, implementation of management protocols like culture-based fisheries and fish stock enhancement protocols has led to considerable improvement in fish production from these resources to the tune of 704.60 kg/ha/year (Borah et al. 2022a) and 455.2 kg/ha/year (Yadav et al. 2021), compared to 206.4 kg/ha/year for unmanaged ones (Yadav et al. 2021). Implementation of culture-based fisheries in Bamuni beel of Assam by ICAR-CIFRI has even resulted in high fish production of 845 kg/ha/year (Borah et al. 2022b).

6.4.4 Ponds and Tanks

Ponds and tanks are one of the most vital fisheries resource of North-eastern region and forms the mainstay of culture fisheries. Total area under ponds and tanks in North-east India is 133,411.16 ha (Anon. 2022), with highest area in Assam followed by Tripura (Table 6.6). The area under ponds and tanks have gradually increased over the years and these resources have contributed immensely to fish production of the North-eastern states.

Table 6.6 Pond and tank resources of North-east Indian states

State	Ponds and tanks (ha) (Anon. 2022)
Arunachal Pradesh	2297
Assam	85,068
Manipur	1747.8
Meghalaya	4320.33
Mizoram	5506.04
Nagaland	3524
Sikkim	1579
Tripura	29368.99
Total	133411.16

Table 6.7 Fish production in North-eastern states of India

State	Fish production (thousand tonnes) (Anon. 2022)				
	2017–18	2018–19	2019–20	2020–21	2021–22
Arunachal Pradesh	4.0	5.0	5.0	5.0	5.0
Assam	327.0	331.0	373.0	393.0	417.0
Manipur	33.0	32.0	32.0	33.0	33.0
Meghalaya	12.0	13.0	14.0	16.0	18.0
Mizoram	8.0	7.0	7.0	5.0	5.0
Nagaland	9.0	9.0	9.0	9.0	9.0
Sikkim	0.4	0.4	0.4	0.4	0.4
Tripura	77.0	70.0	78.0	82.0	82.0
Total	470.4	467.4	518.4	543.4	569.4

6.5 Fish Production

The total fish production from North-eastern states of India during 2021–22 stands at 5.693 lakh tonnes (Table 6.7). The region has witnessed a steady growth in fish production over the years. Compared to 2017 18, fish production in the region has increased by 21.05% in 2021–22. The state of Assam is the leading producer of fish among the 8 states. Total fish production of Assam during 2021–22 stands at 4.17 lakh tonnes (Nandi et al. 2022), followed by Tripura with 0.82 lakh tonnes (Anon. 2022). However, despite possessing rich fisheries resources, contribution of the region to total fish production of the country is quite meagre. During 2021–22, the region contributed 4.70% of the inland fish production and 3.50% of the total fish production of the country (Fig. 6.4). High demand for fish coupled with a large fish-eating population base, the region has to depend on other states of the country to meet its requirements. Among these 8 states, Tripura has the highest per capita fish consumption (25.53 kg), followed by Manipur (18.25 kg), Assam (11.89 kg), Meghalaya (9 kg), which was reportedly higher than the national per capita consumption of 6.31 kg (Fig. 6.5). However, the other states, Nagaland (6.06 kg), Aruanchal Pradesh (3.65 kg), Mizoram (3.38 kg), Sikkim (1.16 kg), had a lower per

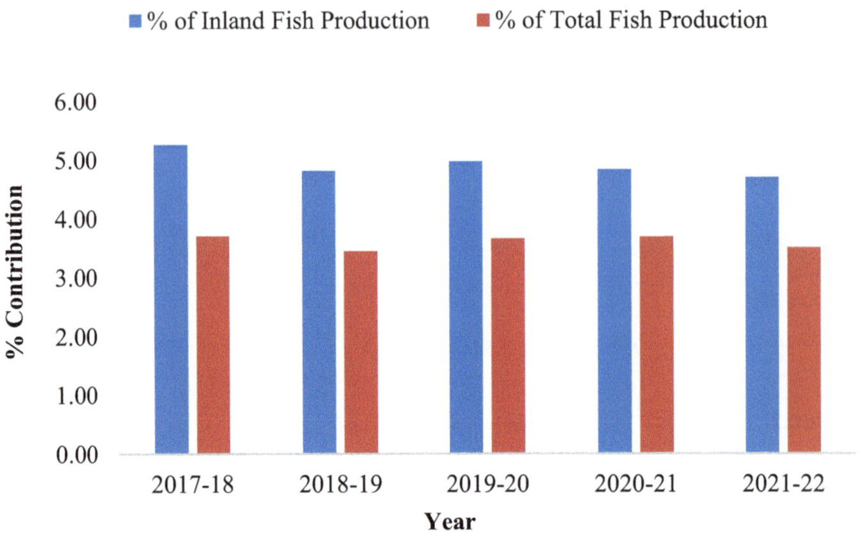

Fig. 6.4 Percentage contribution of North-eastern region to inland and total fish production of India

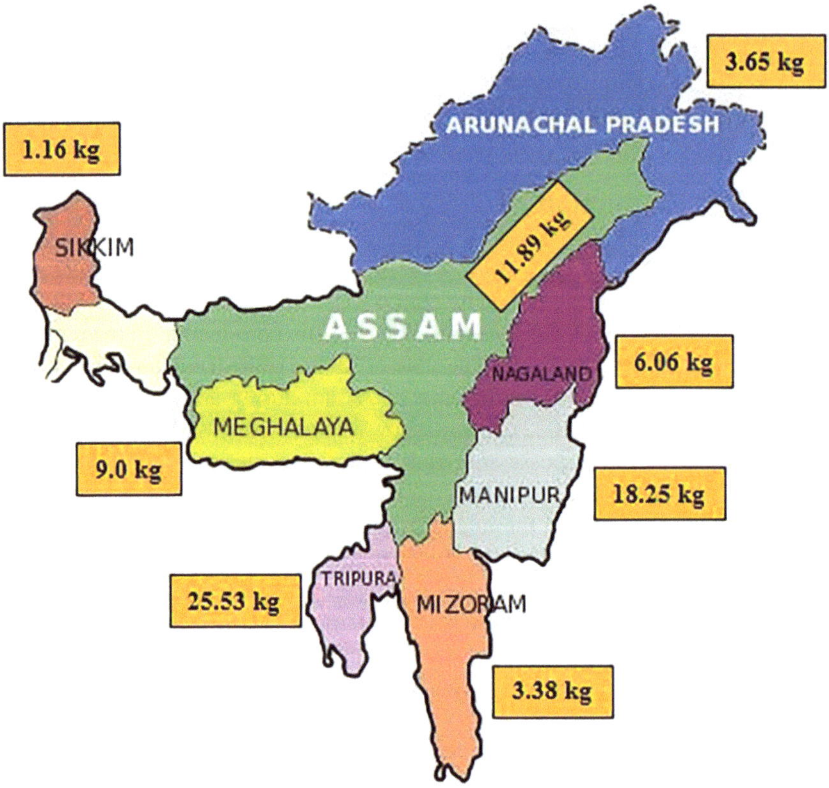

Fig. 6.5 Per capita fish consumption (kg) in North-eastern states

capita consumption in comparison to national average (Anon. 2022). Further, except for the three states of Tripura, Manipur, and Assam, the remaining states had a lower per capita consumption compared to the recommended standard of 11 kg.

6.6 Conclusion

With over 90% of the total population of the region being fish eaters, there lies huge scope for improving fish production and scope for employment opportunities in fisheries sector in North-east India. However, despite being rich in fisheries resources, the states of the region have not been able to fully tap the potential of fisheries sector and have been reliant on other parts of the country to meet the demand for fish. This is primarily due to the absence of a proper strategy and the under-utilization of available resources. The current scenario necessitates responsible utilization of fisheries resources through environmentally sound management programs. Reservoir fisheries along with paddy cum fish farming in hilly states and pond aquaculture along with floodplain wetland fisheries in plains needs greater attention to improve fish production of the region. Sustainable practices, including responsible use of natural resources, promotion of scientific fish farming to boost fish production, support for post-harvest activities, value addition, and the establishment of market linkages can pave the way for the region to transition from being an importer to an exporter in the near future.

Things to Remember

- North-east India is rich in fish diversity with 422 fish species categorized into 133 genera and 38 families, being reported from the region.
- 292 fish species reported from NE India are having ornamental value which is roughly 69.19% of the total fish diversity of the region.
- The region is rich in aquatic resources and floodplain wetlands are one of its most important fisheries resources.
- The region contributes 3.50% of the total fish production of the country, which is very meagre.
- There is huge demand for fish in NE India and there is an urgent need to sustainably utilize the fisheries resources of the region.

References

Anon (2022) Handbook on Fisheries Statistics 2022. Department of Fisheries, Ministry of Fisheries, Animal Husbandry & Dairying, Government of India, p 198

Bhattacharjya BK, Bhaumik U, Sharma AP (2017) Fish habitat and fisheries of Brahmaputra River in Assam, India. Aquat Ecosyst Health Manage 20(1–2):102–115

Bhattacharjya BK, Borah S, Das P, Baruah D, Odyuo NS, Kense K, Kuotsu N, Nakhro K, Das BK (2018) Fisheries resources of Nagaland: status and perspectives. Fish Chimes 38(1&2):66–73

Biswas SP, Singh ASK, Das JN (2015) Conservation and management of ornamental fish resources of North East India. J Aquac Res Dev 6(3):310

Borah S, Bhagawati K, Barman J, Chetia BR, Nath KD (2014) Present status of fisheries in Assam: an overview. Fish Chimes 34(1):58–59

Borah S, Sarkar UK, Das P, Yadav AK, Puthiyottil M, Pame D, Bhattacharjya BK, Das BK (2022a) Managing floodplain wetlands through culture-based fisheries (CBF) for livelihood security and sustainable development: a study from a biodiversity hotspot region of India. Arab J Geosci 15(14):1245

Borah S, Bhattacharjya BK, Das P, Yadav AK, Das BK, Sarkar UK (2022b) Exploring the potential of culture-based fisheries for augmenting income of tribal beel fishers: a success story from Assam. Indian Farm 72(7):23–26

Borah S, Das P, Yadav AK, Das SCS (2023) Sustainable management of Reservoir fisheries in Northeast India. Aqua Post:29–31

Darlington PJ Jr (1957) Zoogeography: the geographical distribution of animals. Museum of Comparative Zoology Harvard University, p 24

Das BK, Bhattacharjya BK, Borah S, Das P, Debnath D, Yengkokpam S, Yadav AK, Sharma N, Singh NS, Pandit A, Ekka A, Mishal P, Karnatak G, Kakati A, Saud BJ, Das SS (2017) Roadmap for development of open water fisheries in north-eastern states of India. ICAR-Central Inland Fisheries Research Institute, Barrackpore, p 120

Das BK, Bhattacharjya BK, Debnath D, Majhi SK (2023) Inland fisheries resources of India with special reference to Northeast India. In: Majhi SK, Debnath D, Yengkokpam S, Yadav AK (eds) Enclosure culture in inland openwaters for production enhancement. ICAR Central Inland Fisheries Research Institute, Barrackpore, pp 1–8

Das P, Landge AT, Nayak BB, Ramteke K, Das BK, Majhi SK, Yadav AK, Kumar R, Saikia P, Akter S, Borah S (2025) Fish community structure and environmental drivers in a tropical river wetland continuum: a study from Brahmaputra Basin in the Eastern Himalayan Region. Ecohydrol 18(3):e70041

Goswami UC, Basistha SK, Bora D, Shyamkumar K, Saikia B, Changsan K (2012) Fish diversity of North-east India, inclusive of the Himalayan and IndoBurma biodiversity hotspots zones: a checklist on their taxonomic status, economic importance, geographical distribution, present status and prevailing threats. Int J Biodiv Conserv 4(15):592–613

ICAR-NBFGR (2022) Annual Report 2022. ICAR-NBFGR, Lucknow, p 180

Nandi B, Priyadarshini P, Debnath R, Borah S, Yadav AK (2022) Fisheries in Assam: status and way forward. Biotica Res Today 4(11):768–770

Ongh SO, Landge AT, Ramteke K, Borah S, Barman J, Akter S, Yadav AK, Das P, Majhi SK, Chakraborty N, Das BK (2025) Fish community structure in accordance with environmental signatures in a tropical river ecosystem in the Eastern Himalayan ecoregion. Aquat Sci 87:40

Raja K, Aanand P, Padmavathy S, Sampathkumar JS (2019) Present and future market trends of Indian ornamental fish sector. Int J Fish Aquat Stud 7(2):6–15

Sarkar UK, Mishal P, Borah S, Karnatak G, Chandra G, Kumari S, Meena DK, Debnath D, Yengkokpam S, Das P, Debroy P, Yadav AK, Aftabuddin M, Gogoi P, Pandit A, Bhattacharjya BK, Tayung T, Lianthuamluaia L, Das BK (2021) Status, potential, prospects and issues of floodplain wetland fisheries in India: synthesis and review for sustainable management. Rev Fish Sci Aquac 29(1):1–32

Sarma D, Singh AK, Baruah D (2018) Checklist of endemic ichthyofauna of North-east India. Indian J Fish 65(3):1–15

Sen N (2003) Fish fauna of North East India with special reference to endemic and threatened species. Rec Zool Surv India 101(3–4):81–99

Sharma N, Sarkar UK, Yengkokpam S, Bordoloi D, Ray BC, Das BK (2022) Exploring potentials of snakehead fishes (Channidae) of Northeast India with emphasis on their ornamental value and conservation status. J Inland Fish Soc India 54(2):104–113

Sinha M (1994) Threatened coldwater fishes of North-Eastern Region of India. In: Threatened fishes of India. Natcon Publication No 4, Utter Pradesh, pp 173–176

WCMC (1998) Freshwater biodiversity: a preliminary global assessment—a document prepared for the 4th meeting of the conference of parties to the convention of biological diversity. World Conservation Monitoring Centre, Cambridge UK

Yadav AK, Das KK, Borah S, Das P, Bhattacharjya BK, Das BK (2021) Impact of fish stock enhancement on fish yield of floodplain wetlands in different agro-climatic zones of Assam India. Aquat Ecosyst Health Manag 24(4):54–63

Role of Primary Fishers' Co-operative Societies in the Management of Small-Scale Inland Fisheries

7

Piyashi DebRoy, D. C. Yateesh, Prateep Kumar Nayak, and Shreya Bhattacharya

Abstract

This chapter explores the structure, functions, and significance of Primary Fishermen Co-operative Societies (PFCS) in managing small-scale inland fisheries in India. It traces the evolution of the co-operative movement, the emergence of fisheries co-operatives, and their institutional foundations within the Indian socio-economic framework. Co-operatives, as one of the oldest forms of collective organization, aim to improve economic and social conditions through democratic governance and shared ownership. In India, PFCSs have become crucial in organizing inland fishers especially those from marginalized communities by facilitating access to credit, training, infrastructure, markets, and welfare schemes. Inland fisheries, which account for about 74.6% of national fish production, face critical challenges including environmental degradation, weak governance, fragmented support, and low investment. Using institutional data and case studies such as the Kalong Kapili Co-operative Society in Assam, this chapter highlights the importance of PFCSs in resource management, community empowerment, and sustainable fisheries governance.

Co-operatives Build a Better World—small-scale inland fishers, often hidden from the eyes of policymakers, are silent warriors against hidden hunger. Through transformative institutions like fishers' co-operatives, their voices must rise and be recognized in shaping sustainable future.

S. Bhattacharya

P. DebRoy (✉) · D. C. Yateesh
ICAR – Central Inland Fisheries Research Institute (ICAR-CIFRI), Barrackpore, Kolkata, India

P. K. Nayak
School of Environment, Enterprise and Development, Faculty of Environment, University of Waterloo, Waterloo, ON, Canada

B. K. Das et al. (eds.), *Small-scale Inland Fisheries in India*, Sustainability Sciences in Asia and Africa, https://doi.org/10.1007/978-981-95-3849-2_7

Keywords

Inland fisheries · Fisheries cooperatives · Rural livelihood · Policy intervention

7.1 Introduction

Co-operatives are one of the oldest socio-economic organizations in the world and were deemed very important from ancient times (Sapovadia 2004). The concept of group dynamism and collectiveness is perhaps the principle behind the successful existence of co-operatives in the world. Co-operative societies facilitate material advancement through united action which turns into self-reliance. Sociologists like Prince Kropotkin emphasized that organizational strength of the weak is the key to survival in a competitive situation. Robert Owen in England initiated several experiments for making co-operation an instrument of equitable economic order. The earliest record of a co-operative comes from Fenwick in Scotland which was found on 14 March 1761 in a local weavers' cottage with oatmeal distribution forming the Fenwick Weavers' Society. Subsequently, Co-operative activities began to expand across Western Europe, North America, and Japan around the mid-nineteenth century. A major milestone in this development was the launch of the modern Co-operative Movement in 1844 by the Rochdale Pioneers in Lancashire, England. This movement emerged as a response to the need for accessible, reliable alternatives to the low-quality and adulterated food and supplies available to ordinary people at the time (International Cooperative Alliance 2025). The surpluses were distributed in the community. The success of the historical movement spread out across the world to all sectors of economy. In 1862, Friedrich Wilhelm Raiffeisen and Franz Hermann Schultz-Delitsch first started the credit union model, which inspired the growth of financial co-operatives across the world (International Cooperative Alliance 2025).

A Co-operative is an autonomous association of persons united voluntarily to meet their common economic, social, and cultural needs and aspirations through a jointly owned and democratically controlled enterprise. The values of co-operatives are self-help, self-responsibility, democracy, equality, equity, and solidarity. The ethical values that are instilled in co-operative principles are honesty, openness, social responsibility, and caring for others. It allows people to take control of their economic future (Food and Agriculture Organization of the United Nations 2012). Co-operatives are not owned by shareholders, and the economic and social benefits of their activity stay in the communities where they are established. The profits are either reinvested in the enterprise or returned to the members. Globally, the Co-operative sector has around 1 billion members. Co-operatives employ around 250 million people in the world. According to the International Co-operative Alliance (2023), the world's top 300 co-operatives have a global turn-over of 2.2 trillion USD. The serious economic plight of many small-scale fishers and fishing communities has been acknowledged since at least the middle of the last century. In the early 1970s, FAO wrote that "The people engaged in these activities and their families continue, with few exceptions, to live at the margin of subsistence and human dignity" (FAO 1974, cited in Béné, 2003). Fisheries is important for the economic development of rural fishing populations. There are fishing co-operative societies around the world since historical times. One such example is the National Federation of Fisheries Co-operative Association (ZENGYOREN) in Japan which is a leading fisheries co-operative in Asia. Fisheries co-operatives established in India in the twentieth century. The subject matter of co-operatives is in the Concurrent List (The Concurrent List is one of the three lists in the 7th Schedule of the Constitution of India, which outlines the subjects on which laws can be made) which means that the Union Government and State Governments are entitled to execute power on fisheries and co-operation. A Co-operative Society in a particular state is registered under State Co-operative Act. However, when the activities of the Co-operative span in a state other than its native state, it is registered under the Central Co-operative Act. It has been studied that out of the many institutional management ways to manage natural resources, co-operative management is the most promising to achieve efficiency, sustainability, equity, and higher level of well-being, and is more acceptable politically and socially (Sapovadia 2004). Inland fisheries are often characterized by small-scale fisheries involving vulnerable and marginalized fishing communities. This chapter explores the role of co-operatives in inland fisheries, providing a historical perspective, identifying key constraints and offering suggestions to address them for the betterment of small-scale inland fishers

7.2 Evolution of the Co-operative Movement in India and the History of Fisheries Co-operatives

The co-operative movement in India originated in the late nineteenth century as a response to the exploitation of farmers by moneylenders, gaining momentum with the enactment of the Co-operative Credit Societies Act in 1904, which established a framework for democratic, member-owned credit institutions. The Co-operative Societies Act of 1912, expanding the scope to include consumer, marketing, and producer co-operatives, and allowing federations to be formed further strengthened this. Influential committees such as the Maclagan Committee (1914) and the Royal Commission on Agriculture (1928) provided key policy recommendations focused on training, structural development, and grassroots outreach. Post-independence, co-operatives were recognized as critical to India's development strategy, with the First Five-Year Plan highlighting their role in rural development. The All India Rural Credit Survey (1951–54) identified major gaps in rural finance and led to comprehensive reforms to strengthen coverage, training, and institutional support. Over time, co-operatives diversified into sectors such as agriculture, dairy, housing, and fisheries. Responding to changing economic conditions, the Government introduced the National Cooperative Policy in 2002 to ensure co-operatives operated as autonomous, transparent, and professionally managed entities, emphasizing governance reforms and legal uniformity.

A significant advancement came in 2021 with the creation of the Ministry of Cooperation, which was established to provide a dedicated administrative, legal, and policy framework to support the sector. Guided by the vision of "Sahkar se Samriddhi" (Prosperity through Cooperation), the ministry works to promote digitalization, enhance ease of doing business, and strengthen coordination with state governments. Today, India is home to the largest co-operative network in the world, with around 8.5 lakh registered societies and a membership of 290 million people, covering nearly 30% of the population, reflecting the movement's enduring importance in driving inclusive and sustainable development.

7.2.1 History of Fisheries Co-operatives

The concept of fisheries co-operatives in India originated as a means to organize and uplift traditional fishing communities, which were often economically marginalized and lacked access to resources and institutional support. The formation of the first fishermen's co-operative society in 1913 in Maharashtra, known as the Karla Machimar Co-operative Society, marked a significant step in applying co-operative principles collective ownership, democratic governance, and equitable benefit sharing to the fisheries sector. This model soon spread to other coastal states, with West Bengal and Tamil Nadu establishing their first fisheries co-operatives in 1918. Over the years, the movement evolved to meet diverse regional needs, resulting in varying structural models across states. Kerala adopting a two-tier system and Maharashtra a four-tier system. A growing consensus now favors a standardized three-tier structure with Primary Fishermen Co-operative Societies (PFCS) at the village level, Central Societies at the district level, and State Federations at the apex. Today, the national fisheries co-operative network includes 17 State Federations, 108 Central Societies, and over 11,000 PFCS, covering around 11.39 lakh members or 21% of active fishers. At the national level, the apex body is the National Federation of Fishermen's Co-operative Limited (FISHCOPFED), formed in 1980 and renamed in 1982. Functioning under the Ministry of Fisheries, Animal Husbandry and Dairying, FISHCOPFED's mission is to promote the fishery co-operative movement, assist in education and development, and act as a voice for co-operative interests. It is supported by schemes from the Department of Fisheries, National Cooperative Development Corporation, and National Fisheries Development Board, playing a vital role in empowering fishing communities and advancing rural development goals.

The presence of fisheries co-operative societies (in numbers) in various states of India is summarized in the following (Table 7.1):

Table 7.1 Geographic distribution of fisheries co-operative societies

State	State-level	Regional level	District level	Primary level	Registered fishers
Andhra Pradesh	1		13	2810	286,410
Arunachal Pradesh				11	230
Assam	1		2	520	90,000
Bihar	1		5	528	410,007
Chhattisgarh	1			1671	55,685
Goa				24	1503
Gujarat	1			701	94,893
Haryana				124	1276
Himachal Pradesh				73	7073
Jharkhand	1			515	32,635
Karnataka	1		2	653	482,115
Kerala	1			888	460,486
Madhya Pradesh	1			2734	96,817
Maharashtra	1	2	36	3775	332,636
Manipur	1	1	3	800	14,258
Meghalaya	1			128	611
Mizoram	1		1	47	1656
Nagaland	1			370	9234
Odisha	1	6		765	154,318
Punjab			1	9	95
Rajasthan	1		1	134	4130
Sikkim				8	230
Tamil Nadu	1		12	1475	761,521
Telangana	1		10	5200	364,357
Tripura	1			308	22,967
Uttar Pradesh	1		23	1125	54,521
Uttarakhand	1		1	167	634
West Bengal	1		20	1433	131,578
Andaman and Nicobar Islands	1		1	129	4149
Daman and Diu				19	3176
Lakshadweep				6	2910
Jammu and Kashmir			1	6	162
Puducherry	1		1	69	64,608
Ladakh				1	21
Total	24	9	133	27226	3,946,881

7.3 National Federation of Fishers Co-operative Limited (FISHCOPFED)

At the national level, the apex co-operative body of fisheries is the National Federation of Fishermen's Co-operative Limited (FISHCOPFED) under Ministry of Fisheries, Animal Husbandry and Dairying. The motto of FISHCOPFED is to promote and develop the fishery cooperative movement in India, to educate, guide, and assist fishers in their efforts to build up and expand the fishery cooperative sector and serve as an exponent of cooperative opinion in accordance with cooperative principles. It was established in 1980 as All India Federation of Fishermen Cooperatives and was renamed as National Federation of Fishers Cooperatives Ltd. in 1982. There are 17 members are from the State Federation of Fishermen's Co-operative Limited. There are also 108 Central Fishermen's Co-operative Societies who are registered under the Central Co-operative Act. FISHCOPFED is well promoted through schemes and funds of the Department of Fisheries of the Ministry of Fisheries, Animal Husbandry and Dairying, Department of Agriculture & Cooperation, Govt. of India, National Cooperative Development Cooperation and National Fisheries Development Board. The main functions of FISHCOPFED (Table 7.2) are as follows:

Table 7.2 Functions of FISHCOPFED

Function	Details
1. Input supply	Fisheries inputs and equipment to co-ops
2. Technology transfer	Through demo-cum-production projects (e.g., in Odisha)
3. Insurance	Implements PMBSY for 30.5 lakh fishers with Center-State cost-sharing
4. Pond/lake leasing	For fish breeding and affordable sale
5. Consultancy	To member co-operatives
6. Training	20,000+ fishers trained in fisheries and co-op subjects
7. Marketing	Via 8 regional offices; includes cold chains, exports, hotel supply
8. Research and studies	Leasing systems, market feasibility, gender integration, etc.
9. Publications	Disseminates scheme details, health info, recipes, news
10. Liaison activities	With Government, ICA, and international organizations
11. Service to members	Includes IT infrastructure and marketing support
12. Database creation	Maintains PFCS data (20,000 units, 31 lakh members)

The status of Fisheries Co-operatives in India is summarized in the following (Table 7.3):

Table 7.3 Status of fisheries co-operatives in India

Level	No. of units	Key functions
National level federation	1 (FISHCOPFED)	Apex policy, coordination, marketing, research, insurance, training
State level federation	21	Promote co-ops, input supply, marketing, technical guidance, export, liaison with State Govts
Regional level federation	7	Intermediate role, often support district-level activities
District level federation	132	Coordinate PFCSs, audit and capacity building
Primary (PFCS)	21,741	Member mobilisation, input supply, fish procurement/sale, training, infrastructure setup
Co-operative Fishermen membership	3,353,115	Ownership and participation, economic empowerment, resource pooling, access to government schemes, capacity building, ensuring fair trade practices

7.4 Structural and Functional Arrangement of PFCS

7.4.1 Primary Fishers Co-operative Society in Wetlands

The fishers of a wetland are registered under a District Fishermen's Co-operative. The managerial hierarchy of the Co-operative Society is shown here (Fig. 7.1).

Some PFCS who initially get audited under the umbrella of a district level co-operative federation gradually develop their own independent auditing also. An introductory statement of an audit report (Table 7.4), which shows the overall performance of the functioning of the co-operative society, in a wetland fisheries co-operative society is shown here.

The other account statements that are audited annually in PFCS include:

(i) Profit and Loss Appropriation Account
(ii) Profit and Loss Account
(iii) Trading Account
(iv) Cash Account
(v) Fish sales statements

The mandatory official activities carried out by PFCS are:

(i) Conducting regular General Meetings of the Office-bearers along with recording of the proceedings of the Meeting
(ii) Conducting an Annual General Meeting of the Office-bearers with the members of the Society and recording of the proceedings of the Meeting
(iii) Showcasing of Balance Sheets
(iv) Showcasing Trading Account
(v) Showcasing Profit and Loss Account
(vi) Showcasing Cash Account
Calculation of Closing Stock of Fish

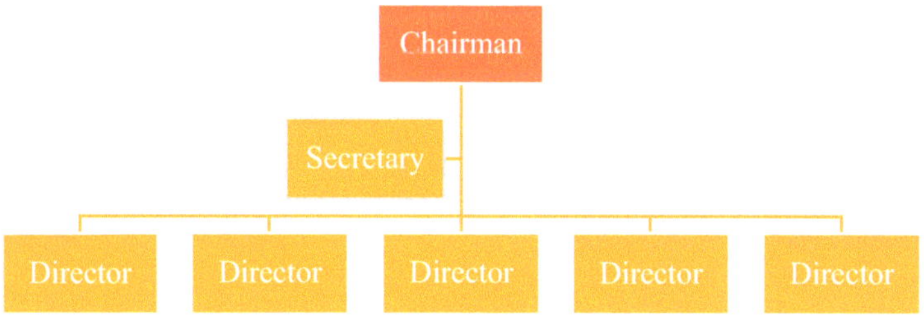

Fig. 7.1 Institutional hierarchy in a District level Fishermen Co-operative Ltd.

Table 7.4 Audit Summary Report of Primary Fishers' Co-operative Society

Introductory statement
Year of Audit:
Period of Audit:
1. Name of Society:
2. Address in full:
3. Registration No. and Date:
4. No. of members:
5. No. of Board Meetings held:
6. Date of last Annual General Meeting (in the presence of how many members):
7. Type of Society:
8. Name of the Affiliated Society Federation Ltd.:
9. Name of Board of Management (along with Designation and Date of Election):
10. Authorized Share Capital:
11. Working Capital:
12. Paid up Share Capital:
13. Statutory Funds:
14. Non-Statutory Funds:
15. Borrowings Refundable:
16. Grant and Subsidy:
17. Advance and Sundry Creditors:
18. Salary payable to Staff:
19. Balance with Banks:
20. Advance and Sundry Debtors:
21. Closing Stock of Fish:
22. Fixed Assets:
23. Gross Profit of Trading:
24. Net Profit:
25. Accumulated Loss:
26. Undistributed Profit:
27. Audit Fee for the Year:
28. Co-operative Education Fund:
29. Audit Classification:
30. Name and Designation of Previous Audit Officer:
31. Name and Designation of Previous Present Officer:
32. Place and Date of Completion of this Audit: Signature of Audit Officer

Source: (i) Audit Reports 2008–09, 2009–10 and 2010–11, 2012–13 to 2017–18, Dooma Primary Fishermen Co-operative Society Ltd., (ii) Audit Reports 2000–01 to 2011–12, Beledanga Barrackpur Fishermen's Co-operative Society Ltd. and (iii) Audit Reports 2017–18 and 2021–22, Sindrani Fishermen's Co-operative Society Ltd.

7.4.2 Role of Primary Fishers Co-operatives in Inland Fisheries

Inland fisheries play a vital role in India's fisheries sector, both in terms of production and employment generation. Accounting for approximately 74.6% of the country's total fish production, inland fisheries have emerged as a cornerstone of India's blue economy, particularly in the context of food security, rural livelihoods, and socio-economic development. The inland fisheries resource base is vast and diverse, encompassing rivers, reservoirs, wetlands, lakes, floodplain systems, estuarine zones, canals, and derelict water bodies, offering enormous potential for sustainable exploitation. According to the Department of Fisheries (2022), the sector provides direct and indirect employment to nearly 23.12 million people, many of whom are from socio-economically weaker sections, including tribal and Scheduled Caste communities. Despite this impressive contribution, inland fisheries often remain underprioritized compared to marine fisheries, facing institutional neglect, poor investment, fragmented governance of transboundary rivers, and degradation of aquatic ecosystems due to pollution, habitat loss, and unregulated development. In this context, fisheries co-operatives become crucial instruments of empowerment and sustainability. These co-operatives help organize small-scale and artisanal fishers, particularly in remote and inland areas, giving them access to credit, training, infrastructure, technology, and welfare schemes. They also enable scientific management of resources through collective stocking, regulated harvesting, and conservation practices. Moreover, co-operatives amplify the voices of inland fishers in policy-making, social inclusion, and strengthen market linkages, helping ensure fair prices and livelihood security.

7.5 Empowering Rural Prosperity Through Inland Fisheries Cooperatives: The Kalong Kapili Success Story

Inland fisheries cooperatives have emerged as powerful agents of change, uplifting not only fishers but also the rural population at large. These cooperative ventures play a crucial role in transforming traditional fishing practices into sustainable and economically viable enterprises.

The following case study delves into the intricate details of a cooperative venture, showcasing how strategic planning, community involvement, and technology adoption have collectively contributed to the upliftment of fishers and the rural population in the Indian context.

The Kalong Kapili Cooperative Society stands as a beacon of rural upliftment through its dedicated efforts in agriculture, allied activities, and fish culture in Assam. Under the visionary leadership of Shri. Jyotish Talukdar, the society has thrived, with 105 members, including 45 empowering fisherwomen. Over three years, their strategic approach and commitment have resulted in a net profit of Rs. 85.68 lakhs, supported by funds from organizations like NABARD.

The society's multifaceted initiatives encompass seed and feed supply, welfare activities, awareness camps, and fish produce marketing. Notably, they've distributed solar pumps, provided insurance to over 1100 fishermen, issued 3740 Kisan Credit Cards, and established a Fish Seed Production Hatchery unit, generating revenue and benefiting thousands. Their strategies focus on sustainable practices, including introducing high-yield fish varieties, locally sourced feed, and integrated farming approaches.

The adoption of technology, such as freshwater prawn culture and innovative carp culture, has significantly increased farmers' income. The society's success has not gone unnoticed, earning them prestigious accolades and recognition from various platforms, including being named the Brand Ambassador of the Department of Fisheries, Assam. Representing the North East Region internationally and receiving accolades from NABARD and the Hon'ble Minister of Fisheries underscore the societal impact and recognition of Kalong Kapili's exemplary contributions.

Insights from the Story: Kalong Kapili's journey illuminates the transformative power of cooperative societies in uplifting rural communities. Their strategic vision, inclusive approach, and adoption of sustainable practices showcase the potential for synergizing traditional knowledge with modern technology. The society's ability to secure funds, drive innovation, and garner recognition highlights the importance of collaboration between government bodies and grassroots initiatives in achieving comprehensive rural development through fisheries cooperatives. The Kalong Kapili Success Story serves as an inspiring template for leveraging local resources, empowering women, and fostering sustainable agricultural practices to create a ripple effect of prosperity in similar regions.

7.6 Challenges Faced by Fishery Co-operatives

Fisher cooperatives across India face multifaceted challenges that hinder their growth, efficiency, and sustainability. Major constraints include institutional bottlenecks, such as delay in accessing timely finance, absence of proper training, lack of technical guidance, and poor infrastructural development (Pandya and Patel 2002; Bhaumik et al. 2005). Many cooperatives are also crippled by internal governance issues, such as dominance by a few individuals, lack of participatory decision-making, and conflict over benefit-sharing (De and Saha 2006; Thakur 2015). Environmental and operational constraints like high water pollution, frequent poaching, weed infestation, and fish diseases further complicate productivity (Bhaumik et al. 2005; Sharma et al. 2018). Additionally, marketing inefficiencies due to middlemen exploitation, low bargaining power, and lack of cold chain infrastructure restrict fair returns to the fishers (Nair et al. 2007; Ekramuddin 2017). Social structures, including caste dominance and limited women's participation, perpetuate inequality and exclusion within cooperatives (Thakur 2015). In areas such as Vasai, Bastar, and Southern Rajasthan, these problems are compounded by low literacy levels, ignorance of rules, and inadequate government support (Nair et al. 2007; Sharma et al. 2018). To ensure long-term sustainability, suggestions have been made to strengthen cooperative marketing, develop local processing units, promote community-based management, and encourage inclusive membership and gender equality within cooperatives (Chandrashekar 2014).

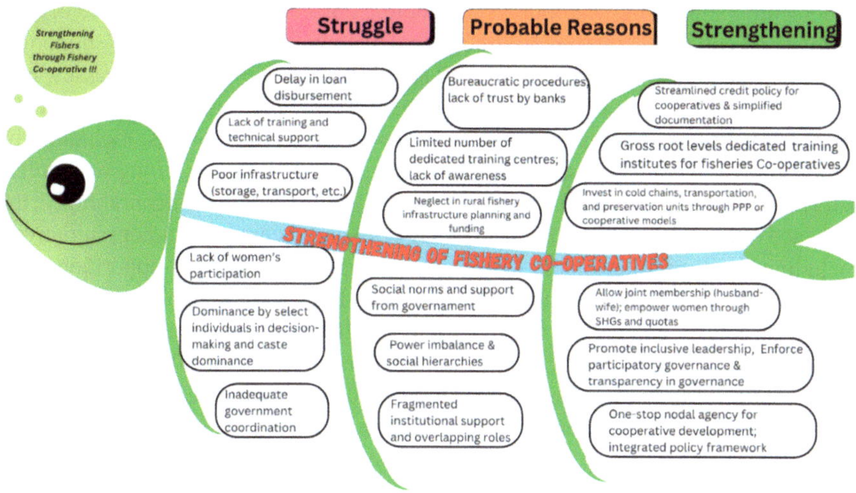

Fig. 7.2 Struggle to Strengthen Primary Fishers' Co-operative Societies

7.7 Future Perspectives for Fishery Co-operatives

Addressing these challenges necessitates strategic interventions and forward-thinking initiatives. Recommendations include organizing techno-financially feasible and professionally managed fishery cooperative societies. Establishing cooperative marketing infrastructure, supported by funding from central, state, and local bodies, is proposed as a crucial step forward. The increase in cooperative credit societies and government intervention to tackle marketing problems is emphasized. Additionally, the adoption of community-based co-management is underscored as a realistic and effective solution for sustaining aquatic resources and livelihoods, outlining a path towards the sustainable growth of fisheries cooperative societies (Fig. 7.2).

7.8 Conclusion

The study of Primary Fishermen Co-operative Societies (PFCS) within the framework of small-scale inland fisheries in India highlights their critical role in promoting sustainable livelihoods, inclusive governance, and participatory resource management. Co-operatives, grounded in principles of collective ownership and democratic decision-making, offer a structured institutional mechanism to organize economically marginalized fishing communities especially those dependent on rivers, reservoirs, floodplain wetlands, and other inland water bodies. While the co-operative model has a long-standing history in India and continues to evolve, it remains a vital tool for addressing systemic issues such as inadequate credit, poor infrastructure, market inefficiencies, and environmental degradation. The findings underscore that PFCSs are not merely economic units but community-based organizations that can integrate conservation with development when properly supported. However, the challenges are significant ranging from governance weaknesses, lack of capacity building, and socio-cultural barriers to environmental and market-related constraints. Addressing these requires targeted interventions, including the establishment of cold chains, training centers, and credit linkages, along with policy reforms that ensure participatory governance and inclusivity. Encouragingly, the presence of apex institutions like FISHCOPFED and recent government initiatives, including the formation of the Ministry of Cooperation, present opportunities to revitalize the co-operative sector. Strengthening PFCSs through a three-tier cooperative structure, enhancing women's participation, and promoting community-based co-management practices are essential for the future of inland fisheries. As India continues to pursue the goals of food security, environmental sustainability, and rural development, fishery co-operatives can serve as transformative institutions for balancing economic growth with ecological stewardship and social equity.

Points to Remember

- Fisheries co-operatives span across 24 states, with 133 district-level and 9 regional-level co-operatives, engaging a substantial 2,722,600 registered fishers.
- FISHCOPFED, established in 1980, plays a pivotal role in promoting and developing fisheries co-operatives in India, providing services from technology transfer to marketing.
- Guided by principles like self-help, democracy, and solidarity, co-operatives prioritize ethical values such as honesty and social responsibility to empower individuals and communities.
- Exemplifying the transformative impact of fisheries cooperatives, Kalong Kapili's journey highlights the synergy of traditional wisdom and modern technology in fostering sustainable rural development.
- Challenges faced by inland fisheries co-operatives, ranging from financial obstacles to marketing issues, can be overcome through strategic interventions, modern technology adoption, and policy shifts for sustainable growth.

References

Audit Report of Beledanga Barrackpur Fishermen's Co-operative Society Ltd (2000–2012) Gopalnagar, Bongaon, North 24 Parganas, West Bengal (Annual reports for 2000–2001 to 2011–2012)

Audit Report of CADP Co-operative Society Ltd (2017–2018) Sindrani, Barojaguly, Haringhata Block, Nadia, West Bengal

Audit Report of Dooma Fishermen's Co-operative Society Ltd (2008–2019) Jhowdhanga, Gaighata, North 24 Parganas, West Bengal (Annual reports for 2008–2009 to 2018–2019)

Audit Report of Sindrani Fishermen's Co-operative Society Ltd (2021–2022) Sindrani, Bagdah Block, North 24 Parganas, West Bengal

Béné C (2003) When fishery rhymes with poverty: a first step beyond the old paradigm on poverty in small-scale fisheries. World Development 31(6):949–975

Bhaumik U, Mittal IC, Das P, Paria T (2005) Constraints in enhancing fish production. Indian J Extn Educ 41:88–90

Chandrashekar BS (2014) Fishery co-operative societies in India: problems and prospects. Glob J Res Anal 3:92–94

De HK, Saha GS (2006) Constraints to community-based aquaculture research. Indian J Extn Educ 42:66–68

Ekramuddin M (2017) An overview of impact of small scale fish farming on socio-economic growth in Birbhum district, West Bengal, India. J Entomol Zool Stud 5:46–52

FAO (1974) Small-scale fisheries: report of the Technical consultation on small-scale fisheries. Food and Agriculture Organization of the United Nations, Rome

International Co-operative Alliance (2023) ICA annual report 2023. ICA

International Cooperative Alliance (2025) History of the cooperative movement. Available at: https://www.ica.coop/en/cooperatives/history-cooperativemovement (Accessed: 25 July 2025)

Nair SR, Pandey SK, Sharma A, Salim SS (2007) Performance of fishery cooperatives in India – an evaluation of primary societies in Thane District of Maharashtra. J Fish Econ Dev 11:16–27

Pandya RD, Patel BM (2002) Constraints in development of inland fish entrepreneurship. In: Abstracts, national seminar on entrepreneurship development in agriculture, p 65

Sapovadia V (2004) Fishermen co-operatives: a tool for socio-economic development. In: IIFET 2004 Japan proceedings, pp 2–9

Sharma P, Mazal B, Rani S, Sharma R, Masoodi Z (2018) Constraints faced by the fishermen in collection of fish and adoption of technology in Southern Rajasthan. J Pharmacogn Phytochem 7:2613–2615

Thakur A (2015) Assessment of status of the fishery cooperatives, SHGs and fishermen groups in Bastar District of Chhattisgarh MFSc (Aquaculture) thesis, Department of Fisheries, Indira Gandhi Krishi Vishwavidyalaya, Raipur

Food and Agriculture Organization of the United Nations (2012) The state of world fisheries and aquaculture 2012. FAO

Gender in Small-Scale Fisheries in India

8

Aparna Roy and Basanta Kumar Das

Abstract

Women play a crucial role in small-scale fisheries (SSF) in India. While men primarily engage in capture from and offshore activities, women contribute significantly through pre- and post-harvest operations, including net mending, fish processing/dry fishing, fish marketing, etc. Even though women play an important role in ensuring household food and nutritional security, they have less access and control over the resources. About 59% of the women in the fishermen community have no access to the extension services because of their ignorance. In decision making, the involvement of women is relatively higher in household affairs than in economic affairs. Women are also facing constraints like poor socio-economic status, poor educational status, poor health status, drudgery, lack of extension contact etc. Addressing these inequities through gender-sensitive policies, institutional support, and empowerment initiatives is crucial for promoting the inclusive and sustainable development of fisheries. Recognizing and mainstreaming women's contributions can strengthen livelihoods, resilience, and social equity in India's SSF sector.

Keywords

Women · Small-scale fisheries · Fishermen community · Livelihoods

Invisible Hands, Vital Roles: Recognizing Women in India's Small-Scale Fisheries

A. Roy (✉) · B. K. Das
ICAR – Central Inland Fisheries Research Institute (ICAR-CIFRI), Barrackpore, Kolkata, India

B. K. Das et al. (eds.), *Small-scale Inland Fisheries in India*, Sustainability Sciences in Asia and Africa, https://doi.org/10.1007/978-981-95-3849-2_8

8.1 Introduction

Small-scale fisheries refer to fishing activities conducted by individuals, families, or small groups using relatively simple and traditional fishing techniques. These fisheries often operate on a local or community scale and are characterized by limited use of capital-intensive equipment and technologies. Small-scale fisheries are crucial for livelihoods, food security, and cultural heritage in many coastal and inland communities around the world. Small-scale fisheries are often embedded in local communities and are closely tied to their social and cultural fabric. Fishing knowledge, practices, and traditions are passed down through generations. It contributes significantly to local economies, providing income and livelihoods for millions of people worldwide, especially in coastal areas and rural communities. Small-scale fisheries also provide fisheries play a vital role in ensuring food and nutritional security for many communities, providing a direct source of protein and essential nutrients. Women play a significant role in small-scale fisheries, participating in various activities such as harvesting, net mending, processing, marketing, and selling fish.

Dr. M.S. Swaminathan once said "Some historians believe that it was women who first domesticated crop plants and initiated the art and science of farming." Infact, women have been at the core of the entire food and nutrition system of society. Traditionally, they have been the backbone of households with multiple responsibilities of production and reproduction. However, with time, the roles and responsibilities of women have changed. The distinction between gender roles has been blurred with time. A woman's role in society is no more restricted to giving birth to children, raising them, and taking care of the family but this change is not equally enjoyed by women from all socio-cultural strata. Most of the women of the farming or fishing communities across the globe belong to the poorer side of the socio-economic continuum and are deprived of their share of rights and opportunities but that does not lessen the workload that they share along with the male members of the family. Moreover, due to the impacts of globalization and some indispensable laws of social evolution, rural men are shifting to urban areas in search of better employment opportunities leaving the women members back home to take care of the food and nutritional system of the household. As a result, the activities related to farming, animal rearing, fishing, and aquaculture are becoming the sole responsibilities of women. This is an emerging phenomenon that developing countries are witnessing today and is often termed as "feminization of the rural workforce." This trend is very much visible in the fisheries sector across the globe.

Even though data are not available on a sex-disaggregated basis, however, case studies indicate that women may make up almost 30% of all employment in fisheries, including primary and secondary activities. About 5.4 million women were employed in the primary sector of fisheries in 2008, according to data provided by FAO of 86 different countries. In major fish-producing nations, i.e., China and India women made up 21% and 24% respectively, among all fishermen and fish farmers (FAO 2010). Women make up nearly half (46%) of the fisheries workforce, and it is even higher in countries like Cambodia (57%), India (72%) and Nigeria (73%) it was reported by FAO, 2010. As per the Handbook of Fisheries Statistics, 2020 about 5.4 million people in India actively fish, including 1.6 are women. Although women frequently play important positions and are held accountable for a variety of duties in small-scale inland fisheries, they are frequently left out in the decision-making process (Moser 2013; Fitriana and Stacey 2012). Women carry fishing gear and boats, transfer fish caught to shore, and also do several other crucial tasks that enable their husbands to fish (Moser 2013). According to Fitriana and Stacey (2012), women are largely responsible for processing and selling fish at markets and devote 40–50% of their time to these tasks. However, because women's fishing job sometimes entails aiding their husbands or providing food and money to their families, it is frequently regarded as domestic or household work rather than fishing. Women are often not given a role in decisions about fisheries management despite their active involvement (Moser 2013; Fitriana and Stacey 2012).

Moser's Triple Role

Reproductive roles: Rarely considered as 'real work,' rarely paid, unrecognized and unappreciated (childbearing and caring, domestic tasks).

Productive roles: Activities that produce goods and services for consumption or trade.

Community roles: Participation in community and social activities (Participating in common interest groups, self-help groups, training and development programmes etc.)

Since women comprise a large section of the total fishing population and extend their share of responsibilities in certain activities more than the male counterparts of their communities, they are to be made an integral part of the development system. Without paying attention to their gender-specific needs and opportunities, full utilization of human resources would not be possible in the inland fisheries sector. Their multi-dimensional empowerment is required to channelize their potential to contribute to the food, nutritional, and livelihood security of the fishing communities. Therefore, in-depth studies with a gender analytical approach are needed in the inland fisheries sector. These studies would shed light on a number of topics, including gender roles, empowerment, obstacles, and potential future strategies for mainstreaming women in the fisheries sector. Some of the research projects and development activities conducted at CIFRI so far have explicitly focused on the gender arena. This chapter attempts to bring out some of the major issues of women in the inland fisheries sector in light of the findings and investigation made by ICAR-Central Inland Fisheries Research Institute, and thus presents a holistic scenario of the gender roles, challenges, and opportunities. The picture that we attempt to present through this document is expected to be of use to the policymakers and development agents working in this arena.

"Gender Role" in Inland Fisheries

The phrase "Gender Role" refers to the tasks that men and women carry out in various contexts, at various times, and within various cultures, classes, castes, ethnic groups, etc. Various factors influence men's and women's roles, and they differ from one socio-cultural system to the next. Moser's Triple Role framework assumes particular relevance in this context. This paradigm, created by Caroline Moser, aids scholars in comprehending the division of labor and time within the family and community. The concept of women's "triple role," or their triple function in production (agricultural/livestock/fisheries work), reproduction (home duties and childcare), and community affairs, is introduced by Moser.

Women have traditionally held several roles in India's inland fisheries sector. In order to maintain families and communities and secure food security through farming, fishing, and other related endeavours, the responsibilities of women are crucial. However, fishing is typically regarded as a man's occupation. Making and repairing nets and gear, assisting in the planning of fishing trips, and unloading and sorting the fish are all productive roles that women play in the fishing sector and that have a significant impact on household incomes and regional economies. They actively participate in catching small indigenous fish, clams, crabs, fish fry by using artisanal fishing gears such as trap and small mesh net such as mosquito net, push net and hand picking in various states of India like West Bengal, Assam, Bihar, Andhra Pradesh, Orissa, and the North Eastern states. Women in the fisherfolk community have played and will continue to play a significant part in the growth of inland fisheries, but their contributions are barely acknowledged.

Study on Gender Impact Assessment in Inland Open Waters Conducted in the Hooghly Estuary of West Bengal

- ICAR-CIFRI conducted a study on Gender Impact Assessment in Inland Open waters during 2012-14.
- 400 randomly selected fisher women in the Hooghly estuary of West Bengal were interviewed personally
- Findings of the study reflected on the socio-economic status and role of fisher women in livelihood security with special focus on their time allocation pattern, financial contribution in livelihood earning, roles in fisheries activities, etc.

The findings of the studies by ICAR-CIFRI, which directly or indirectly addressed the gender issues in inland fisheries, are presented in the document along the lines of this "triple role" concept. The major findings regarding the time allocation pattern of fisher-women, their productive and community roles, participation in fishery-related activities to earn a livelihood, and their access and control over resources are presented in the following paragraphs.

The study of demographic composition revealed that the overall male-female ratio in the upper stretch of the Hooghly estuary was 1000:1100, whereas in the lower stretch the same was 1000:1080. The Schedule Cast made up the majority of women (56%) while the Schedule Tribe made up around 18% of the population. About 15% of the population was made up of immigrants who came from Bangladesh. According to an analysis of the age distribution of fisherwomen, 66% of the women in this age range were involved in various fishery-related activities. The results showed that, because of diminishing returns and societal taboos, young women were not as interested in fishing activities (Roy and Bhaumik 2012).

It was found that women of the fishermen community in the Hooghly Estuary participate in many different kinds of occupations relating to fishing and other related fields. Women participated less in inland fishing activities in the upper stretch of the Hooghly estuary than in the lower stretch. Women were observed to actively participate in seed gathering, seed selling, fish grading, fish vending, fish drying, and fish processing in the lower stretch. They also worked as laborers and engaged in other activities that generate income. Roy and Bhaumik (2012 reported that altogether, 62% of women participated at a medium level in various fishing and income-generating activities. However, just 16% of women had high levels of participation, and the remaining 22% had low levels.

Livelihood diversification has become imperative in the present era to ensure and resilient livelihood of rural people. Dependence on a single source of living often becomes risky in the face of changes in natural and as well as man-made environments (Fig. 8.1). In Hooghly estuary they generate revenue from dressmaking, working as house maids, livestock rearing, wool knitting, nursing etc. The average monthly income earned by the fisherwomen was Rs.1450 and Rs. 1550 in freshwater gradient area and marine areas, respectively (Roy et al. 2017).

A study was conducted in the Sunderbans area of fishermen with special emphasis on women's role in dry fish making. In addition to taking care of their family or working in other fields, women in Sundarbans are involved in both commercial and artisanal fishing. Due to social stigma, women often conduct other fishery-related tasks to meet their home needs instead of going out to sea to fish. The fisheries activities of women include fish seed collection, labor work like fish sorting and grading, aquaculture in the household ponds, capturing or harvesting of fish from the water channels, canals, paddy fields, creeks with gill nets and various indigenous traps like Baki, Chokhia, Ghuni etc., fish vending/marketing, Net mending, etc.

8.2 Women Involved in Fish Drying Activities in the Sunderbans

Fisherwomen of Sundarbans have a major contribution to fish drying. The winter migratory bagnet fishery contributes a lot to the dry fish business. The temporary fishing camps are usually established for drying fish, which are known as Khuties. The women of the fisherfolk community are active in these camps to earn their livelihood. Sun drying is the simplest and most ancient method of preserving fish and is extensively practised in the lower Sundarbans area. This activity is from October to the middle of February. It was found in the study that by fish drying activities alone,

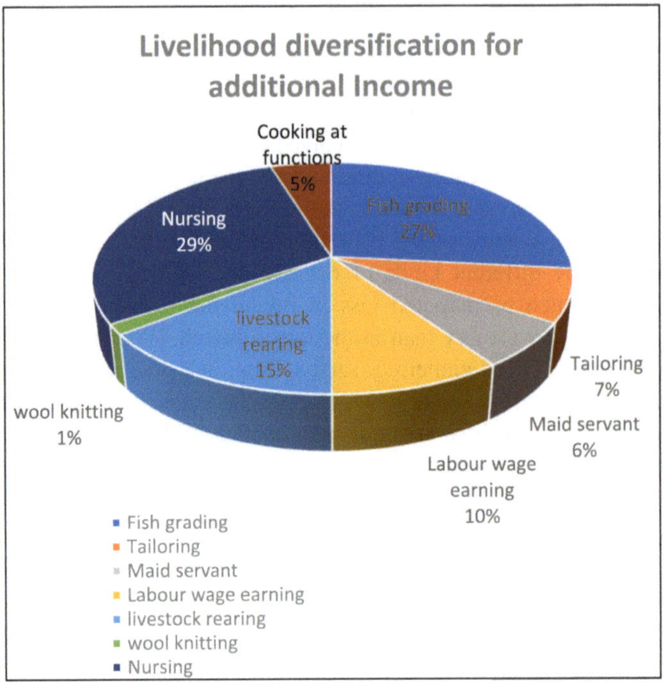

Fig. 8.1 Livelihood diversification for additional income

women fisherfolk of Sunderbans contribute 23% to the family income on average. With an hour-long lunch break, they work roughly eight hours a day. The majority of dry fish camp employees (63%) believe that their socioeconomic status is improved by working there (Roy et al. 2017).

For fish drying, men are mostly responsible for washing and cleaning the fish. Women have higher participation in fish drying and grading. Women also keep watch and guard during the drying of the fish. But both men and women are responsible for packaging the dry fish. In the Sunderbans, women also work as fish vendors.

In a project on "Population characteristics of small indigenous fishes (SIFs)," the roles of women fishermen were thoroughly analyzed during 2012–17. In the northern part of West Bengal in the Torsa River region (Totopara to Madarihaat), the involvement of women in fishing activities was higher. It was found that roughly 52% of the women in the fishing community are highly engaged in fish catching, while 48% are highly engaged in fish marketing (Fig. 8.2).

For earning a livelihood, women were working in tea gardens and raising and selling livestock. The major percentage of a fisherman's family's average income (62%) comes from fishing and fish harvesting on the Torsa River. This is followed by farming (7%) and livestock husbandry (17%) (Sinha et al. 2023) (Fig. 8.3).

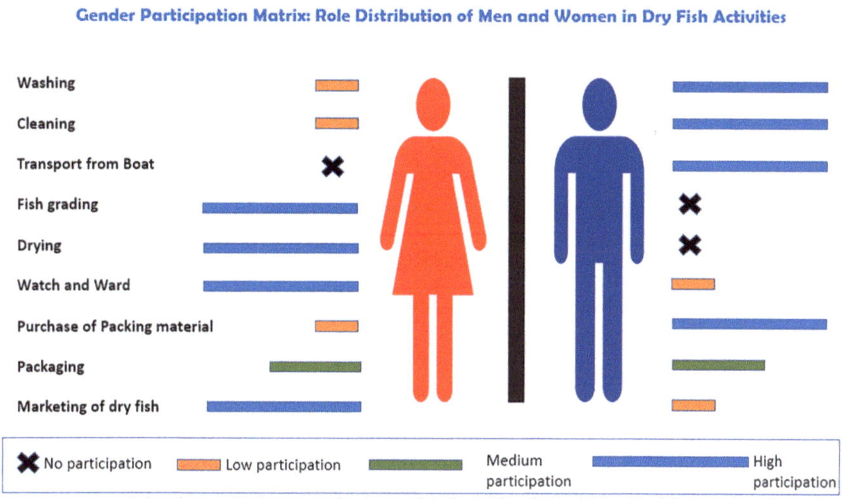

Fig. 8.2 Gender participation matrix

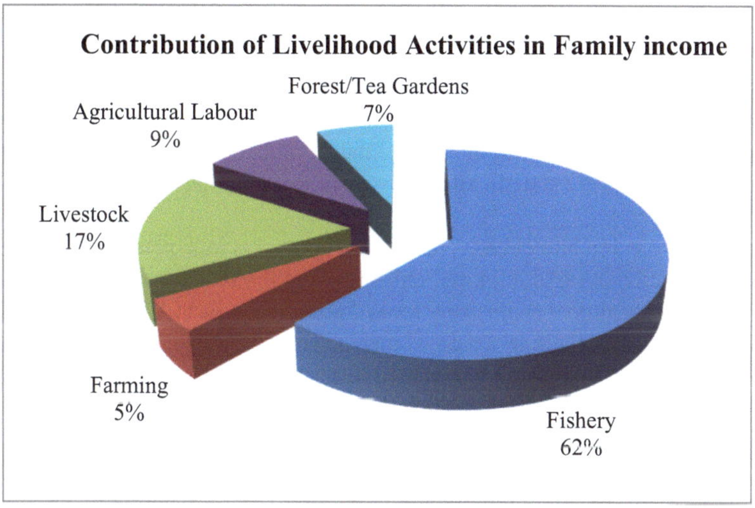

Fig. 8.3 Contribution of livelihood activities in family income

8.3 Status of Women's Empowerment

Empowerment in the true sense refers to the ability of an individual to take his or her own decisions by having enough control over own life, rights and opportunity to use the resources as per one's need, and power to decide on the use of the resources. Various studies focusing on the Gender aspect conducted at ICAR-CIFRI focused on this area and investigated fisherwomen's status regarding their access and control over rights, resources, assets, and decision-making in day-to-day life. In a patriarchal country like India even many of the elite class women do not have their say when it comes to major decisions in their lives. The scenario is even poorer for rural women of poor socio-economic strata. Nonetheless, the condition of fisherwomen is no exception. Fisherwomen often have limited control over resources.

Fisher Women's Access and Control Over Resources

In Sunderbans area of West Bengal, fisherwomen do not have equal access and control over resources. Girls are often denied access to general education. The fisherwomen do not have absolute control over most of the productive resources, like land, capital, mechanized equipment etc. Women of the fishermen's community do not have access and control over fishing craft, because of social taboos. Women (59%) of the fishermen community have no access to the extension services, because of their unawareness, but fisherwomen have short-term access and control over the natural resources like water and forest. Fisherwomen involved in fish seed collection have absolute control over the gears they use for fish seed collection.

A study conducted at the Lalganj and Devnibas villages of Frezarganj area of the Sunderbans area of West Bengal, where decision-making was studied by Roy and Bhaumik (2013) in detail. Two hundred women selected at random were interviewed personally. To ascertain the data regarding women's decision-making, process participatory tool like a decision matrix was used (Fig. 8.4)

Focus-group discussion and a community meeting were also conducted to collect general information. The ranks given by the respondents were then converted into percentage positions with the help of the formula Garrett's Ranking Technique. The obtained data were statistically analyzed. Tabular and percentage analyses were employed.

Women's role as a homemaker, though supportive, is indispensable for men to go for fishing. Activities like cooking, childcare, children's education, family health, and sanitation are almost exclusively looked after by women (Ashaletha et al. 2002). These are the home affairs in which women actively participate in decision-making. Eleven aspects are taken into account to describe the role of women in decision-making regarding home affairs. The highest participation of women was on the matter related to the preparation of food items (67%) and giving children supplementary food items (67%). They consult their husband regarding the age of children to send to school (68%), immunization of children (63%), whether loans should be taken (45%) and reproductive decisions (41%) (Fig. 8.5). Important areas which require attention and timely decision-making are the education level of children and the marriage of children women are hardly consulted. Her husband plays a major role in taking decision in this matter. For membership in any organization, the respondents also consult (40%) relatives, neighbours, and others.

The investigation shows that there are five ways fishermen generally take decisions about various practices related to fisheries. The majority of fishermen decide their own regarding fish catching of fish (44%), how fish should be marketed (36%),

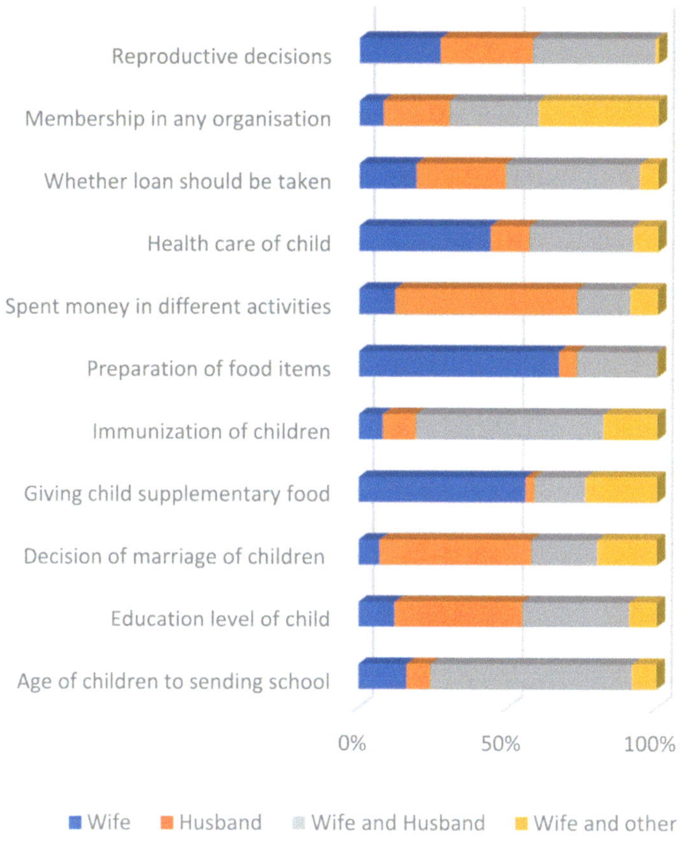

Fig. 8.4 Pattern of decision making regarding different areas related to home affairs

and where fish should be marketed (34%). In the case of fisherwomen, the majority decide to select the variety of pickle to be prepared (56%), followed by net repairing (46%). Since pickles are used for domestic consumption, women are conversant whether fish pickle or prawn pickle has better marketing scope or consumer preference. Again, repair the net on their own, so that the fishing program of the next day by the male members is not affected. With regard to decisions taken by husband and wife together, the highest percentage of responses was found in the case of fish seed collection (51%) and fish/prawn seed segregation (54%). From these findings, it is clear that the involvement of rural women of Sunderbans is relatively higher in those practices which are easier to housewives to perform.

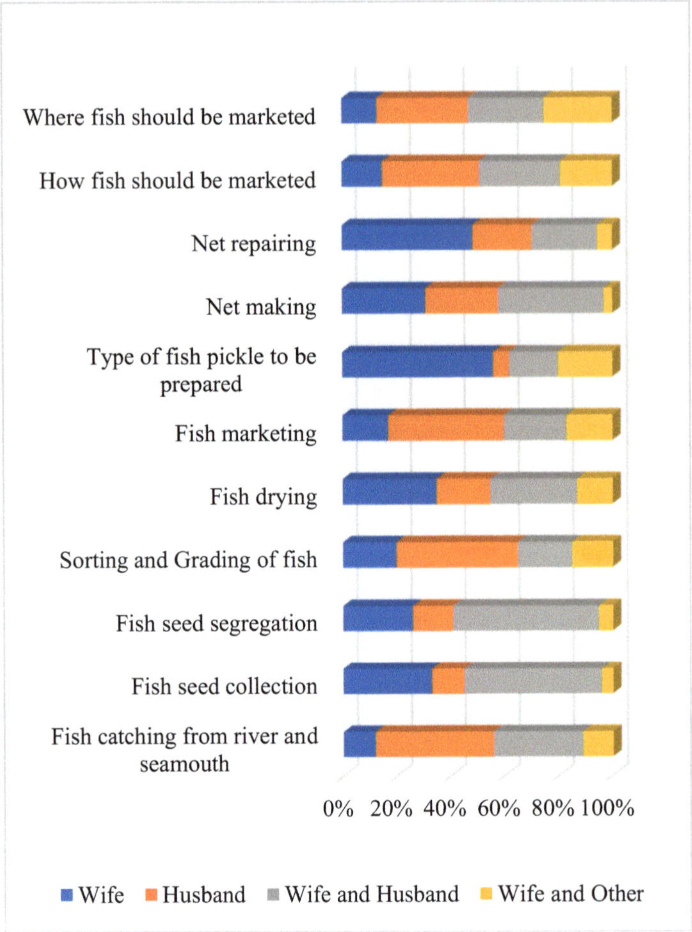

Fig. 8.5 Pattern of decision making regarding different areas related to fishery activities

8.4 Issues

Gender roles and dynamics play a significant role in small-scale fisheries in India, as in many other parts of the world. While the extent of gender influence can vary based on factors such as geographical location, culture, and socioeconomic conditions, certain general trends and issues are often observed in the context of gender and small-scale fisheries in India.

- ***Division of Labor***: Women in fishing communities are typically involved in various aspects of the fisheries value chain, including activities such as fish harvesting, processing, sorting, cleaning, and selling. Men tend to be more involved in fishing operations such as boat handling, net mending, and fishing.

- ***Access to Resources***: Access to fishing resources, credit, and technology can be unequal between genders. In some cases, women may have limited access to fishing gear, boats, and credit facilities. This hinders their ability to participate in fishing activities.
- ***Property Rights***: Traditional norms and legal frameworks sometimes result in unequal property rights, with men having more control over fishing assets and decision-making related to resource use.
- ***Decision-Making***: Decision-making power within fishing households and community-level management structures skewed towards men. This affects the allocation of resources and the prioritization of activities
- ***Vulnerabilities and Risks:*** Women in fishing communities may face specific vulnerabilities and risks, such as limited access to healthcare, exposure to harsh weather conditions, and a lack of social security measures.
- ***Climate Change Impact***: Climate change may disproportionately affect women in fishing communities, as they often have responsibilities related to securing food, water, and shelter for their families.

8.5 Constraints Perceived by Women of the Fishermen Community

Several challenges lie for rural women of India belonging to poor socio-economic classes. Their participation in productive work is often not recognized and ends up adding to their cumulative drudgery in day-to-day life. The situation of fisherwomen is no better. As Marxist theorists have stated that women's participation in agriculture and allied sectors is often distress-driven and not by their choice, the problems become even more severe. They are often pushed into these responsibilities without commensurate power and self-dependence. This has given rise to a discrepancy that is creating a problematic situation for women involved in several livelihood activities. Increased participation without consistent opportunities is leading to more workload and discrimination against women. Constraints for women in the fisheries sector are manifold. Some are directly visible like poverty, lack of education, knowledge and competency, wage difference, limited access to facilities and resources, limited access to input and credit, etc. Some of the constraints are deep-rooted in the collective psyche of the society and thus are less visible to the naked eye. These problems are poor participation in decision-making and a lack of control over assets and resources. Their roles in livelihood earning often go unnoticed and unrecognised. They end up providing labor without getting paid for it thus constituting a large portion of marginal workers in the country. Despite the significant roles that they play in livelihood earning, they hardly have control even over their own earnings. In the inland fisheries sector, women are thought of as "economically inactive" or merely play a supportive role as fishermen's wives. Their effort in both domestic and economic activities is not only financially unrewarding but also thankless. Different studies undertaken by ICAR-CIFRI have addressed the perceived constraints of women in fisheries sector which are presented under different heads in the following paragraph.

Poor Socio-economic Status:

The poor socio-economic status of women is often intertwined with issues of gender equity and inequality. In many societies, including India, gender norms and biases can exacerbate the challenges faced by women living in poverty.

Poor Educational Status

The lack of education of rural women is another obstacle that prevents them from taking part in decision-making, having control over resources, claiming their rights *etc.* In the study conducted on the Hooghly estuary, it was found that Illiteracy among fisherwomen was very prevalent. About 38% of the sample fell into the illiterate group, and only 12% had education up to a secondary level but the scenario is changing. Almost all the fisher women (94%) are sending their children to school irrespective of their gender but most of the fisher women perceived that less number of secondary schools in nearby areas discouraged them for sending their daughters for higher education.

Lack of Credit Facilities

One big problem of the fisherfolk community is indebtedness. The fishermen purchase their nets and boats by borrowing money from moneylenders cum fish traders. To pay back, the fisherfolk would sell their catch to the traders at throw-away prices. Fisherwomen often pay lease amounts to the government by debt against gold and silver ornaments of women are common in Bihar (Chaurs and Mauns) (Kumari 2016). The difficulty in availing credit from institutionalised sources is often rooted into lack of women's right and access to assets. Moreover, in most cases, fisherwomen were not aware of any type of facilities offered by the State as well as Central Government.

Lack of Extension Contact

The lack of extension contact in the context of gender equity refers to the insufficient or limited outreach, engagement, and support provided by agricultural extension services to address gender-specific needs and promote gender equity within the agriculture and rural development sector. Agricultural extension services are designed to provide farmers with information, resources, and guidance to improve their agricultural practices and livelihoods. However, when gender-specific considerations are overlooked, it can perpetuate inequalities and limit the potential of women farmers and rural women.

Technological and Market Constraints

Women are also less aware of technologies and market mechanisms, which makes them lag in modern fishery-related activities and utilizing the full market potential of pursuing entrepreneurial ventures. The information gap is wide between men and women, as women in particular find it difficult to access the existing socio-cultural structure (Ababa 2009). The women dry fish workers of Sunderbans undertake fish processing activities like drying fish, prawns and salt curing using traditional know-how. Modern methods like the use of chemicals to delay the process of decay in dry fish, mechanical drying, hygienic handling, and preparation of diversified fish products are unknown to the women working in the fish processing field in Sunderbans. Women in are Sundarbans do not have formal training in aquaculture. So, they are getting less produce the desired. The technological know-how of ornamental fish, carp, and prawn has not come to the access of rural women of Sunderbans (Roy et al. 2017).

Drudgery

Drudgery is physical and mental strain, agony, and hardship experienced by human beings (Moharana et al. 2016). The long hours of work in saline water for fish seed collection, women suffer from psychological tensions of being taken away by tidal waters or crocodiles. Most of the fisher women suffer from skin diseases (48%), followed by leukorrhea (41%), due to working in saline waters for long hours. A shortage of drinking water is a serious issue in the Sundarbans. Women of the fishermen community are largely responsible for water collection. They have to walk a long distance to collect drinking water.

Domestic Issues

Some domestic issues are deep-rooted in the socio-cultural system of the fisherwomen. These are often in the form of alcoholism and gambling among fishermen, unemployment, school dropout among children, child marriage, and dowry. The fishermen often spend all the money on liquor, leaving nothing for health and education expenses (Kumari 2016).

Poor Health Status

In the study conducted in the Hooghly estuary, health status of the fisherwomen was studied by collecting specific information on the mortality rate of mother and child during delivery, the distance of health care services, and the nutritional status of the fisherwomen. During the study period, 29 and 45 delivery cases were reported from fresh freshwater gradient stretch and the marine stretch, respectively. In the marine stretch, the mortality rates were 2.22% and 6.66% for mother and child, respectively (Roy et al. 2015). Malnutrition as well as unhygienic and unscientific delivery processes at home may be some of the reasons behind these deaths. However, no death case was reported from the freshwater gradient.

8.6 Gender Fairness in Fisheries Policies

The gender fairness of fisheries policy in India is an important consideration for ensuring the equitable and sustainable development of the sector. While efforts have been made to integrate gender considerations into fisheries policies, there are still areas that require further attention and improvement. Here's an overview of the gender fairness of fisheries policy in India:

- *Inclusion of gender in policies*: In recent years, there has been increased recognition of the importance of gender in fisheries policies. Some policies and programs acknowledge the roles of women in fisheries and highlight the need for their empowerment.
- *Capacity building*: Some policy initiatives include provisions for training and capacity-building programs targeted at women in fishing communities. These programs aim to enhance their skills and promote their active participation in various aspects of the sector.
- *Women's collectives*: Some policies support the formation of women's self-help groups and cooperatives, which provide women with a platform to collectively address their concerns, access resources, and engage in decision-making.
- *Gender-Disaggregated Data*: There is growing awareness about the need for gender-disaggregated data to better understand the roles, contributions, and challenges faced by women in the fisheries sector. This information can inform policy development and implementation.

8.7 Challenges and Areas for Improvement

- While gender considerations are mentioned in policies, the actual implementation of gender-responsive measures is limited. There's often a gap between policy intent and on-the-ground realities.
- Despite policy intentions, women still face challenges in accessing resources such as credit, training, and technology. Policies need to address these barriers effectively.
- Women are often underrepresented in decision-making bodies related to fisheries governance and management. Policies should focus on increasing women's participation in these forums.
- Some policies lack a comprehensive gender analysis that considers the specific needs, constraints, and opportunities of women in different segments of the fisheries value chain.
- Policies lack robust mechanisms for monitoring and evaluating the gender impacts of their implementation. Without proper evaluation, it's difficult to assess the effectiveness of gender-responsive measures.
- Cultural norms and gender stereotypes can hinder the effective implementation of gender-fair policies. Policies need to take these factors into account and address them sensitively.
- Gender considerations should be integrated into multiple levels of fisheries policy, including resource management, livelihood development, and community engagement.

8.8 Conclusion

In this chapter, the sharing of roles, responsibilities, decision-making between men and women, and constraints were thoroughly analyzed and which has brought out the present scenario of women in the fisheries sector. It reflected upon the fact that despite women's significant contribution to livelihood and nutritional security, their roles are often unrecognized, leading to unfair distribution of benefits between men and women. By default, most of the policies and technologies have been designed in favour of men. Even the women-centric policy interventions focus only on the more visible aspects of development and overlook the subtle dimensions of true empowerment like women's control over their life decisions and equitable access to resources and opportunities. Findings of several studies conducted at ICAR-CIFRI indicate that various practical and strategic constraints that women in the inland fisheries sector face which range from poor socioeconomics, education, and health status to inadequate social participation and mass media exposure, nutritional insecurity, lack of institutional linkage, and lack of access to credit facilities. These challenges, if left unattended, would give rise to deeper gender discrimination and devaluation of human resources. A lot of untapped opportunities lie for rural fisherwomen in enterprise building and catering to the overall livelihood improvement of the fishing community, along with the development of the fisheries sector. These opportunities must be harnessed to their fullest potential. Gender-sensitive extension approaches, training methodologies, and technologies are to be designed to extend the reach of research and development to the rural fisherwomen of the remotest parts of the country. Apart from that the research findings of the gender analytical studies are expected to guide the policy makers to design and implement gender mainstreaming policies in the inland fisheries sector. To achieve the ultimate goal of improving the fishing community's overall standard of living, programs must go beyond the fundamental requirements of women and focus on their sociocultural and economic empowerment.

Things to Remember

To enhance the gender fairness of fisheries policy in India, the following steps can be taken:

- *Strengthen Implementation*: Ensure that gender-responsive measures outlined in policies are effectively implemented, with clear guidelines and mechanisms for accountability.
- *Capacity Building*: Expand capacity-building programs for women in fisheries to improve their skills, knowledge, and leadership capabilities.
- *Data Collection:* Improve the collection and analysis of gender-disaggregated data to inform policy decisions and monitor progress.
- *Participation and Representation*: Promote the active participation of women in decision-making processes and increase their representation in fisheries governance bodies.

- *Awareness and Sensitization*: Conduct awareness campaigns and training to challenge gender stereotypes and promote gender equity within fishing communities and among policymakers.
- *Monitoring and Evaluation*: Develop mechanisms for continuous monitoring and evaluation of policies to assess their gender impacts and make necessary adjustments.
- *Collaboration*: Foster collaboration between government agencies, NGOs, research institutions, and fishing communities to ensure a holistic and gender-responsive approach to policy development and implementation.

By addressing these challenges and building on existing achievements, India can develop fisheries policies that are more equitable, inclusive, and supportive of gender fairness in the sector.

References

Ababa A (2009) Economic and social council. Africa Review Report on Mining (Summary). Economic Commission for Africa, p. 20

Ashaletha S, Ramachandra C, Sheela I, Diwan AD, Sathiadas R (2002) Changing roles of fisherwomen in India: issues and prospective. In: Kohli MP, Singh T, Ratna (eds) Women in fisheries. Indian Society of Fisheries Professionals, Mumbai, pp 21–43

FAO, GSO, MoP (2010) National gender profile of agricultural households, 2010. Report based on the 2008 Cambodia Socio-Economic Survey. Food and Agriculture Organization of the United Nations/General Statistics Office and Ministry of Planning, Rome/Phnom Penh

Fitriana R, Stacey N (2012) The role of women in the fishery sector of Pantar Island, Indonesia. Asian Fish Sci Spec Issue 25:159–175

Handbook of Statistics (2020) Department of Fisheries, Ministry of Fisheries, Animal Husbandry & Dairying, Government of India, New Delhi, 176. https://dof.gov.in/sites/default/files/2021-02/Final_Book.pdf

Kumari B (2016) Decreasing trend in participation of women in fish marketing in Patna, Bihar (India). Asian Fish Sci S 29:205–212

Moharana PC, Santra P, Singh DV, Kumar S, Goyal RK, Machiwal D, Yadav OP (2016) ICAR-Central Arid Zone Research Institute, Jodhpur: erosion processes and desertification in the Thar Desert of India. Proc Indian Natl Sci Acad 82:1117–1140. https://doi.org/10.16943/ptinsa/2016/48507

Moser C (2013) Participation by women fishers in Sekong River fishery management in Lao PDR. Interim report. CGIAR Challenge Program on Water and Food–Mekong, Vientiane. http://mekong.waterandfood.org/wp-content/uploads/Interim-Report-11-Mar-2013-Participation-by-Women-Fishers-in-Fishery-Management-in-SekongBasin-Lao-PDR.pdf

Roy A, Bhaumik U (2012) Women's role in fisherfolks' communities of Hooghly estuary. J Interacad 16(4a):1100–1106

Roy A, Bhaumik U (2013) Participation of women of Sundarbans in fisheries decision making. J Inland Fish Soc India 45(2):23–27

Roy A, Pandit A, Sharma AP, Bhaumik U, Majunder S, Biswas, DK (2015) Socioeconomic status and livelihood of fisher women of hooghly estuary. J Inland Fish Soc India 47(1):49–56

Roy A, Sharma AP, Bhaumik U, Pandit A, Singh SRK, Saha S et al (2017) Socio-economic features of womenfolk of Indian Sundarbans involved in fish drying. Indian J Ext Educ 53(2):142–146

Sinha A, Koushlesh SK, Kumari K, Chanu TN, Roy A, Gogoi P (2023) Diversity and threat to cold-water fishes of the Torsa River at the Terai Region of West Bengal India. In: Fisheries and aquaculture of the temperate Himalayas. Springer, Singapore, pp 61–79

Climate Change and Small-Scale Inland Fisheries (SSIF) in India: Impacts, Vulnerabilities, and Resilience

9

Mishal P., Gunjan Karnatak, Suman Kumari, Lianthuamluaia, U. K. Sarkar, and Basanta Kumar Das

Abstract

Climate change poses a significant threat to small-scale inland fisheries (SSIF) in India, impacting both aquatic ecosystems and the livelihoods of millions dependent on them. This chapter assesses the ecological and socio-economic vulnerabilities of SSIF across rivers, floodplain wetlands, and reservoirs, highlighting key indicators such as rising temperatures, altered rainfall patterns, increased frequency of extreme events, and changes in species composition and fish productivity. Case studies from the Ganga basin and other inland systems reveal significant declines in fish diversity, shifts in species distribution, habitat degradation, and increased fisher migration. The chapter discusses vulnerability assessment frameworks and methods, with emphasis on exposure, sensitivity, and adaptive capacity. It also outlines major constraints in SSIF, including limited awareness, weak institutional support, and lack of climate-resilient infrastructure. Adaptation and mitigation strategies such as enclosure culture, ranching, use of climate-resilient fish species, habitat restoration, and livelihood diversification are reviewed. The chapter underscores the need for policy attention, capacity building, and integration of climate-smart approaches to enhance the resilience of SSIF and ensure long-term sustainability in the face of increasing climate variability.

Assessing Climate Change Impacts and Vulnerability in Small-Scale Inland Fisheries of India

Mishal P. (✉) · G. Karnatak · S. Kumari · Lianthuamluaia · U. K. Sarkar · B. K. Das
ICAR – Central Inland Fisheries Research Institute (ICAR-CIFRI), Barrackpore, Kolkata, India

B. K. Das et al. (eds.), *Small-scale Inland Fisheries in India*, Sustainability Sciences in Asia and Africa, https://doi.org/10.1007/978-981-95-3849-2_9

9.1 Introduction

Climate change is considered as one of the most crucial environmental challenges in the present century with enduring consequences on aquatic resources, fisheries, and food security. It effects especially the poor and developing countries which are more vulnerable and least capable to deal with the adverse impacts. A vast majority of rural populations in India are dependent on climate sensitive sectors (including agriculture and fisheries) and natural resources viz. freshwater, mangroves, coastal zones, grasslands for their income and livelihoods. Climate change is a recent addition to the list of crucial factors threatening the sustainability of fisheries. According to FAO, around 120 million people depend on fisheries for their livelihood, of which 97% are living in poor countries and more than 90% are contributing in small-scale fisheries (SSF). In India, the inland capture fish production is 1.46 million tonnes (FAO 2020), in which majority (>90%) come from SSF sector. SSF make a considerable contribution to overall fish and food output and provide livelihoods to millions of people in the fish value chain. SSF in inland sector is expected to generate more livelihood opportunities than the marine counterparts and will play a crucial role in local nutrition and food security (Welcomme et al. 2010; Funge-Smith 2018). The workforce in SSF is vividly more diverse than anticipated with women accounting for half of all small-scale fishers worldwide (World Bank 2012).

India is blessed with vast inland open water resources viz. rivers (29,000 km), estuaries (0.3 Mha), reservoirs (3.5 Mha), and floodplain wetlands (0.51 Mha) which play an important role in fish production. A total of 9.45 million fishers are dependent on inland fisheries for their livelihood. The SDG 13 calls to take urgent action to combat the climate change impacts. The SSF in open water sector are threatened due to climatic change and associated factors including anthropogenic activities. SSFs can be used as tools to alleviate poverty and increase the food security (Nayak and Berkes 2019). Although SSFs are generally considered vulnerable, they have their strengths too, several diversifications and adaptive strategies are employed by certain small-scale fishing communities to counter the effects of climate change (Galappaththi et al. 2021) and the resilience shown by the SSFs in the advent of globalization and "developmental decades" in India is encouraging. Although neglected for long years, recent decades witnessed a marked increase in importance of SSF and it received great attention from planners and policy makers as means of attaining sustainable development goals (SDGs). In this background, the present chapter deals with climate change impact on small scale inland fisheries, vulnerabilities, and various mitigation and adaptation strategies.

9.2 Climate Change Issues and Indicators

The increase in the global surface temperature was +0.99 °C throughout the last two decades of the twenty-first century (2001–2020), but during just one decade (2011–2022), it reached +1.09 °C (IPCC 2023). Climate change directly or indirectly poses major threats to ecology and fisheries which has deleteriously impacted the aquatic environment, biota, fish diversity, fish production, and thereby the dependent communities and their livelihood in inland sectors. The frequencies of extreme weather events including droughts, cyclone, flood, heat waves etc., are increasing and are considered to have considerable impacts on SSF. According to India's climate change forecasts, the surface air temperature has risen by 0.4 °C in the last century. Analysis of climate data along India's major river basins has shown a warming trend, decreased precipitation, and fluctuating seasonality of rainfall occurrence. According to studies, the annual rainfall in the Ganga basin has reduced by 257 to 580 mm and the mean air temperature has risen by 0.20 to 0.47 °C over the past three decades. Increased of water temperature, decreased in rainfall of the Ganges River systems basin indicate the prevalence of climate change. On the other side, along Cauvery River, the span of the monsoon season and a rise in yearly rainfall were observed. The erratic weather and climatic pattern have impacted the ecology and fisheries of the river system of India (Das et al. 2019).

9.3 Impact Pathways and Vulnerability

The impact of climate change on fisheries is direct as well as indirect effects of a number of physical and chemical factors (Brander, 2010). The millions of fisherfolk dependent on fisheries and allied fields for their livelihood are affected by climate variation in the world. The poor economies which are extremely reliant on fish as a cheap source of protein are vulnerable to the impacts of climate change (Allison et al. 2009). It is predicted that the tropical fish stocks are expected to decline by 40% by 2050 (Lam et al. 2020). The effects are clearly visible in India as evidenced by changes to physiological behavior, altered geographic distribution, and loss of species. There have been several inland fisheries research achievements in India over the years.

The coastal regions and low-lying coastal wetlands are being severely impacted by increased global temperature, sea level rise, ocean acidification, and other effects. Losses of fish stocks and production infrastructure as a result of extreme weather events like floods and cyclones, as well as a rise in disease and parasite incidence and harmful algal blooms, are only a few of the short-term effects of climate change. Other impacts are loss of biodiversity, threat to genetic resources, reproductive failures, reduced availability of wild seed, altered species distribution increasing competition for freshwater effecting SSF in open waters in long term. Among the inland open waters, floodplain wetlands are becoming more vulnerable to climate change and associated factors. Loss of biodiversity, decreased fish production, impaired hydrological regimes, siltation, eutrophication, weed proliferation, loss of river connectivity, depth reduction, and shrinkage of water area are the major threats in floodplain wetlands due to climate change (Fig. 9.1).

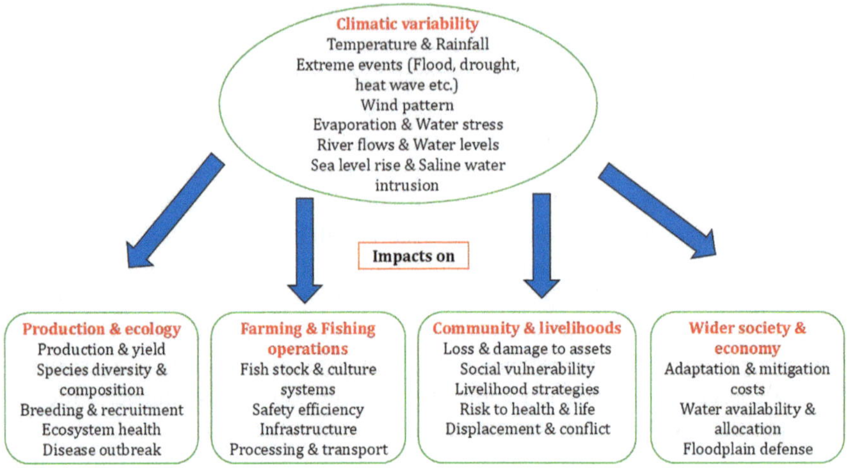

Fig. 9.1 Climatic variability and its impacts. (Adapted and modified from Badjeck et al. 2010; FAO 2018; Sarkar and Das 2021)

9.4 Ecological and Social Vulnerability

A species, system, or resource's susceptibility to the adverse impacts of climate change is known as vulnerability and it has three components viz. exposure, sensitivity, and adaptive capacity (IPCC). Emerging methods like vulnerability assessments can be employed as the first stage of the adaptation planning procedure. It assists in determining which systems, habitats, or species are most at risk from climate change. There is no single standard method/framework to perform the vulnerability assessment, and instead diverse methods are used at various levels, and consequently the interpretation has to be done carefully.

The decreased availability of the fish spawn or seed abundance of the Indian Major Carps (IMC) in the middle stretch of River Ganga was due to the decrease of rainfall during the monsoon and shift of monsoon seasons which is the breeding period of the IMCs. The shift of the geographical distribution of some warm water fishes including *Glossogobius giuris*, *Puntius ticto*, *Xenentodon cancila* and *Mystus vittatus* to the colder area in the Ganga River system clearly indicated the colder area of the River has transformed to be congenial for warm water fishes due to climate change. The floodplain wetlands are important habitat not only to support biodiversity but also play an important role to provide enormous ecosystem services and important fisheries resources to support livelihood of many people but due to climate change and associated factors the floodplain wetlands are becoming vulnerable. In the Ganga basin, significant reductions in depth and area (between 37.20 and 57.68%), diversity of native fish fauna in wetlands (between 22.85 and 54%), and transition from mesotrophic to eutrophic trophic states have been noticed (Sarkar and Das 2021) (Fig. 9.2).

A case study in a seasonally open floodplain wetland in West Bengal, India, which became almost dry due to the shortfall of precipitation which in turn resulted

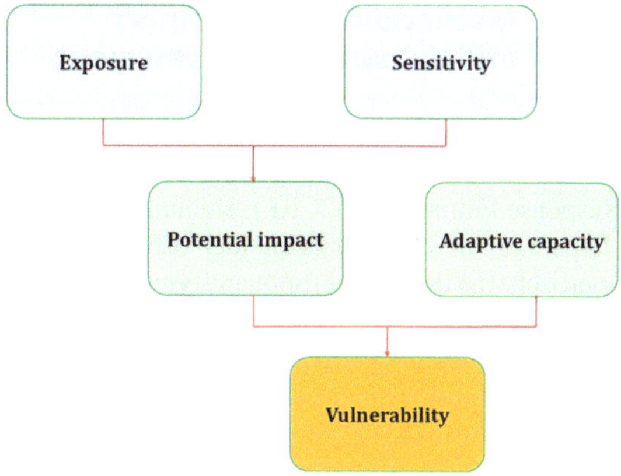

Fig. 9.2 Climate change vulnerability assessment. (Adapted from Glick et al. 2011)

in drastic reduction (41%) of the fish production, loss of fish (17%), plankton (54%), and benthic (31%) diversity and increased of macrophytes coverage (80%). Shrinkage of Gangetic wetlands due to climate change and other factors including environmental variations and anthropogenic activities were also reported. The study conducted by ICAR-CIFRI indicated that the abundance of tiny planktons was influenced by the climate and environmental anomalies in the Gangetic floodplain wetlands. Dominance of a plankton population by a fewer plankton species in higher temperature and erratic temperature fluctuation conditions has been reported (Rasconi et al. 2017). Due to the impact of extreme climatic events on the fisheries, many fishers have opted alternate livelihood to maintain their livelihood. The shift to alternate livelihood options evidenced the impact of climate change on SSF and socioeconomic conditions of the stakeholders.

Heat Waves (HWs) are impacting the inland open water ecosystem. The study conducted by ICAR-CIFRI indicated that HWs has an impact on the biotic community and environmental variables, particularly primary producers in the aquatic ecosystems of lower Ganga basin. Assessment on the diet composition and feeding intensity of selected fishes from the Ganges River basin during summer months indicated lower Gastro-Somatic Index (GaSI) values evident the impact of heat-wave on food composition and feeding intensity important fish species. It also significantly reports that catastrophic events like heat waves may become the major climatological drivers to induce harmful cyanobacterial bloom in aquatic ecosystems in recent climatic scenario.

In recent years, the occurrence of extreme climatic events such as depressions, cyclonic storms, and severe cyclonic storms has increased over the coastal areas of West Bengal. Due to extreme climatic events (i.e., Amphan in 2020 and Yaas in 2021), many rivers associated bheries in the coastal area of West Bengal were destroyed which has resulted in loss of fish production due to the escape of many stocked fishes (*Lates calcarifer*, *Mugil cephalus*, *Macrobrachium rosenbergii*, *Penaeus monodon* etc.) to adjacent river. Based on study, it is observed that the fish diversity was decreased by 14.33% in the selected wetlands. Due to this situation, many fishers have to quest for alternative livelihood opportunities. The migration of workers in pursue of works was increased by 50.88% from the study area.

9.5 Methodologies for Impact and Vulnerability Assessment

The most popular methodologies for climate change vulnerability assessment are Nature Serve Climate Change Vulnerability Index (CCVI), Climate Change Response Framework (CCRF), Habitat Vulnerability Model, Expert opinion workshops and surveys. A stakeholder driven methodology alternative to data driven approach was developed for quantifying climate induced threat to wetland fisheries. The vulnerability index (0.0–1.0) was used based on inland fisheries related indicators. A methodology based on the combination of stakeholders' perception and ecological conditions was also developed by for vulnerability assessment. An innovative tool for assessing species vulnerability and predicting the spawning success in the changing climatic scenarios using GAM was also developed (Fig. 9.3). The method

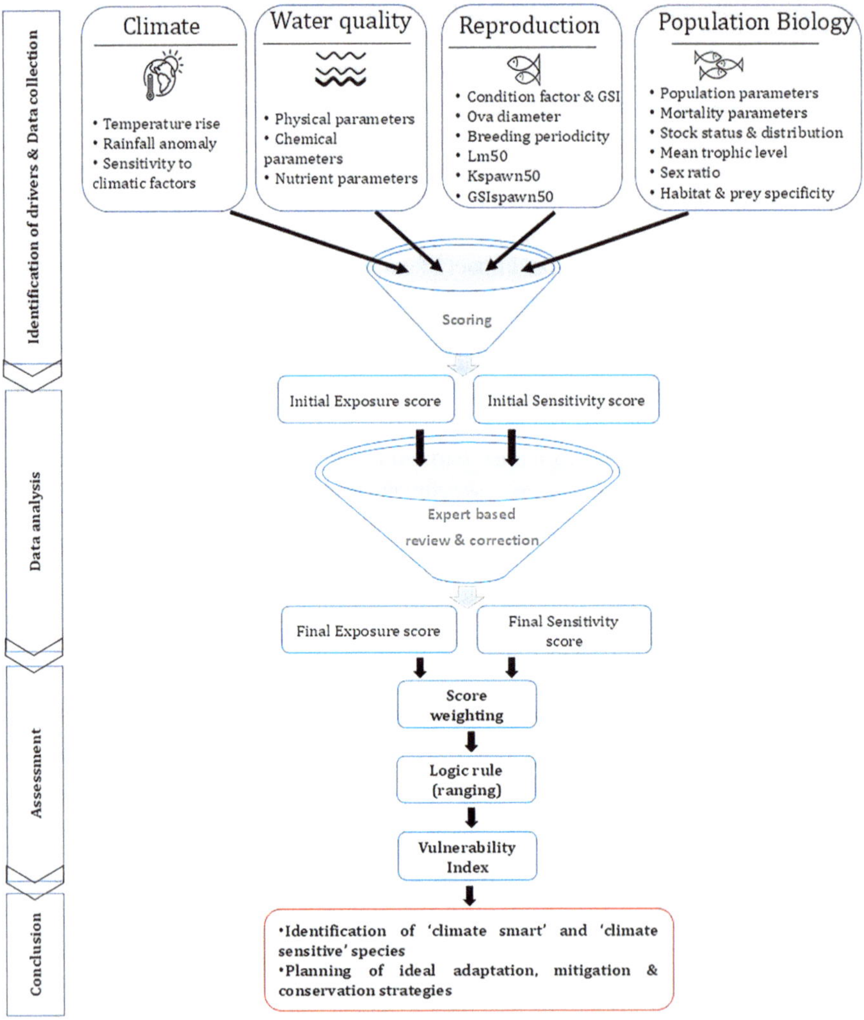

Fig. 9.3 Framework for vulnerability assessment. (Adapted from Lianthuamluaia et al. 2023)

uses novel parameters viz. Threshold GSI, Kspawn 50 (pre-spawning fitness), Girth spawn50, and associated climate optima and limnological parameters. Thermal tolerance of inland freshwater fishes collected from resources with varied water temperature was estimated using CTmax and CTmin. Most of the methodologies developed for assessing the vulnerability due to climate change are based mainly on the secondary data as the collection of primary data for several years are difficult which is the major limitation for most of the developed methodologies. The delineation of other drivers of vulnerability from climate change is also a major challenge in climate change impact studies. Forecasting and prediction models have to be developed for inland open water fisheries with respect to climate change.

9.6 Constraints and Challenges in SSIF

The inland SSF in India is largely unorganized and lacks robust database on production and its contribution to income, livelihood, and nutritional security. Although impact of climate change on inland fisheries has been attempted, the information of the vulnerability to climate change has not been adequately disseminated to the stakeholders. Many fishers are unaware of the concept of impact of climate change on SSF. Certain policies and guidelines are made for climate change mitigations; however, monitoring and implementation of such guidelines and policies are a major challenge. There is also dearth of comprehensive database on vulnerability of inland open water fisheries to climate change and forecasting/prediction models. Limited technically skilled manpower is available to study the influence of climate change and to devise climate smart inland SSF. This necessitates more capacity building and climate literacy programmes across the country. Additionally, small-scale artisanal fishers confront a variety of socioeconomic difficulties and limitations. A few of them include uneven distribution, various management regimes, and poor governance; a lack of credit support; frail or inactive cooperative societies; competition between parties for the lease of water bodies; a lack of alternative means of subsistence; a declining share of fishermen in consumer rupee; and restricted access to educational opportunities and primary healthcare.

9.7 Adaptation Strategies and Resilience

The SDG 13 aims to take urgent action to combat the climate change impacts and it also calls for promotion of mechanisms to increase the adaptive capacity for effective climate–change related planning and adoption of management measures. The initiatives to adapt to and mitigate climate change should be implemented with due consideration to vulnerability status through new approaches and methods, available resources and stakeholder's participation. As fishers are vulnerable to the impacts of climate change, special focus must to be given while designing the adaptation measures and meeting the targets of poverty alleviation, food security, and

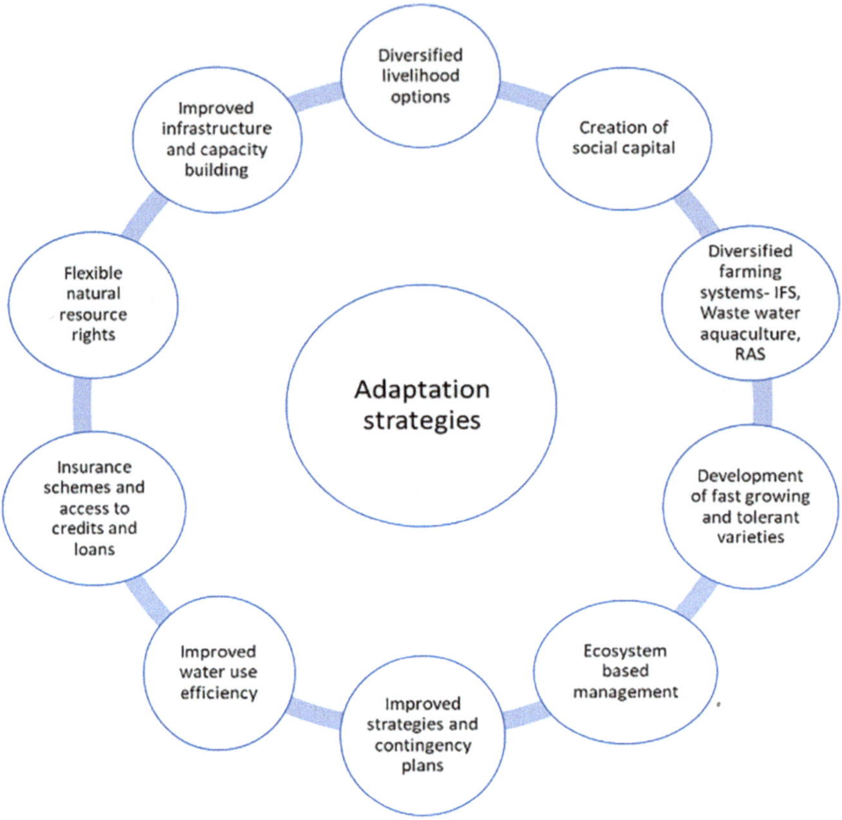

Fig. 9.4 Major climate adaptation strategies and measures. (Source: Badjeck et al. 2010; FAO 2018)

empowering women (Fig. 9.4). Increased awareness on climate change and focus on developing adaptation strategies are emerging in developing countries in recent years. However, the local wisdom and knowledge of the fishers regarding the climate change adaptation is not well documented, refined, and implemented.

9.8 Climate Smart Reservoir and Wetland Fisheries

The climate adaptation and mitigation strategies suggested and adopted in wetlands and reservoirs are ranching for replenishment of fish stocks and biodiversity conservation, enclosure culture, pre-summer enclosures, deep pool refuge system, and periphyton-based fish farming, sensitization of the stakeholders, and incorporation of climate resilient species in culture-based fisheries. Fast growing tolerant fish species with short duration farming may be prioritized. The following strategies are recommended for climate smart reservoir and wetland fisheries:

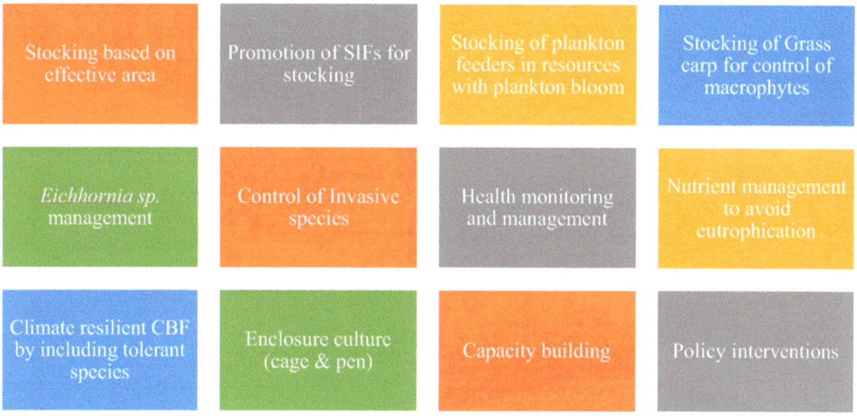

Fig. 9.5 Recommended strategies for climate smart reservoir and wetland fisheries

Desiltation, maintenance of link channels, provision of deep pools, minimization of anthropogenic pressures, control of exotic species, integrated-ecosystem based approach, early warning system, sensitization of stakeholders, improved farming and diversified livelihood options, improvization of fishing implements, and selection of climate resilient fish species are a few mitigation strategies for wetland fishers (Fig. 9.5) Potential climate resilient management strategies identified for wetlands are *temporary pre-summer enclosure*, *deep pool refuge*, *autumn stocking*, *submerged branch pile refuge*, and *floating aquatic macrophyte refuge-based fishery*. The popularization, widespread adoption, and optimization of these pre-existing traditional fishery practices along with enclosure/pen culture as climate change mitigation strategies would help the fishers to cope up with climate induced water stress in wetlands. Integrated farming, community approach, diversification of livelihood and farming systems, ranching, and judicious exploitation are some of the recommended adaptation strategies for SSF to climate change.

Black clam, *Villorita cyprinoides*, was found to be a resilient species which could withstand the deluge in Kerala. In Assam and West Bengal, IMCs along with minor carp, *Labeo bata* were successfully cultured in pens. Natural spawning of the SIFs, *Amblypharyngodon mola* and *Gudusia chapra*, was observed in the pens indicating effectiveness of CRPS in conservation through auto recruitment. The cage culture with resilient fish species viz. *Labeo bata* and *Puntius sarana* along with IMCs is a viable option as an alternative livelihood option for fishers (Fig. 9.6). The climate smart model villages must be developed through demonstration of climate smart adaptation strategies and capacity building programs. National climate campaign, mass awareness, and farmers sensitization programs can help to increase the responsiveness of farmers to climate-induced threats in SSF and skill development to cope up with these threats.

Fig. 9.6 Resilient fish species for climate smart SSF

9.9 Conclusion

The climate change impacts on fish production and ecology, farming and fishing operation, community, and livelihoods and wider society and economy. Strengthening of Primary Fisheries Cooperative Societies (PFCS), maintaining sustainability of the fisheries resources including rational exploitation, restore, conserve, protect and manage ecosystems, creating wealth and alternative sources of income, having more control over systems and robust regulatory regimes, acknowledging the importance of SSFs communities, and including small-scale fishing communities in the decision-making and conflicts resolution mechanisms for sustaining livelihood of SSF are recommended. It is necessary to expand the established climate resilient adaption measures into different regions to raise awareness among fishers about climate change. The vision should be to secure sustainable resource use and access right to the fishers, ensure good catch from the resources. It is high time to strive for social, economic, and human rights of small-scale artisanal fishers and climate smart SSF in inland waters.

Points to Remember

- Climate change is a major threat to small-scale inland fisheries, affecting fish production, biodiversity, livelihoods, and food security—especially in vulnerable communities.
- Rising temperatures, altered rainfall, floods, and droughts are disrupting riverine ecology, fish breeding cycles, species distribution, and wetland productivity.
- Floodplain wetlands are highly vulnerable, with observed shrinkage, biodiversity loss, siltation, eutrophication, and changes in trophic state due to climate stress and anthropogenic activities.
- Fisher communities, especially women and artisanal fishers, are socio-economically impacted and often forced to migrate or shift to alternate livelihoods due to declining fish resources.
- Adaptation and mitigation strategies, including enclosure culture, deep pool refuge, climate-resilient fish species, ranching, and integrated farming, offer viable climate-smart solutions.
- Challenges include poor policy implementation, lack of climate literacy, limited data and manpower, and inadequate access to credit, infrastructure, and health/education services for SSF communities.
- Future focus should be on strengthening cooperatives, promoting ecosystem-based management, and building capacity through awareness, skill development, and inclusive decision-making in SSF governance.

References

Allison EH, Perry AL, Badjeck MC, Adger WN, Brown K, Conway D, Halls AS, Pilling GM, Reynolds JD, Andrew NL, Dulvy NK (2009) Vulnerability of national economies to the impacts of climate change on fisheries. Fish Fish 10(2):173–196

Badjeck MC, Allison EH, Halls AS, Dulvy NK (2010) Impacts of climate variability and change on fishery-based livelihoods. Mar Policy 34(3):375–383

Brander K (2010) Impacts of climate change on fisheries. J Mar Syst 79(3–4):389–402

Das BK, Sarkar UK, Roy K (2019) Global climate change and inland open water fisheries in India: Impact and adaptations. In: Climate change and agriculture in India: Impact and adaptation. Springer, Cham, pp 79–95

FAO (2020) The state of world fisheries and aquaculture 2020: Sustainability in action. FAO, Rome

Funge-Smith S (2018) Review of the state of the world fishery resources: Inland fisheries, FAO Fish Aquac Circ No. C942. FAO, Rome

Galappaththi EK, Ford JD, Bennett EM, Berkes F (2021) Adapting to climate change in small-scale fisheries: Insights from indigenous communities in the global north and south. Environ Sci Policy 116:160–170

Glick P, Stein BA, Edelson NA (2011) Scanning the conservation horizon: A guide to climate change vulnerability assessment. National Wildlife Federation, Washington, DC

Harrod C, Ramirez A, Valbo-Jorgensen J, Funge-Smith S (2018) In: Barange M, Bahri T, Beveridge MCM, Cochrane KL, Funge-Smith S, Poulain F (eds) Impacts of climate change on fisheries and aquaculture: Synthesis of current knowledge, adaptation and mitigation options, FAO Fish Aquac Tech Pap No. 627. FAO, Rome, p 628

IPCC (2023) Summary for policymakers. In: Core Writing Team, Lee H, Romero J (eds) Climate change 2023: Synthesis report. Contribution of Working Groups I, II and III to the Sixth Assessment Report of the IPCC. IPCC, Geneva, p 36

Lam VW, Allison EH, Bell JD, Blythe J, Cheung WWL, Frölicher TL, Gasalla MA, Sumaila UR (2020) Climate change, tropical fisheries and prospects for sustainable development. Nat Rev Earth Environ 1(9):440–454

Lianthuamluaia L, Kumari S, Sarkar UK, Borah S, Mishal P, Karnatak G, Das BK, Ghosh BD, Das A, Debnath S, Ray A (2023) Improving approaches and modeling framework for assessing vulnerability of Asian leaf fish in the major river basin floodplains of India in changing climate. Ecol Inform 73:101926

Nayak PK, Berkes F (2019) Interplay between local and global: Change processes and small-scale fisheries. In: Jentoft S, Chuenpagdee R, Barragán-Paladines MJ, Franz N (eds) Transdisciplinarity for small-scale fisheries governance. Springer, Cham, pp 203–220

Rasconi S, Winter K, Kainz MJ (2017) Temperature increase and fluctuation induce phytoplankton biodiversity loss: Evidence from a multi-seasonal mesocosm experiment. Ecol Evol 7(9):2936–2946

Sarkar UK, Das BK (2021) Research advances in climate and environmental change impacts on inland fisheries of India: Status, vulnerability and mitigation strategies. Aquat Ecosyst Health Manag 24(3):7–17

Welcomme RL, Cowx IG, Coates D, Bene C, Funge-Smith S, Hall A, Lorenzen K (2010) Inland capture fisheries. Philos Trans R Soc Lond B Biol Sci 365(1554):2881–2896

World Bank (2012) Hidden harvest: The global contribution of capture fisheries. World Bank, Washington DC. http://openknowledge.worldbank.org/handle/10986/11873

Policy and Guidelines for Regulating Small-Scale Inland Fisheries in India

10

Basanta Kumar Das, Liton Paul, Piyashi DebRoy, and Pranaya Kumar Parida

Abstract

Small-scale inland fisheries in India employ over 2.3 million people and underpin food security and rural livelihoods, yet they face mounting pressures from overfishing, habitat loss, climate change, and weak regulation. These fisheries depend on rivers, lakes, reservoirs, ponds, and floodplains, and rely on traditional techniques and limited capital. State policies are crucial for their management, but their effectiveness varies: Kerala, Andhra Pradesh, and West Bengal have adopted community-based management, tenure rights, and sustainability measures, while other states lack up-to-date, comprehensive frameworks. This patchwork leads to uneven resource stewardship, poor enforcement, and the marginalization of fishing communities. Many laws remain disconnected from national priorities and overlook traditional knowledge, emerging threats such as pollution and market integration, and gender equity. Strengthening governance will require co-management models, secure customary rights, ecosystem-based approaches, and better coordination with the National Fisheries Policy 2020 and Sustainable Development Goals. A unified, inclusive, and science-driven state policy framework is essential to safeguard livelihoods, preserve ecosystems, and build climate-resilient rural communities.

Keywords

Inland fisheries · Policymaking · Sustainability · Livelihoods · Co-management

Managing small-scale inland fisheries involves more than rules; it's about nurturing rivers, enhancing livelihoods, and preserving a prosperous heritage for the future.

B. K. Das (✉) · L. Paul · P. DebRoy · P. K. Parida
ICAR – Central Inland Fisheries Research Institute (ICAR-CIFRI), Barrackpore, Kolkata, India

B. K. Das et al. (eds.), *Small-scale Inland Fisheries in India*, Sustainability Sciences in Asia and Africa, https://doi.org/10.1007/978-981-95-3849-2_10

10.1 Introduction

Small-scale inland fisheries in India encompass fishing activities conducted in inland water bodies like rivers, lakes, reservoirs, ponds, and floodplains, using traditional or simple technologies and limited capital investment. Over 2.31 million inland fishers constitute approximately 80% of the total workforce in the fisheries sector of the country (Ministry of Fisheries, Animal Husbandry & Dairying 2021). Its contributions to fish production have been significant. For example, inland fisheries contribute around 65% (7.5 million tonnes) to India's total fish production (11.4 million tonnes) (Ministry of Fisheries, Animal Husbandry & Dairying 2021). Having wide geographical distribution, inland fisheries are spread across diverse water bodies in all Indian states, with significant concentrations in Andhra Pradesh, West Bengal, Tamil Nadu, and Assam. Small-scale inland fisheries in India are overwhelmingly family-run ventures, where knowledge of riverine habitats and seasonal rhythms is handed down across generations. Operating with simple boats, nets and traps, these fishers sustain household food security and supply fresh fish to local markets, supporting the diets and livelihoods of millions of people, especially women and marginalized groups, who have few alternative income sources. These communities also act as informal custodians of aquatic ecosystems, drawing on traditional stewardship practices to protect habitats and spawning grounds yet today their resilience is undermined by overfishing and indiscriminate gear use, habitat degradation from pollution and water extraction, and the growing impacts of climate variability on river flows. These ecological threats are compounded by a fragmented legal framework marked by unclear licensing, unenforced closed-season rules, and insecure resource access, as well as significant post-harvest losses due to inadequate cold-chain and transport infrastructure. In this context, a comprehensive policy and guideline framework is essential: one that formally recognizes community access rights, enforces science-based quotas, mesh sizes and seasonal closures, integrates pollution control with watershed management, strengthens market logistics, and fosters participatory co-management to improve both fishery sustainability and community well-being (FAO 2024; National Fisheries Policy 2020).

Fisheries governance falls under both the Union and the State domains, depending on the geographical scope. The Union List gives authority to the central government over inter-state rivers and fisheries beyond territorial waters, which primarily refers to marine fisheries. In contrast, inland fisheries are governed primarily by state legislatures. The State List includes subjects such as water, land rights, irrigation, and fisheries. Entry seventeen covers water supply, irrigation, canals, drainage, embankments, and water storage, while entry twenty-one addresses inland fisheries. This arrangement gives states the legal authority to make rules and regulations related to the management, conservation, and use of inland fishery resources (Constitution of India, art. 246, Sch. VII, List I, Entry 1).

In addition to this federal structure, the Constitution also includes important environmental responsibilities. Article 48A under the Directive Principles of State Policy obliges the State to protect and improve the environment and safeguard the country's forests and wildlife (Constitution of India, 1950, art. 48A). Similarly, Article 51A outlines the fundamental duties of citizens, including the duty to protect the natural environment, rivers, lakes and wildlife, and to show compassion for all living creatures (Constitution of India, 1950, art. 51A). Conflicts over water resources, especially in the context of Inter-state Rivers, are addressed under Article 262, which empowers Parliament to establish legal mechanisms for the adjudication of disputes over the use and control of Inter-state Rivers and to exclude such matters from the jurisdiction of the courts (Constitution of India, 1950, art. 262). At the national level, several laws provide indirect support to inland fisheries through environmental protection and conservation. The Indian Fisheries Act of 1897 is one of the oldest pieces of legislation and penalizes harmful practices such as poisoning or blasting water bodies to kill fish (Indian Fisheries Act, 1897). The Environment (Protection) Act of 1986 is a more comprehensive framework that addresses pollution and degradation of aquatic habitats. Other laws such as the Water (Prevention and Control of Pollution) Act of 1974, the Wild Life Protection Act of 1972, the Biological Diversity Act of 2002, and the Wetlands (Conservation and Management) Rules of 2010 play significant roles in maintaining ecological balance and protecting freshwater biodiversity.

This chapter aims to an examination of the legal policies adopted at the state level and central level. The analysis focuses on selected states to understand how inland fisheries are regulated and to identify gaps and good practices in current governance models. The study is based entirely on secondary data, collected from official sources such as policy documents, legislative acts, government reports, and fisheries department publications. By comparing these state-specific legal instruments with central laws and guidelines, the chapter aims to assess the coherence and adequacy of current regulatory structures. The goal is to propose more inclusive and effective policy that strengthen sustainability, promote equity, and recognize the vital role of small-scale inland fisheries in rural India. To achieve this, the chapter draws on a detailed review of secondary data, including legislative documents, policy briefs, and official reports from central and state fisheries departments. Acts, rules, and regulations applicable to inland fisheries were systematically collected and analyzed for selected states, along with relevant national schemes. By mapping and comparing the regulatory landscape across regions, the chapter identifies both best practices and critical shortcomings in existing approaches and outlines strategic recommendations for more robust, community-centered governance of India's small-scale inland fisheries.

10.2 Policy Framework for Regulating Small-Scale Inland Fisheries in India

The Government of India recognizes the vast potential of inland fisheries to contribute significantly to rural livelihoods and to double the income of the fishing community. According to the National Fisheries Development Board (NFDB 2025), the inland fisheries sector holds an estimated potential of 17 million metric tons. With the country's population steadily increasing, the demand for fish as a source of protein is also rising, highlighting the urgent need for the sustainable development and management of inland aquatic resources. Governance of small-scale inland fisheries in India remains fragmented a patchwork of policies and legal frameworks that vary in their relevance and effectiveness, due to this is important to examine both central and state-level policies to understand their scope, implementation gaps, and implications for the development of inland small-scale fisheries.

10.2.1 State Level

State-specific fisheries acts and rules: Each state has its legislation with varying degrees of detail and effectiveness. Some states, like Kerala and Andhra Pradesh, have pioneering policies for inland fisheries management (Table 10.1).

Table 10.1 Legal instruments and key policy features for inland fisheries at the state level in India

State	Act/regulation/policy framework	Key policy features
West Bengal	West Bengal Inland Fisheries Act, 1984	Prohibits dams without fish passes, sewage-fed fishery limits, lease restrictions, and protected area rules
Madhya Pradesh	Madhya Pradesh Fisheries Act, 1948	Prohibits explosives and harmful gear; enforces mesh sizes, license requirements, and fish sale standards
Bihar	Bihar Jalkar Management (Amendment) Act & Fish Seed Certification Act, 2007, 2018	"Jalkar" settlement; free jalkar zones; exotic species ban; seed certification committees
Punjab	Punjab Fisheries Act & Rules, 1914, 1965	Mandatory licensing; banned during closures; area restrictions; mesh and minimum size limits; fines
Chhattisgarh	United State of Gwalior, Indore and Malwa Fisheries Act (Samvat 2006)[a]	Requires permission for dams/weirs; prohibits destructive substances; seasonal bans; mesh-size enforcement
Gujarat	Gujarat Fisheries Act, 2003	Bans lime, poisons, explosives; invasive species control; gear standards; effluent restrictions; public leasing
Uttar Pradesh	Uttar Pradesh Fisheries Act, 1948	Gear specifications; dam construction limits; brooder/juvenile hilsa protection; minimum size; protected areas rules
Kerala	Kerala Inland Fisheries & Aquaculture Act, 2010	Govt. sole conversion power; coastal aquaculture ban; protected area restrictions; size/weight limits; effluent controls
Himachal Pradesh	Himachal Pradesh Fisheries Act, 1976	License requirement; bans on destructive gear; import/export control; mesh-size standards; price regulation
Manipur	Manipur State Fisheries Act, 1988	Prohibits low mesh gear; poison bans; size standards; penalizes illegal fishing and cross-border transport
Mizoram	Mizoram Fisheries Act, 2002	Ban seasons; fish sanctuaries; effluent and diversion restrictions; size/weight minimums; penal provisions
Rajasthan	Rajasthan Fisheries Act, 1953	Prohibits dams/weirs in public waters; protected area no-fishing; mesh standards; mandatory licenses

[a]Samvat 2006 corresponds to 1949 CE. Data drawn from official state fisheries legislation, departmental notifications of Department of Fisheries websites and state-level policy documents (1953–2024)

10.2.2 Insights of These Policies

The review of state-level policies underscores several critical insights emerging from the initial characterization of small-scale inland fisheries and their governance gaps. First, the ubiquity of preferential leasing mechanisms, across West Bengal, Madhya Pradesh, Gujarat, and other states, reflects a shared effort to redress historical marginalization of traditional fishing communities. By granting longer lease tenures, reduced fees, and priority for PFCS or SHGs and Scheduled Castes/Tribes, these policies aim to secure resource access and strengthen livelihoods.

Second, conservation and resource stewardship surface as central policy objectives. States such as West Bengal, Odisha, and Punjab mandate sanctuary zones, closed seasons, habitat mapping via GIS, and mesh-size and size-limit regulations to marine ecosystem not inland waterbodies. These measures respond directly to overfishing and habitat degradation identified at the outset, translating ecological knowledge, both scientific and traditional, into enforceable rules.

Third, modernization of inland fisheries infrastructure is evident in multiple approaches. Andhra Pradesh and Odisha leverage remote sensing and cluster-based hatchery models to ensure year-round seed supply, while Tamil Nadu and Uttar Pradesh invest in cold-chain linkages, market infrastructure and advanced culture techniques like cage and pen systems. Such interventions address post-harvest losses and market access constraints highlighted in earlier sections.

Fourth, the socio-economic and gender dimensions of governance are increasingly recognized. Welfare schemes, skill training, and reserved membership quotas for women in cooperatives (e.g., Madhya Pradesh's 33% reservation) underscore the policy focus on inclusive growth and human resource development. Similarly, policies across states converge on extension services and capacity building to integrate fishers into co-management arrangements.

Fifth, the persistence of colonial-era regulatory frameworks, prominently in Punjab's Fisheries Act of 1914, illustrates a contrast between top-down enforcement models and emerging participatory regimes. Punjab's strong licensing, penalty, and seasonal closure regimes offer valuable enforcement insights but lack the community-driven co-management and infrastructural support emphasized elsewhere.

Collectively, these insights reveal a policy landscape that is simultaneously fragmented and innovative. While individual states have pioneered targeted interventions, the absence of a cohesive national guideline, one that harmonizes leasing protocols, conservation mandates, socio-economic safeguards and modern infrastructural standards, remains the principal governance gap. Addressing this gap will require synthesizing best practices, reconciling enforcement with empowerment, and fostering adaptive frameworks that can respond to ecological and livelihood challenges of small scale fishers in India's inland fisheries.

10.2.3 Access, Tenure, and Property Rights in Inland Fisheries

Inland water bodies are common property resources, but the rights to use them are not always equal or consistent. Property rights are the foundation for determining who can access and use these resources. These rights can range from highly controlled and privately held to open access, particularly in rivers where little regulation exists. The form of property rights depends on many factors including the size of the water body, traditional and customary use, physical geography, ease of access, and river connectivity (Chandra 2011a, b; Chandra and Das 2019; Kelkar and Arthur 2022a, b).

Governments and local authorities manage inland fisheries with multiple objectives in mind, like ensuring economic benefits, protecting livelihoods, sustaining biodiversity, promoting equity, and maintaining the ecosystem. The establishment of well-defined property rights helps build trust among users and authorities and provides users with motivation to manage the resources sustainably. These arrangements must balance public responsibility with local participation. The public trust doctrine permits governments to lease or grant access to water bodies so long as these actions do not harm the collective public interest (Henquinet and Dobson 2005).

Property rights in fisheries are generally described as a bundle of rights which include access, withdrawal, and exclusion. Access means the ability to enter and use a defined area. Withdrawal refers to the ability to harvest resources, and exclusion is the right to limit access to others. Leasing or tenure rights involve transferring the right to use a water body from the government or another legal owner to a fisher, group, or cooperative. This can be done through auctions, direct negotiations, or as prescribed by law. In most states, leases range from five to ten years, and the lease fee is usually calculated based on factors like fish catch potential, total water area, and the effective area suitable for fishing (NABARD 2018a, b).

In practice, access to fisheries is most often governed by leaseholders. However, traditional access rights are sometimes recognized, particularly for small-scale or subsistence fishers. Typically, leaseholders control fishing for larger or commercially valuable species, while access to smaller fish is often allowed for specific groups or under regulated conditions.

Access and management of inland water bodies, especially wetlands and reservoirs, pose significant challenges. Various state agencies and institutions have been authorized to manage fishing rights, and their approaches can vary widely. As a result, the outcomes of such management also vary and have different implications for fishers (Huppart 2005).

Fishing leases represent a shift in property rights, from common access to more structured arrangements. These leases can be granted to individuals, cooperatives, or entire communities. While many state governments aim to use leasing as both a revenue generation tool and a means to support fisher welfare, this dual goal often results in conflicting priorities when it comes to fair access and sustainability.

Historically, fishers had only the exclusive right to the fish they caught. Other rights, such as access to fishing grounds or management decisions, were usually held in common. As fish resources have become scarcer and conflicts over access have grown, states introduced regulatory mechanisms like gear restrictions and seasonal closures (Chandra 2011a, b; Chandra 2014; Chandra and Bhattacharyya 2016).

Management of inland water bodies is complex and involves many layers. Stakeholders include not just fishers but also government agencies, cooperatives, and local institutions. Among these, fishers remain the most directly affected as their livelihoods depend on how these resources are managed (Chandra and Das 2019). Creating policies that fairly distribute rights and responsibilities, while supporting conservation and livelihood objectives, remains a critical challenge for sustainable inland fishery development. The details of fishing rights tenure policies in different states are given in Table 10.2.

Table 10.2 Fishing rights tenure policies in different states

State	Type of water bodies	Tenure of fishing right (leased period)	Ownership	Right to	Lease to
Madhya Pradesh	100–1000 ha 1000–2000 2000–above	5 years 5 years 5 years	ZilaParishad Fisheries Dept MP fish cooperative federation	Fish Fish Mainly sale of fishes	Fishermen, Coop, SHG Contractor
Uttar Pradesh	All reservoirs	5 years	Dept of Fisheries UPFDC, Fish Fed	Fish	Fishermen, Coop, SHG, Contractor
Assam	Wetlands	7 years	Dept of Revenue, Fisheries	Fish	Individual Fisherman and Cooperative
Bihar	All water bodies	7–10	Fisheries Department	Fish	Fishermen and Coop.
West Bengal	All water bodies	5–12	Dept of Fisheries, Irrigation Dept	Fish	Fish Coop. society
Chhattisgarh	200–1000 ha 1000–5000 5000–above	5 years 5–7 years 10 years	Dept of Fisheries CG fish coop fed Fisheries Department	Fish Fish Fish	Fishermen, Coop, SHG Fishermen, Coop, SHG FCS, Contractor
Gujarat North Gujarat Tribal area	200 and above 20–1000 1000–above	5–10 years 5–10 years 5–10	Fisheries Department	Fish Fish Fish	Fishermen,FC Contractor Tribal Coop S Tribal Co Federation

Source: This table is compiled from official state-level fisheries policies, departmental notifications, and government circulars accessed through respective Departments of Fisheries websites. Information is cross-verified with secondary literature including Chandra (2011a, b), Kelkar and Arthur (2022a, b), NABARD (2018a, b), and relevant documents available on the National Fisheries Development Board and FAOLEX repositories

10.2.4 Central Level

At the national level, India's fisheries governance rests on a collection of statutes and policies that while ambitious in their overall aims, offer only limited and indirect support to inland small-scale fishers. The Fisheries Act of 1971 establishes the legal framework for regulating and developing the sector but contains no measures specifically tailored to the distinctive ecological and socio-economic realities of inland waters. The National Fisheries Development Board has driven aquaculture expansion and marine initiatives, yet its programs have historically given scant attention to capture fisheries in rivers, lakes, and wetlands. Meanwhile, the National Water Policy of 2012 provides a strong mandate for sustainable and equitable management of the country's water resources but does not explicitly incorporate inland fisheries as a legitimate water-use sector, creating a critical oversight in the integration of fisheries into broader water governance.

In 2020, the Central Government introduced its first comprehensive *National Fisheries Policy,* which seeks to harmonize resource conservation with food and nutritional security, enhance fisher welfare, and boost India's global competitiveness in fish products. Grounded in constitutional principles and a people-centered approach, the policy promotes equity, gender inclusion, entrepreneurship, and subsidiarity and outlines a ten-year roadmap for the sector (NFP 2020). It highlights the importance of community welfare by extending credit facilities, exploring insurance for fishing assets, strengthening cooperatives and producer organizations and providing safety nets for both resident and migrant fish workers. It also advocates for responsible fishing practices, diversification of livelihoods to ease pressure on over-exploited stocks, and improvements in vessel design and navigational training to protect fishers from hazards (Table 10.3).

Despite these initiatives, the broad scope and generalized provisions of existing national instruments leave inland small-scale fishers insufficiently recognized and ground level implication has not been done yet. To meet the urgent needs of these communities, who depend on vulnerable ecosystems and face complex governance challenges, India requires a more finely tuned policy framework that explicitly addresses the unique dynamics of inland fisheries and integrates them into the nation's wider water and rural development agendas.

Table 10.3 Inland fisheries policy highlights under NFP 2020

Focus area	Key provisions
Ecological Restoration	Water flow regulation, habitat restoration, pollution control, fish passes, estuarine health
Regulation and Management	Updated leasing, traditional rights, closed seasons/areas, gear/method regulation
Stock Enhancement	Ranching, stock replenishment with endemic species
Governance	Central-state cooperation, interstate water management, national policy adoption
Community Welfare	Climate adaptation, disaster response, livelihood and nutritional support

10.3 Assessing the Effectiveness and Deficiencies of Indian Policies Concerning Small-Scale Inland Fishers

Despite some progress in inland fisheries development, existing Indian policies remain largely ineffective in addressing the specific needs of small-scale inland fishers (SSF). At the national level, policy attention continues to favor marine fisheries and commercial aquaculture, resulting in a systemic neglect of the inland SSF sector. This mismatch is further compounded by an outdated and fragmented legal framework that lacks clear, enforceable provisions tailored to the realities of small-scale fishers operating in rivers, lakes, reservoirs, and wetlands. While schemes and subsidies have been introduced in some states, their implementation is often patchy and fails to reach the most marginalized communities due to weak institutional capacity and poor targeting. Small-scale fishers are rarely included in policy formulation or governance processes, which lead to the exclusion of their traditional knowledge and erode local stewardship. Moreover, the absence of reliable data on inland fish stocks and the socio-economic status of fishers hamper evidence-based policymaking and limit the scope for meaningful intervention. Conflicting priorities across sectors, such as agriculture, water management, and infrastructure, often result in policies that directly harm inland fisheries by degrading aquatic ecosystems or restricting access to fishing grounds. These deficiencies, taken together, illustrate that while some policy mechanisms exist; they are insufficient, misaligned, or poorly executed, making the overall framework ineffective for ensuring the sustainability and well-being of India's inland small-scale fishers.

10.4 The Need for a Comprehensive and Integrated Policy Framework for Small-Scale Inland Fisheries

A comprehensive and integrated policy framework for small-scale inland fisheries (SSF) refers to a coherent set of legal, institutional, and strategic interventions that address the ecological, economic, and social dimensions of inland fisheries in a unified manner. In the Indian context, such a framework would prioritize the distinct realities of SSF operating in rivers, reservoirs, floodplain wetlands, lakes, and tanks; systems that differ fundamentally from marine or large-scale aquaculture environments. It would combine updated legal provisions, inclusive governance mechanisms, reliable data systems, intersectoral policy coherence, and climate-responsive strategies into a single, coordinated structure. The objective is not merely to regulate resource use, but to promote sustainable livelihoods, social equity, and ecosystem resilience across diverse inland water bodies.

India urgently requires such a framework (Fig. 10.1) to bridge the existing policy gaps and institutional fragmentation that undermine the sustainability of inland SSF. The current landscape is marked by outdated legislation, conflicting mandates across departments, limited community involvement, and weak enforcement. A robust framework should begin by formally recognizing the strategic importance of inland SSF for food security, employment, and biodiversity conservation. Legal

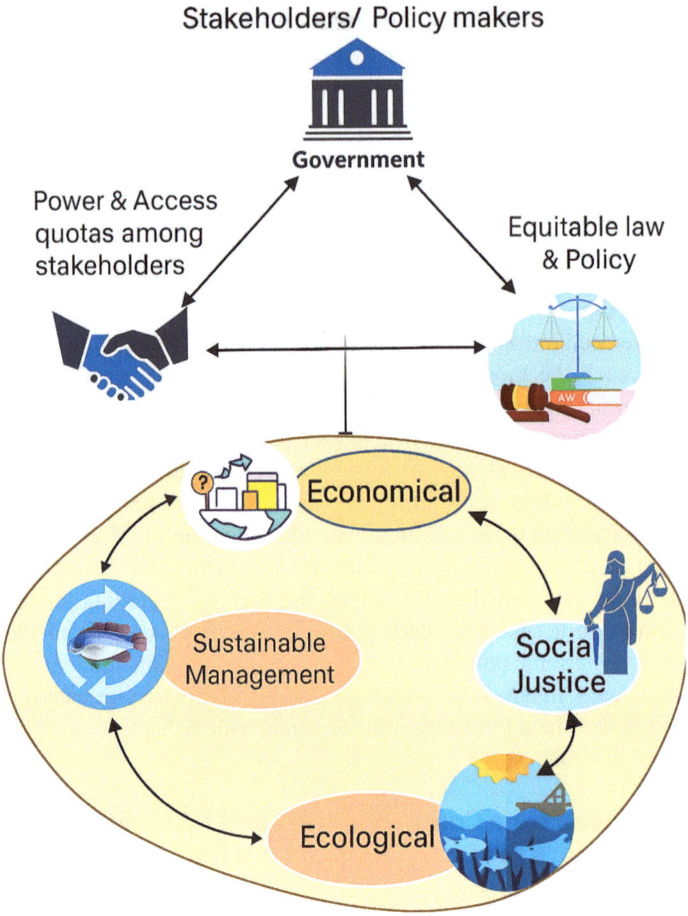

Fig. 10.1 Governance framework for sustainable and equitable resource management

foundations must be strengthened through the consolidation and modernization of fisheries laws, with explicit provisions for inland systems. Empowering fishing communities through participatory governance and co-management is essential to ensure local stewardship and accountability. The development of robust and regularly updated data systems, covering fish stock assessments, socio-economic profiles, and catch trends, will enable evidence-based policymaking. Cross-sectoral coordination among water, agriculture, industry, and environment departments must be institutionalized to avoid contradictory policies that harm fisheries. Moreover, the promotion of responsible fishing practices, sustainable habitat management, and equitable market access will ensure long-term ecological and economic viability. Climate adaptation strategies must also be mainstreamed to enhance the resilience of SSF-dependent communities in the face of changing hydrological patterns and extreme events. In summary, a comprehensive and integrated policy framework is not just a technical necessity but a strategic imperative for safeguarding the future of India's small-scale inland fisheries. Such a framework will protect livelihoods, improve nutrition and food security, and contribute to inclusive rural development while ensuring the ecological integrity of inland aquatic ecosystems. The following (Table 10.4) outlines the key pillars and components of a comprehensive and integrated policy framework tailored for small-scale inland fisheries in India.

This table integrates research-backed recommendations and innovations, ensuring a robust, evidence-based policy framework for India's small-scale inland fisheries.

Table 10.4 Proposed policy framework for small-scale inland fisheries in India

Policy pillar	Key components	Expected outcomes
Legal and Institutional	Update and consolidate inland fisheries laws (Das and Roy 2022), Clearly define access and tenure rights for SSF (Kumar et al. 2019), Promote co-management and decentralized governance structures (Arthur 2020)	Clear legal foundation (Das and Roy 2022), Reduced conflicts (Kumar et al. 2019), Improved enforcement and compliance (Arthur 2020)
Community Empowerment	Recognize and integrate traditional knowledge (Arthur 2020), Facilitate participatory governance (Arthur 2020), Provide capacity building and leadership development (Bhattacharya et al. 2024)	Stronger community ownership (Arthur 2020), Increased social equity and accountability (Bhattacharya et al. 2024)
Ecological Sustainability	Implement ecosystem-based management (FAO 2024), Conserve and restore inland aquatic habitats (FAO 2024), Regulate destructive fishing practices (Das and Roy 2022)	Sustainable resource use (FAO 2024), Healthier ecosystems and fish populations (FAO 2024)
Knowledge and Data Systems	Develop regular stock assessments (Bhattacharya et al. 2024), Monitor socio-economic conditions (Das et al. 2022a, b), Establish integrated data platforms (Bhattacharya et al. 2024), Leverage digital monitoring tools (Bhattacharya et al. 2024)	Evidence-based policymaking (Bhattacharya et al. 2024), Targeted, adaptive interventions (Das et al. 2022a, b), Reduced data gaps (Bhattacharya et al. 2024)
Livelihoods and Markets	Strengthen fisheries value chains (Arthur 2020), Improve infrastructure and cold storage (Das and Roy 2022), Ensure fair pricing and access to markets (Das and Roy 2022), Facilitate direct market integration (Bhattacharya et al. 2024)	Better income security (Arthur 2020), Enhanced market competitiveness (Das and Roy 2022), Fairer value distribution (Bhattacharya et al. 2024)
Cross-Sectoral Coordination	Align fisheries with agriculture, water, and environment sectors, Establish inter-departmental coordination mechanisms (FAO 2024)	Reduced policy conflicts (FAO 2024), Harmonized sectoral planning (FAO 2024)
Climate Resilience	Integrate fisheries into climate adaptation plans (Das et al. 2022a, b), Promote risk mitigation (e.g., insurance, livelihood diversification) (Das et al. 2022a, b)	Increased resilience to environmental shocks (Das et al. 2022a, b), Long-term viability of SSF (Das et al. 2022a, b)

(continued)

Table 10.4 (continued)

Policy pillar	Key components	Expected outcomes
Distress and Safety Nets	Establish 24/7 fisheries call centers for real-time distress alerts and advisory services (Bhattacharya et al. 2024), Implement distress response mechanisms (e.g., SOS features, insurance) (Das et al. 2022a, b)	Enhanced fisher safety and well-being (Das et al. 2022a, b), Rapid response to emergencies (Bhattacharya et al. 2024), Reduced vulnerability (Das et al. 2022a, b)
Information and Communication	Expand access to information through mobile advisory services and digital literacy programs (Bhattacharya et al. 2024), Disseminate weather, market, and regulatory updates via SMS and apps (Bhattacharya et al. 2024)	Improved decision-making, Reduced information asymmetry, Increased uptake of best practices (Bhattacharya et al. 2024)

Source: Arthur (2020); Bhattacharya et al. (2024); Chuenpagdee and Jentoft (2018); Das and Roy (2022); Das et al. (2022a, b); FAO (2024); Henquinet and Dobson (2005); Jena and George (2018); Johnson (2013); and Kumar et al. (2019)

10.5 Conclusion

India's small-scale inland fisheries, which employ around eighty percent of the nation's fisheries workforce and play a vital role in rural food security, are under mounting pressure from overfishing, habitat degradation and an outdated legal framework. Although concern about these challenges has grown and a diverse group of stakeholders is eager to contribute to sustainable solutions, the existing array of central, state, and local policies remains fragmented and inadequate in enforcement and community engagement. To protect the livelihoods of millions and ensure the long-term health of inland aquatic ecosystems, India must embrace a holistic policy reform that emphasizes responsible resource management, guarantees secure fishing rights, and embeds genuine co-management with local communities. By bridging regulatory gaps, encouraging fisher participation in decision-making and adopting adaptive, evidence-based strategies, the country can turn its inland fisheries into a resilient and productive sector. The growing awareness and willingness to act among researchers, policymakers and fishers alike make this an opportune moment to unite efforts and forge a sustainable future for India's inland fisheries.

Things to Remember

- Small-scale inland fisheries in India employ over 2.3 million people and significantly contribute to food security and livelihoods.
- These fisheries face threats from overfishing, habitat degradation, and outdated regulations.
- The current patchwork of policies lacks focus, enforcement, and community participation.

- A comprehensive policy overhaul is needed to prioritize sustainable resource management, secure fishing rights, and empower small-scale fishers.
- This will ensure the sustainability of fisheries, livelihoods, food security, and healthy ecosystems.

References

Arthur RI (2020) Small-scale fisheries management and the problem of open access. Mar Policy 115:103867

Bhattacharya S, Das BK, Roy A, Nayak PK, Saha A, Parida PK, Lianthuamluaia L, Mondal K, Chakraborty S (2024) Socio-economic profile and vulnerabilities assessment in small-scale fisheries of Lower Gangetic flood plain wetland: an initiative towards achieving wetland ecosystem sustainability and community well-being. Sustainability 16(21):9583

Chandra R (2011a) Institutional and legal dimensions of inland fisheries in India

Chandra R (2011b) Property rights and inland fisheries in India: issues and perspectives. CIFRI Technical Paper

Chandra R (2014) Customary rights and local governance in Indian inland fisheries

Chandra R, Bhattacharyya A (2016) Fisheries, institutions, and sustainability: India case studies

Chandra R, Das B (2019) Policy and governance in inland wetland fisheries: a review

Chuenpagdee R, Jentoft S (2018) Transforming the governance of small-scale fisheries. Marit Stud 17:101–115

Constitution of India (Art. 246, Schedule VII, List I, Entry 1) Government of India (1950). https://www.mea.gov.in/images/pdf1/S7.pdf

Constitution of India, Article 48A. Government of India (1950). https://indiankanoon.org/doc/871328/

Das BK, Roy A (2022) Small-scale inland fisheries: an answer to livelihood and nutritional security in India. Retrieved from https://www.icsf.net/wp-content/uploads/2022/11/Small_Scale_Inland_FIsheries_BK-DAS_AParna-Roy.pdf

Das BK, Parida PK, DebRoy P, Roy A (2022a) A situational analysis of small-scale inland open water fisheries in India: from vulnerability to viability

Das BK, Parida PK, DebRoy P, Roy A (2022b) A situational analysis of small-scale inland open water fisheries in India: from vulnerability to viability. Work Paper V2V:2022–2028

FAOLEX (2023) Legal and policy database on fisheries and aquaculture. Food and Agriculture Organization of the United Nations. Retrieved from https://faolex.fao.org

Food and Agriculture Organization of the United Nations (2024) A review of the inland fisheries of India. Retrieved from https://openknowledge.fao.org/items/38d9d218-7178-45d9-ba81-bf6b137adde9

Government of Andhra Pradesh (2015) Fisheries Policy 2015–2020

Government of Bihar (2008) Fisheries Policy

Government of India (1897) The Indian Fisheries Act. Ministry of Agriculture and Farmers Welfare

Government of India (1974) The Water (Prevention and Control of Pollution) Act. Ministry of Environment, Forest and Climate Change

Government of India (1986) The Environment (Protection) Act

Government of India (2002) The Biological Diversity Act

Government of India (2010) Wetlands (Conservation and Management) Rules

Government of Madhya Pradesh (2008) Fisheries Policy

Government of Odisha (2015) Odisha Fisheries Policy

Government of Punjab (1914, 1965) Punjab Fisheries Act & Rules. Department of Fisheries, Punjab

Government of West Bengal (2020) State Inland Fisheries Policy Documents

Henquinet JW, Dobson T (2005) The public trust doctrine and sustainable ecosystems: a Great Lakes fisheries case study. NYU Environ Law J 14:322

Huppart J (2005) Leasing systems and fisher welfare in inland waterbodies

Jena JK, George G (2018) Small-scale fisheries in India: An appraisal

Johnson D (2013) Livelihoods in the context of fisheries – a governability challenge. In: Chuenpagdee R (ed) Governability of fisheries and aquaculture: theory and applications, pp 67–86

Kelkar N, Arthur RI (2022a) Access and tenure in South Asian inland fisheries

Kelkar N, Arthur RI (2022b) Reframing inland fisheries governance: tenure, access, and equity, FAO Fisheries and Aquaculture Circular No. 1183. FAO, Rome

Kumar RS, Kumar DK, Prakash S, Krishnan M, Ojha SN (2019) Status of small-scale inland fisheries in Bihar, India. SSRG Int J Econ Manag Stud 6(8):224–229

Ministry of Fisheries, Animal Husbandry & Dairying (2021) About the department. Department of Fisheries, Government of India. Retrieved from https://dof.gov.in/about-us/about-department

NABARD (2018a) Financing inland fisheries and aquaculture: Policy, potential and opportunities. National Bank for Agriculture and Rural Development

NABARD (2018b) Status of fisheries and lease practices in Indian states. National Bank for Agriculture and Rural Development

National Fisheries Development Board (2020–2024) State fisheries policies and schemes. Government of India. Retrieved from https://nfdb.gov.in

Perspective of Small-Scale Fisheries in Chilika Lagoon

11

Arun Pandit, Anjana Ekka, Basanta Kumar Das,
L. Chakraborty, and S. K. Mohanty

Abstract

Chilika Lagoon, Asia's largest brackish water lagoon and a designated Ramsar Site, supports a rich biodiversity and sustains the livelihoods of approximately two lakh fishers residing in 141 shoreline villages. Small-scale fisheries (SSF) dominate the lagoon's fishery landscape, with fishing serving as the primary occupation for over 70% of fisher households. This chapter presents a comprehensive socio-economic assessment of Chilika's SSF, examining fishing practices, income patterns, institutional arrangements, livelihood diversification, and vulnerability to natural disasters. The study highlights the significant dependence of marginalized communities on fishing, despite limited access to assets such as boats and modern gear. Institutional frameworks include a multi-tiered governance structure involving FISHFED, Primary Fishermen Cooperative Societies (PFCS), the Chilika Development Authority (CDA), and local administrative bodies. However, challenges such as informal credit systems, market dependency on middlemen (Mahajans), and conflicts over leasing rights, particularly between traditional and non-traditional fishers which undermine the financial stability of fisher households. The study also documents the sectoral differences in fishing intensity, boat ownership, and gear types, reflecting the ecological diversity across northern, central, southern, and outer channel sectors of the lagoon. The resilience assessment framework identifies major gaps in financial

---

Beneath Chilika's Waters: Stories of Resilience and Reform in Small-Scale Fisheries

---

A. Pandit (✉) · A. Ekka · B. K. Das · L. Chakraborty
ICAR – Central Inland Fisheries Research Institute (ICAR-CIFRI), Barrackpore, Kolkata, India

S. K. Mohanty
Chilika Development Authority, Bhubaneswar, Odisha, India

B. K. Das et al. (eds.), *Small-scale Inland Fisheries in India*, Sustainability Sciences in Asia and Africa, https://doi.org/10.1007/978-981-95-3849-2_11

and infrastructural capital among fisher households. The chapter concludes with policy recommendations emphasizing livelihood diversification, cooperative strengthening, sustainable exploitation, and inclusive governance to ensure long-term socio-ecological sustainability. A new fisheries policy under development must balance economic aspirations, ecological conservation, and social justice for all stakeholders, particularly the traditional fishing communities whose heritage and well-being are intimately linked to the health of Chilika Lagoon.

Keywords

Chilika Lagoon · Small-scale fisheries · Livelihood · Resilience · Governance · Cyclone impact

11.1 Introduction

In Asia, Chilika, the largest brackish water lagoon, lies between 19°28′ and 19°54′ N and 85°05′ and 85°38′ E. It is famous for its vast water spread area, scenic beauty, variety of biodiversity and is considered as an eco-wonder. The lagoon became a Ramsar Site in 1981 due to its rich biodiversity and importance in socio-economic life of riparian communities. Ministry of Environment and Forests, Government of India included this lagoon in the list of wetlands for conservation and management. The maximum area of the lagoon is 1165 km^2 during monsoon and minimum 906 km^2 in dry season. The length (north-South) is 64.3 km and the mean width is 20.1 km (Ghosh and Pattanaik 2005). The lagoon is an example of a unique assemblage of marine, brackish and fresh water ecosystems. The lagoon harbors high biodiversity and it is the abode of several endangered and endemic species. The Chilika can be divided into four ecological sectors. The vast water body falls in eight blocks of three districts (Ganjam, Khurda and Puri) of Odisha. In Odisha, the fishing activities are mainly concentrated around the Chilika lagoon (Nayak et al. 2008). It also serves as the largest wintering ground for water fowls.

141 villages of fishers reside in the shoreline of Chilika. Around 2 lakh fishers are drawing livelihood from this lagoon (Mohapatra et al. 2014). The major chunk of the fishers is very poor and belong to the lowest strata of the society. The livelihood of these fishers largely depends on the Chilika wetland system. Rich fishery, in the form of fish, invertebrate and crustacean species of this lagoon generates over 6% of Odisha's international export earnings.

11.2 Stakeholders, Institutional Arrangements, and Governance

The Chilika covers the Puri, Khordha, and Ganjam districts to the adjoining district of Odisha. There are many stakeholders in managing the fisheries of the lagoon. The district collectors of Ganjam, Puri and Khordha grants, lease the fishery resource of

Chilika to the FISHFED for sub leasing the areas in favor of the PFCS. He/she also issues specification of fishing gears. At the lowest level of institutional arrangement, there are Primary Fishermen Cooperative Societies. There were 158 active PFCS in 2018 (The Hindustan Times 2018). The Registrar of Cooperative Societies implement the policies and programmes of the Central and State Governments. Their major aim is to make available schemes and benefits to the farmers and other economically weaker sections. The administration, monitoring, and supervision of the cooperative societies are done by the Directorate. The Asst. Register of Cooperative Society (ARCS) has been established during the year 1959 to registrar the Fishermen Co-operatives in the capacity of Asst. Registrar of Co-op. Societies (Fy) Chilika Circle, Balugaon with the help of sub- Asst. Registrar of Co-op. Societies & Inspector of Co-op. Societies. It manages and supervises all the PFCSs of the Chilika Circle. The main aim and objective of the ARCS is to promote socio-economic condition the poor fishermen.

The Revenue Dept of the Odisha Government collects the revenue from FISHFED. FISHFED gets 10% of the lease value. District Fishery Office, Chilika Wildlife Division is the implementation agency of Forest Act and Wildlife Act in the lagoon. They also manage the Nalbana Wildlife Sanctuary. Asst. Director of Fisheries, Balugaon register and issue licenses to all the fishing boats. Eviction of illegal and un authorized fish culture was the major task of the 'Task force' at the district level. The Asst. Director of Fisheries is also the member of the Task Force. The ADF also collect fish catch data along with CDA, monitor physico-chemical parameters of the lagoon, and implement OMFR Act and Rule. Overall management, conservation, and research of the lagoon are the responsibility of the Chilika Development Authority (CDA). CDA, which was established in 1992, has been executing the activities of restoration and sustainable management of the lagoon. To uplift the economic condition of the community, the authority also does various developmental activities either on its own or through some other agencies. In May 2017, the CDA, with the help of Odisha Marine Fishery Regulation Act 1982, undertook the job of eviction drive to clear unauthorized gheris for prawn culture. The Authority also banned the use of zero nets for conservation of fishery resources. The Police is an another stakeholder, which maintains the law and order in and around Chilika lagoon.

11.3 Perspective of SSF in the Chilika Lagoon

The fishery in the Chilika is predominantly small scale in nature. Fishing is the major occupation of large percentage of households providing more than 50% of their livelihood. They are socially and economically marginalized people with poor resource base. Even many of the fishers could not afford to buy the essential crafts and gears required for fishing. They act as fishing labor or "crew member." The ICAR-CIFRI study during 2019 showed that around 65–70% of the fishermen households owned boats across different sectors (Table 11.1). Overall, among the boat owners, around 36% households possessed motorized boats and 64%

Table 11.1 Boat ownership by the fishers (%)

Sector	% household owned boat	Among boat ownership	
		% motorized	% non-motorized
Northern	68.53	42.5	58.57
Central	62.50	38.62	61.00
Southern	72.56	31.44	68.66
Outer channel	78.21	30.12	70.16
Total	70.24	36.63	63.75

Source: ICAR-CIFRI study, 2019

Table 11.2 Fishing gears operated by the fisher households (% of households)

Net/gear	Overall	Northern	Central	Southern	Outer channel
Khonda	73.11	60	83	73	83
Dubi	28.44	18	29	43	29
Zero net	17.22	23	15	13	15
Menjia	16.67	14	17	20	17
Khepa	2.44	–	–	–	11
Patua	6.22	14	7	–	–
Dosti	2.22	–	–	10	–

possessed non-motorized boats. For single or two fishermen fishing, small boats were used, while for multi-fishers operations larger boats were used. Both non-motorized boats as well as motorized boats with out-board engine were found to be used by the fishers. However, Das and Choudhury (2013) found less boat possession by the fishers. They reported that in all the sample villages, less than 35% respondents had either motorized or non-motorized boat. However, in Arakhakuda village, their study found that 47.3% of sample respondents have non motorized boat.

A variety of nets were possessed and used by the fishers (Table 11.2). *Khonda*, a traditional box net, was the predominant gear possessed and used by the fishers of Chilika lagoon. It was found that in general, each family had 1 to 3 members involved in fishing activities. The fishermen typically spent between 5 and 12 h a day fishing, with the duration varying according to the season.

11.3.1 Fish Catch, Fishers' Income, and Livelihood

The average annual fish landing from Chilika during 2011–12 to 2015–16 was 12,781 tonnes, of which brackish water prawn comprised 36.5% of the total catch and 55.7% of the total value (Table 11.3). At the 2018 farm-gate price level, the economic value of this landing was Rs. 234.5 crores. If this value is averaged and spread over the fishers, each fisher household would get an annual income of Rs. 70,420.

The survey of ICAR-CIFRI (2019) estimated the overall yearly income from all sources at around Rs. 73,000 across the sectors (Table 11.4). The fishing in Chilika

Table 11.3 Group-wise landing of fish and shell fishes (average of 2011–12 to 2015–16)

Fish group	Avg landing (tonnes)	Value (Rs. lakh)
BW Prawn	4665.6	13,064
FW Prawn	436.9	677
Mullets	901.0	1351
Clupeoides	1519.9	1292
Perches	798.3	2994
Beloniformes	290.2	116
BW catfishes	1324.0	927
FW catfishes	27.2	19
Sciaenids	544.9	504
Threadfins	252.5	341
Cichlids	365.7	183
Elopiformes	3.9	4
Elasmobranchs	2.2	5
Tripod	179.3	201
Featherbacks	344.4	344
Murrels	182.8	128
Eels	101.4	81
Carps and minnows	266.4	293
Misc.	230.9	35
Crab	343.8	894
Total	12,781.3	23,452

Table 11.4 Annual income of the fisher households (in Rs.)

Sector	Total income	Income from fisheries	Share of fisheries in total income (%)
Northern	64,784	49,139	75.85
Central	67,025	50,153	74.83
Southern	83,406	65,518	78.55
Outer channel	73,469	58,884	80.15
Overall	73,182	56,857	77.69

contributed around 78% of the total income. The estimated income is very close to the revenue obtained from total value of fishes. The income was significantly lower for "fishing labor" group who did not possess any boats. Among all the sectors the income was higher in southern sector, contributed by more income from fisheries.

According to Das and Choudhury (2013), 55% of the respondents in the Chilika region had an annual income of less than Rs. 50,000. Kumar et al. (2011) highlighted that over 70% of the households were living below the poverty line, significantly higher than the state average of 52%. Moreover, only 16% of the population had access to essential social services such as clean drinking water and proper sanitation facilities. Formal economic infrastructure was also dismal. They also found that 76% of the population took loan from moneylender who often charges exorbitant interest rates. Mohanty (2012) in a village of Chilika lagoon found that there were 65% BPL and 35% APL families. The highest amount of expenditure was on food, 53%. About 90% of the village people were depend on fishing and only 10% on agriculture and other business.

Hence, all the studies show that the income is much below as compared to the state average (Rs. 101,550 per capita per annum in 2017–18 at current prices; https://statisticstimes.com/economy/india/odisha-economy.php). The ICAR-CIFRI found that migration and occupational diversification were also practiced by the fishers to supplement the livelihood. The study identified as many as 19 sources from which one household or the other derived the livelihood. The investigation further revealed lack of sufficient institutional credit systems and a poor asset base of the fishers and hence fishermen are compelled to take loans from the middlemen (*Mahajan*). As observed by Kumar et al. (2011) the share of fisheries in total income has declined over the period. They found that around 42% of the households derived incomes from sources other than fishing which was nearly nil in the past.

11.3.2 Disposal of Catch

There were 18 landing centers registered by CDA. Kalupadaghat, Bhusandapur and Sorona and Balugaon are the major four among the lot. These 4 centers handle 70% of the landing. Many fishers land their catches on the village itself. Many fish collection boats, purchase fish from the contracted fishers. They also supply food, water to the fishers. These contract fishers operate for a longer period of time. Some operate 2–3 weeks per fishing voyage.

The fishing and fish marketing system operated in a manner that placed fishermen at a disadvantage. One of the main financial challenges they faced was the lack of access to formal credit institutions and their limited asset ownership. As a result, fishermen often relied on loans from middlemen (Mahajans), who imposed conditions requiring them to sell their entire catch exclusively to them. These Mahajans typically set the purchase prices 10–12% lower than prevailing market rates. Over 65% of fishing households were found to be in debt, primarily relying on informal credit sources.

Kumar et al. (2011) observed that a complex trading structure had developed in the Chilika lagoon, involving three key actors: fishers, commission agents, and Mahajans. Mahajans acquired the full catch through informal agreements made with commission agents and, in return for offering interest-free loans, secured exclusive rights to the harvest. Commission agents often entered into similar arrangements with the fishermen. This system created an intricate web of informal financial dependencies among the three levels, as described by Shimpei and Shaw (2009).

Kumar et al. (2011) reported that over 90% of the total catch from approximately 33,300 fishers was controlled by around 500 middlemen and 800 local traders. Consistent with the findings of Shimpei and Shaw (2009), their study also revealed that fishermen often borrowed money or received advances from middlemen at high

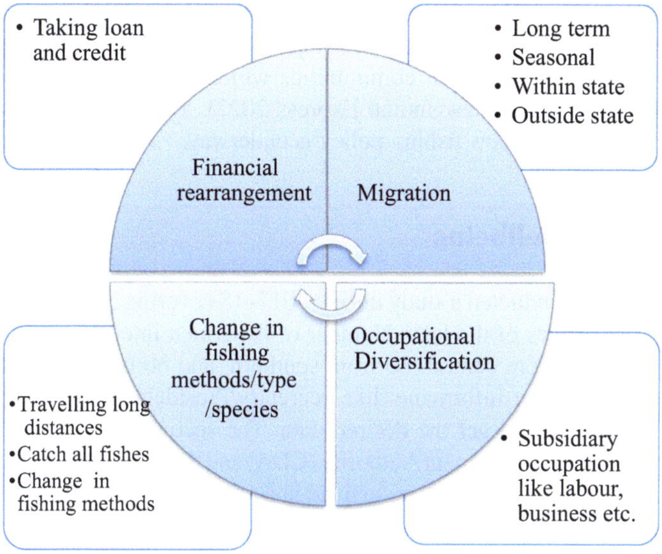

Fig. 11.1 Livelihood strategies adopted by the Chilika fishers

interest rates. The availability of basic amenities among the fishing community was significantly lower than the state average. A 2007 survey showed the literacy rate among fishers was just 42%, compared to the state average of 63%, while access to drinking water and sanitation facilities stood at only 14.75% and 14.6%, respectively.

Das and Choudhury (2013) recommended that both central and state governments should introduce various welfare programs aimed at improving the living standards of communities residing in the Chilika lagoon region (Fig. 11.1).

11.3.3 Conflict Among Fishers-Non-fishers

Shrimp farming was introduced in the Chilika lagoon during 1984–85 as a means to boost the income of local fishers. The promise of higher profits from prawn cultivation drew many individuals to take up aquaculture in the region (Pradhan and Flaherty 2007; Pattanaik 2008). As a result, the area under aquaculture in and around Chilika expanded significantly—from 3856 km^2 in 1994 to 5831 km^2 by 2004 (Mishra and Griffin 2010).

The entry of non-fishing communities, often with better resources and modern fishing gear, created pressure on traditional fishing practices and equipment. This led to the loss of traditional fishing grounds for local communities, resulting in occupational displacement. Such developments triggered tensions and conflicts between the traditional fishers and incoming groups (Dujovny 2009). However, intervention by the Court and Government policy did a favor to the fishermen. FISHFED gets the leasing rights from the District collectors every year. Subsequently, the FISHFED sub-leased to PFCSs. The lease value is increased 10% every year.

Capital-affluent non-fishers also get these sub-letting rights. The illegal prawn culture continued for long, however, were being dismantled. There has been a resentment among the non-fishermen communities which demanded the protection of their livelihood too (The New Indian Express 2022). The Hon'ble High Court of Orissa intervened and a new fishing policy is underway.

11.4 Fishers' Wellbeing

The ICAR-CIFRI conducted a study during 2017–18 covering 300 fishermen spread across the four sectors of the lake. Number of fishermen interviewed was 80 from northern sector, 80 from central, 80 from southern, and 60 from outer channel. In addition to that 18 key informants like secretary/President of the Coop. Society were also interviewed to get the desired data. The secondary data were also collected from Chilika Development Authority (CDA) and the Dept. of Fisheries, Odisha.

The study found that each sub caste lived in a particular hamlet and generally did not live with other sub castes. There are few fishing villages where a sub caste group keep good relationship with other sub caste group in the village. As per the Fishery Dept, there were around 7500 fishing boats in the lagoon. Keuta (also known as Kaibarta or Khatia) was the predominant (68%) fishing caste in the sampled villages. Kandara (14.2%), Tiara (6.3%), Nolia (6.7%) were the other major fishing castes. Das and Choudhury (2013) also found that fisher caste was divided into five sub caste groups. Those are Keuta, Kandara, Nolia, Tiara, and Bengali refugee. Further, similar to the present study, Keuta and Kandara were predominant in the sample as they jointly contributed 71.67% of the sample (Table 11.5).

There were 23,115 fishermen households in and around the Chilika lagoon and their population was 1.46 lakh (2011 census). Around 30% of fishers household possessed agricultural land in northern, central and southern sectors. Fishing was the primary occupation to around 73% of the households (Table 11.6).

Fishing was the primary occupation to around 73% of the fisher respondents; it ranged from 55 in central to 90% in outer channel sectors. Agriculture farming was the second most significant avenue of livelihood, particularly to northern and central sector fishers. In some areas like Arakhakuda, Satapada, tourism is also a sector where many youths of the fishing families are engaged. The sector is booming: the annual number of tourists visiting the wetland during 2000–2006 averaged 378,370, an increase of over 60% as compared to arrivals during 1994–1999 (Kumar et al. 2011).

Table 11.5 Household characteristics of the fishermen

Particular	Northern	Central	Southern	Outer channel
Avg. age of respondent (years)	48	46	48	43
Avg. family size	6	6.7	6.3	5.5
Literacy % (members above >6 years)	71.3%	72.3%	79.68	77.4%
% HH owned agri. land	39.9%	44.5%	27.6%	6.6%

Table 11.6 Primary occupation of the fisher households (HH)

Population	Northern	Central	Southern	Outer channel	Total/ average
No. of fishers HH	8288	4027	7024	3776	23,115
Fisher population 2011	60,571 (41.5%)	22,866 (15.6%)	40,163	22,431	146,031
Primary occupation (% of HH)					
Fishing	67	55	81	90	72.6
Fish trading	1	3	6	–	2.3
Middleman	0.7	2.1	~1	<0.5	0.97
Agriculture farming	20	23.5	4	3	13.4
Petty Business	4	10.3	3	4	5.2
Govt./private service	1.5	2.3	~1	1	1.4
Others	5.8	3.8	5	3	4.6

11.5 Impact of Disasters on Chilika Fisheries

Since the fishers family are resource poor, the natural disasters have profound negative impact on their livelihood. As a case study, an investigation was carried out by the ICAR-CIFRI in Chilika lake to assess the impact of severe cyclonic storm, Fani, on livelihood of fishers in 2019. The study found that the storm created a havoc in the rural economy of the affected areas of the lagoon. The fishers households lost considerable assets and livelihoods as a result of the storm. In Chilka outer channel and Chilka north sectors, 55% and 22% of the semi-*pucca* houses were partially damaged. Almost all the *kachha* houses in the Chilika outer channel were either partially or severely destroyed. Among them around 38% were fully damaged. Overall, across all the sectors around 28% of the *kachha* houses were severe damaged. Electrical installations, together with other household items, were badly hit by the cyclone.

Many fishers lost the livestock and poultry. In Puri district, the stored agricultural produces were damaged to a large extent. Loss of fishing nets was highest in Chilika outer channel in Puri district, where 75% fishers reported net loss. Around 68% of the boats in the outer channel were severely damaged. The corresponding figure for the central sector was 6%. Damages in nets and boats not only caused loss of a important asset but also caused loss of fishing opportunities. Overall, fishermen could not go for fishing for 7–22 days starting 2–3 days before the storm.

11.6 Resilience of the Fishermen Families

Study of resilience of the household is crucial to understand how effectively a system can respond to various challenges (Pandit et al. 2021). Assessing household resilience is crucial to understanding how effectively a system can respond to various challenges (Pandit et al. 2021). Such evaluations also enhance managerial capabilities, as identifying the sources of resilience supports improved decision-making. This allows stakeholders to pinpoint areas with low resilience and implement necessary corrective actions. Ultimately, this process contributes to building a stronger management framework and reducing the long-term vulnerability of fishing households. Household resilience plays a key role in enabling affected individuals to recover from damage and return to normalcy. As noted by Collins (2014) and Rawlani et al. (2014) in the context of Bangladesh, coastal communities are particularly susceptible to cyclones and storm surges driven by high winds.

The ICAR-CIFRI conducted a study (Pandit et al. 2021) during 2019 in the coastal state of Odisha. The assessment was based on a set of indicators developed with slight modifications to the framework proposed by Béné et al. (2009). The study considered nine resilience indicators grouped under four types of livelihood capital: natural, physical, financial, and human. Findings revealed that only three of the nine indicators fell into the "desirable" category, while all indicators under financial capital were categorized as being in "crisis." Fishers had limited savings—only 7.5% of their income—and lacked sufficient income diversification. In terms of infrastructure, just one indicator was found to be "desirable." Overall, out of 12 indicators assessed, six were in the "crisis" category, two were "stable," and only four were "desirable." The study concluded that with 50% of the indicators falling under the "crisis" category, urgent interventions are necessary to shift them toward the "desirable" level.

11.7 Suggestions and Policy Recommendations

A longstanding conflict exists between traditional fishermen and non-fishermen. The traditional fishing livelihood has faced challenges, particularly due to the excessive use of motorized boats (Ray and Garada 2018). The Hon'ble Orissa High Court had directed the concerned department to develop a new fishing policy on Chilika fisheries by the end of 2023 (The Times of India 2023). The new policy should incorporate the demand of traditional fishermen, demand of the professional fishermen (non-fishermen), and the sustainability of the lagoon fisheries.

There are many fishermen cooperatives numbering approximately 200 in 2016; however, many of them are not properly active. These cooperatives should be revived and made active. Additionally, FPOs, SHGs may also be formed to express their grievances, highlight their struggles, and suggest remedies to address their challenges (Das and Choudhury 2013).

Special efforts may be made for upliftment of education status of the fishers and their family members. Education and awareness may help them to be freed from the clutches of the commission agents and other middlemen. Fishermen will be able to sell their catch at proper price. They should be made aware of the different government schemes available for them. Easy and hassle-free institutional loan may be provided at the landing centre itself. Awareness and incentives may be provided to the fishermen to exploit the lagoon sustainably.

The government should ensure the provision of adequate infrastructure, such as clean and cemented landing centers, well-constructed jetties, cold storage, and transportation facilities. Ray and Garada (2018) emphasized the need for efficient management of motorized boats and strict regulation of fishing rights to protect the lake's rich biodiversity and support the sustainability of its fisheries.

It has been seen that the present level of exploitation in the lagoon is higher than the MSY (11,376 MT as estimated by ICAR-CIFRI), it would be much higher if the production from illegal prawn gherries are taken in to account. There should be rationalization of the exploitation level. The authorities may create opportunities for supplementary income generating activities. Eco-tourism, ice factory, fish processing facilities, marketing infrastructure, self employment schemes like tailoring, carpentry, masonry work, e-rickshaws, and small business opportunities may ease off the fishing pressure. The upcoming new fishing policy should consider the interest of all the stakeholders, particularly of the traditional fishermen, and the same may implemented as soon as possible.

11.8 Conclusion

The small-scale fisheries of Chilika Lagoon are intricately woven into the ecological fabric and socio-economic livelihood of over two lakh marginalized fishers. Despite being the backbone of the lagoon's vibrant fishery economy, these communities continue to face persistent challenges—ranging from natural disasters and shrinking fishing rights to socio-political conflicts and exploitation by informal credit networks. The increasing vulnerability, inadequate infrastructure, and weak institutional support have constrained the resilience of fisher households. However, with appropriate policy reforms, cooperative revival, sustainable fishery practices, and inclusive governance, Chilika can become a model of equitable fishery development. Future strategies must center around empowering traditional fishers, diversifying incomes, conserving biodiversity, and ensuring participatory management to preserve both the ecology of Chilika and the dignity of those who depend on it.

Points to Remember

- Chilika Lagoon, Asia's largest brackish water body, supports over 2 lakh fisherfolk and contributes significantly to Odisha's fishery-based economy.
- Fisheries are predominantly small-scale, with limited asset ownership—only 36% boats are motorized, and many fishers operate as laborers or under middlemen.
- The fishing governance system involves multi-level stakeholders, including PFCSs, FISHFED, Chilika Development Authority, and local administrative bodies.
- Middlemen (Mahajans) dominate the trade, offering informal credit and controlling over 90% of fish sales, often at unfair prices.
- Fishers face multiple vulnerabilities—from climate disasters like cyclone Fani to income instability and poor access to education, health, and sanitation.
- Incomes are low, with average annual household earnings around ₹73,000—well below the state average—and 78% of that coming solely from fisheries.
- The chapter emphasizes urgent policy reform, sustainable exploitation, financial inclusion, and diversified livelihoods (e.g., eco-tourism, SHGs, cold chains) for long-term resilience.

References

Béné C, Kodio A, Lemoalle J, Mills D, Morand P, Ovie S, Sinaba F, Tafida A (2009) Participatory diagnosis and adaptive management of small-scale fisheries in the Niger River Basin. CGIAR CPWF Project Report 72. www.waterandfood.org

Collins J (2014) A rising tide in Bangladesh: livelihood adaptation to climate stress. Aust Geogr 45:289–307

Das S, Choudhury J (2013) Fishermen of Chilika Lake of Odisha: an enquiry into their socio-economic conditions. Srusti Manag Rev 6(2):54–60

Dujovny E (2009) The deepest cut: political ecology in the dredging of a new sea mouth in Chilika Lake, Orissa, India. Conserv Soc 7:192–204

Ghosh AK, Pattanaik AK (2005) Chilika lagoon: experience and lessons learned brief. GEF Lake Basin Management Initiative. http://www.worldlakes.org. Accessed 28 Apr 2010

ICAR–Central Inland Fisheries Research Institute (2019) Annual report 2019. ICAR–CIFRI

JICA (2009) Socio-economic condition of fishers in and around Chilika. Report submitted to Chilika Development Authority

Kumar R, Horwitz P, Milton GR, Sellamuttu SS, Buckton ST, Davidson NC, Pattnaik AK, Zavagli M, Baker C (2011) Assessing wetland ecosystem services and poverty interlinkages: a general framework and case study. Hydrol Sci J 56(8):1602–1621

Mishra S, Griffin AL (2010) Encroachment: A threat to resource sustainability in Chilika Lake, India. Applied Geography 30(3):448–459

Mohanty AK (2012) Socio-economic condition of fishermen of Samala Village of Chilika Lagoon, Odisha

Mohapatra S, Sathyan N, Afsal VV, Thomas JV (2014) Empowerment of fishermen of Chilika lagoon, Odisha: a NETFISH-CDA joint venture. Fish Chimes 34(2&3):40–44

Nayak SK, Swain P, Nanda PK, Dash S, Shukla S, Meher PK, Maiti NK (2008) Effect of endotoxin on the immunity of Indian major carp. Labeo rohita, Fish and Shellfish Immunology 24(4):394–399. https://doi.org/10.1016/j.fsi.2007.09.005

Pandit A, Saha A, Chakraborty L, Swain HS, Sharma SK, Mohanty BP, Das BK (2021) Resilience of inland fishers against nature's fury: a study on the effect of extremely severe cyclone Fani on socio-economy and livelihood of fishers' households in Odisha, India. Aquat Ecosyst Health Manag 24(4):75–84

Pattnaik AK (2008) Management of coastal lagoon ecosystems: The Chilika experience. Chilika Development Authority

Pradhan D, Flaherty M (2007) National initiatives, local effects: Trade liberalization, shrimp aquaculture, and coastal communities in Orissa, India. Society & Natural Resources 21(1):63–76

Rawlani AK, Sovacool BK, Ahammad R, Nandy P, Hussain P, Collins J, Islam MM, Sallu S, Hubacek K, Paavola J et al (2014) Building responsiveness to climate change through community-based adaptation in Bangladesh. Clim Dev 124:733–746

Ray S, Garada R (2018) Boat automation and fishery livelihood: a case of Chilika Lake in Odisha. Environ Dev Sustain 20(5):2399–2414

Shimpei I, Shaw R (2009) Linking human security to natural resources: perspective from a fishery resource allocation system in Chilika Lagoon, India. Sustain Sci 4(2):281–292

The Hindustan Times (2018) Odisha officials advise against seaplane service in Chilika lake over ecological concerns. https://www.hindustantimes.com/india-news/odisha-officials-advise-against-seaplane-service-in-chilika-lake-over-ecological-concerns/story-EHPUeeZBMXwqP1MNjU9uwK.html

The New Indian Express (2022) Odisha govt directed to take policy decision on non-fishermen in Chilika. Published 7 Apr 2022. https://www.newindianexpress.com/states/odisha/2022/apr/07/odisha-govt-directed-to-take-policy-decision-onnon-fishermen-in-chilika-2439010.html

The Times of India (2023) Odisha's Chilika lake development under scrutiny. https://timesofindia.indiatimes.com/articleshow/100264834.cms

Small Scale Fisheries in Wetlands of West Bengal and Bihar: Approaches and Way Forward

12

Suman Kumari, Gunjan Karnatak, Mishal P., and Basanta Kumar Das

Abstract

Wetlands in the states of West Bengal and Bihar play a crucial role both ecologically and economically, serving as key hubs for one of India's largest small-scale fisheries (SSF) sectors. These freshwater ecosystems support biodiversity, contribute to food security, and sustain rural livelihoods. However, despite their significance, the SSF sector in these regions grapples with various challenges such as environmental degradation, weak institutional frameworks, and suboptimal productivity. This chapter examines the nature, relevance, and present condition of SSF in the floodplain wetlands of eastern India. It underscores the socio-economic contributions of artisanal fishing communities while identifying persistent issues like limited access to modern technologies, fragmented markets, and insufficient policy support. The discussion incorporates the vulnerability-to-viability (V2V) framework, supported by field observations and case studies from selected wetlands in Bihar and West Bengal. To address these challenges, the chapter explores promising interventions such as culture-based fisheries, cage and pen aquaculture, and stock enhancement practices. These technologies offer scalable pathways to boost fish yields, enhance sustainability, and strengthen community resilience. The chapter advocates for cohesive institutional mechanisms, effective habitat and fisheries governance, and empowerment of local fishers to close the gap between the potential and actual productivity of wetlands. It concludes by proposing a strategic framework that integrates sustainable prac-

---

Empowering Small-Scale Fisheries through Wetland Sustainability: Challenges, Interventions, and Opportunities in Bihar and West Bengal

---

S. Kumari · G. Karnatak · Mishal P. · B. K. Das (✉)
ICAR – Central Inland Fisheries Research Institute (ICAR-CIFRI), Barrackpore, Kolkata, India

B. K. Das et al. (eds.), *Small-scale Inland Fisheries in India*, Sustainability Sciences in Asia and Africa, https://doi.org/10.1007/978-981-95-3849-2_12

tices, community engagement, and policy coherence to ensure the long-term ecological, social, and economic viability of small-scale fisheries.

Keywords

Small-Scale Fisheries (SSF) · Wetlands · Floodplain Ecosystems · Culture-Based Fisheries (CBF) · Biodiversity

12.1 Introduction

Wetlands are among the most productive and biologically diverse ecosystems on Earth, playing a vital role in supporting the small-scale fisheries (SSF) sector across the globe. These ecosystems cover an estimated area of 5.3 to 12.8 million square kilometers, accounting for approximately 9% of the planet's land surface. Wetlands offer a wide array of ecological services, including functioning as breeding and feeding grounds for numerous fish species, including shellfish. Additionally, they serve critical functions such as filtering water, storing and sequestering carbon, acting as carbon sinks, conserving biodiversity, and regulating climatic variables—alongside supporting SSF and generating livelihoods for millions of fishers. Small-scale fisheries are essential for ensuring food and nutritional security and play a key role in the livelihoods of local communities dependent on wetland resources. Although the term "small-scale fisheries" has traditionally been associated more with marine environments, its application to inland water bodies has gained wider acceptance and is now considered highly appropriate. SSF, often referred to as artisanal fisheries, typically involve fishers operating small, often non-motorized or wooden boats with low capital investment and intensive physical effort to harvest fish from inland waters for subsistence and income.

In recent times, global attention has increasingly acknowledged the critical contributions of SSF to food security, nutrition, and livelihoods at both local and national levels. This sector provides employment to over 1.2 million individuals globally, with about 97% of them based in developing countries. Fishers in the SSF sector are involved in various related activities such as net mending, gear repair, and low-cost fish processing. Many engage in seasonal or occasional fishing operations to earn short-term income. The majority of small-scale fishers are self-employed and rely on fishing to meet household food requirements and generate income. Global inland fish production has shown a steady upward trend, reaching 12 million tonnes in 2019, accounting for nearly 12% of the total global capture fisheries production. India was a leading contributor to this figure, producing approximately 1.8 million tonnes. However, a slight decline in global inland fish output to 11.5 million tonnes was recorded in 2020, attributed to the disruptions caused by the COVID-19 pandemic. Asia continues to dominate inland fish catch statistics globally, with small-scale fisheries being the primary contributors to this production, mainly supplying fish for local consumption. In the Indian context, around 16.1 million fisher families are engaged in various fisheries-related activities, with over 65% of them being men (Fisheries Statistics of India 2023). The fisheries sector contributes significantly to the national economy, accounting for 1.1% of India's Gross Domestic Product (GDP) and 6.72% of the agricultural GDP. Additionally, the sector has been experiencing a robust annual growth rate of 10.34% in total fish production.

The Government of India has set the target to achieve fish production from 14 to 22 million metric tonne by 2025 including marine, inland fish capture and aquaculture. Fish production from capture fisheries of marine and inland has reached to the stagnant phase there is scope for increasing production from wetlands, reservoirs, and ponds. An inland fisheries resource includes rivers, streams, reservoirs, lakes, ponds, canals, mangroves, estuaries coastal and floodplain wetlands. Inland fish production is steadily increasing globally due to diversification of culture system from inland resources. This sector provides livelihood to millions of fishers and fish farmers at the primary level (Govt. of India). In spite of huge water resource in Bihar gap in fish demand and supply is wider. West Bengal is highest fish producing states still underutilized their resources.

12.2 Small Scale Fisheries (SSF) in India

India is bestowed with vast inland fisheries resources comprises of the rivers and canals (1.95 lakh km), reservoirs (3.54 million ha), floodplain wetlands and derelict water bodies (0.798 million ha), ponds and tanks (2.4 million ha), mangroves (3.56 Lakh ha), and estuaries (2.85 Lakh ha) are major contributor to SSF. River Ganga and Subernrekha are two main river flowing through the West Bengal. Five main rivers such as Ganga, Ghagra, Kosi, Gandak, and subernrekha are flowing through the various district of Bihar. These rivers are main source of water for floodplain wetlands, reservoirs, ponds, and tanks. Small Scale Fisheries play vital role in sustaining livelihood and nutritional security in India. Almost entire inland fisheries of India may be considered as Small-Scale Fisheries because engagement of fishing activity by using traditional or environmentally friendly fishing methods. In artisanal fisheries, fishers' uses non-motorized, hand operated country craft, small vessels for short fishing trip with low input cost, fishing near shore for own consumption or small income. Most of the produce of tradition fisheries are un-reported and not considered in the actual valuation or catch estimation, so there are considerable chances of less reliable data available in public domain on SSF. Approximately 81% of total fisheries sector in India is under SSF. Small Scale Fisheries apart from nutrition and livelihood security also contribute in poverty alleviation, employment generation, and wider distribution in supply chain. The concept of by-catch in SSF is almost insignificant, environmentally friendly fishing practice, socially and culturally rich but there is huge gap in the production and productivity. Wetlands of West Bengal and Bihar state have huge potential of SSF. Ichthyofaunal diversity rich in wetlands due to its feeble connection with parent rivers either seasonal or perennial. Major Small-Scale Fisheries contributing to the livelihood are from family cyprinidae.

12.2.1 Characteristics of Small-Scale Fisheries

- Use of environmentally friendly or less harmful fishing gear such as gill net, cast net, push net, on shore encircling net (especially in reservoirs).
- The gears used in SSF are less intensive, operated by 1–10 people without using motorized boat.
- The stakeholders/fishers have good understanding about the resources and awareness on resources should pass from one generation to other generation.
- Majority of fishers are work as labor in fishing operation, few are working as owner or lessee of resources.
- Unorganized marketing channel due to scattered human habitation on bank of river or wetlands.
- Fishing is restricted only for few days in breeding seasons, fishing holiday is not uniform thought-out Indian states for inland resources shifting monsoon pattern.
- Marketing channel is underdeveloped, most of time bulk catch is sold on site to the wholesaler or vender.
- Lack of marketing infrastructure due to interrupted roads connectivity.
- The fish is highly perishable commodities so fishers have financial dependency on middlemen, lessee of water bodies for their harvest.
- Fishers are highly skilled in manual operation of fishing gears and craft.
- Limited expenditure on taking care of safety measures on boat or off boat.

12.2.2 Importance of Small-Scale Fisheries

- It is estimated that 90% of fisheries work force in India is directly or indirectly employed in fisheries work. Small-Scale Fisheries play active role in economic and social growth of our country.
- SSF plays significant role in addressing challenges like nutritional deficiency in women and children, local to global nutritional and livelihood security.
- Important contributor to poverty alleviation and national rural economy.
- SSF is low expenditure venture and less harmful to the environment.
- In Indian wetland management system, wetlands are allotted to the individual/group of fishers/lessee of fishermen community for certain period of time. Lessee allows fishing community to fish small indigenous fish/unstocked fish/ self-recruiting population of fish throughout the year without paying taxes.
- Small Scale Fisheries give tremendous opportunity to sustain their livelihood during population displacement due to extreme natural events for additional source of income due to most of the inland fisheries resources are open access in nature.
- Traditionally low value SIFs contributes major share in animal protein consumption among the local low income group people.
- Local consumption of SIFs are commonly in fresh, salted, and dry form in different parts of India especially in Bengal and North East parts of India.

- In India especially in Inland fisheries sector, all the vessels are operated under Small Scale Fisheries.
- SSF is backbone of Indian rural economies for large section of society and deep-rooted in its regional cultural traditions.

12.3 SSF in Wetlands of Bihar and West Bengal

India possesses an extensive network of floodplain wetlands, characterized as low-lying areas adjacent to major rivers that become seasonally inundated due to overflow from the main river channels (Sugunan et al. 2000). The eastern states, particularly Bihar and West Bengal, contribute significantly to inland fisheries, accounting for 35% of the country's inland fish production and supplying 26% of the total national fish demand, including fish seed requirements (Das et al. 2017). These wetland ecosystems offer a range of ecological goods and services, supporting local communities through the provision of food, fuel, fodder, and microclimatic regulation. The wetlands of Bihar and West Bengal are seasonally flooded by river water and are commonly covered with aquatic macrophytes of varying density. These water bodies also serve as critical breeding and nursery habitats for a variety of small indigenous fish species (SIFs).

Fish, as a vital ecosystem service, plays a crucial role in ensuring both livelihoods and nutritional security. However, achieving the full potential of these resources for sustainable fisheries development requires coordinated efforts from government bodies and community-based institutions. In practice, the sector faces multiple barriers, including limited access to scientific knowledge and modern technologies, inadequate fisheries infrastructure, weak financial support mechanisms, lack of cooperation among stakeholders, poor coordination between relevant departments and user groups, and unorganized institutional frameworks. These challenges are particularly evident in the states of Bihar and West Bengal, where the existing potential of floodplain wetlands for fish production remains largely underutilized.

12.3.1 Bihar

Bihar, located within the Ganga river basin, comprises 38 districts: 21 in North Bihar and 17 in the southern part of the state. It is richly endowed with diverse aquatic resources, including rivers and their associated ecosystems such as floodplain wetlands (including oxbow lakes, meanders, and seasonal floodplains), reservoirs, ponds, and tanks. The state has approximately 3200 km of rivers, 2.4 lakh hectares of chaurs and floodplain wetlands (Fig. 12.1), 9500 hectares of oxbow lakes or *mauns*, 36,811 hectares of reservoirs, and 93,296 hectares of ponds and tanks. Despite this abundance, the current fish supply falls short of meeting the state's growing demand. Nevertheless, Bihar has witnessed a steady rise in fish production, which has reached an annual output of 6.83 lakh metric tonnes (Table 12.1). The number of fishermen engaged in fisheries activity full or part time are 4,088,096

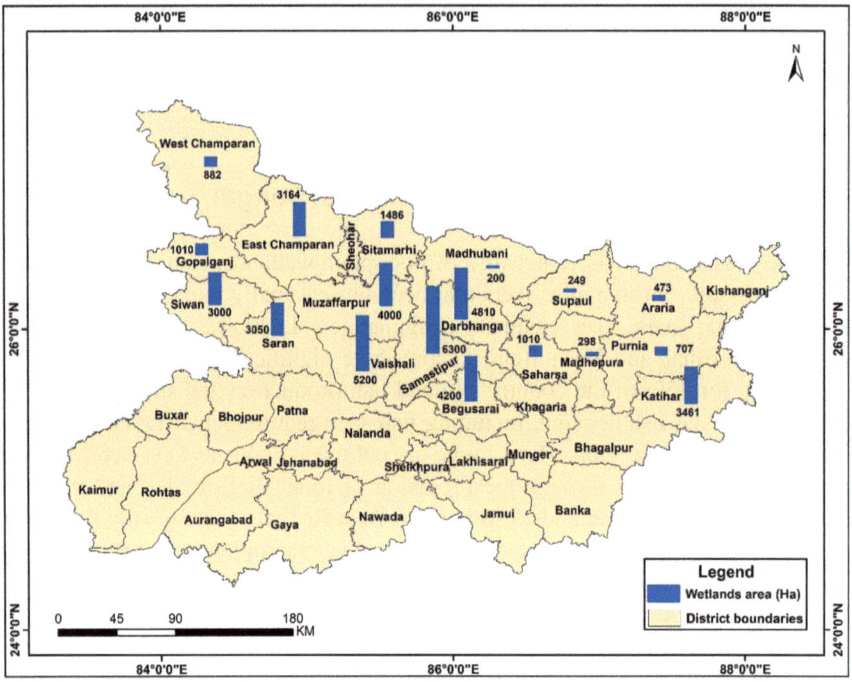

Fig. 12.1 Wetland fisheries resources of Bihar in GIS map. (Source of data: Das et al. 2017)

of which 52.33% are male (Handbook of Fisheries Statistics 2022). The livelihood of these fishermen solely/ partly depends on these fisheries. Apart from Ganga river system, 15 rivers and rivulets such as Koshi, Gandak, Budhi Gandak, Kareh, Kamla, Bagmati etc. A total 88 species of fish have been reported from the state. The commercially important fishes such as *Labeo catla, Labeo rohita, Cyprinus carpio, Cirhhinnus mrigala, Ctenopharyngodon idella, Hypophthalmichthys molitrix, Labeo bata, Labeo gonius, Mystus tengara. Tor tor, Notopterus notopterus, Notopterus chitala, Heteropneustis fossilis, and Clarius batrachus* comprise major fisheries in the Bihar.

12.3.2 West Bengal

West Bengal is bestowed with huge inland aquatic resources such as rivers, beels, ponds and lakes, reservoirs and tanks, brackish water bheries and canals etc. apart from marine and coastal fisheries resources (Table 12.3). The state has a geographical area of 88,752 sq. Km. The state has two principal divisions viz. the Terai Region in foothills of Himalayan in the North and the Alluvial plain in The South. The southern most part is deltaic region created by river Bhagirathi-Hooghly and their tributaries. The southern region occupied by largest mangrove forest such as

Table 12.1 Inland fisheries resources and fish catch composition of Bihar (2021–22)

Inland resources of Bihar				
Reservoirs	Area (ha)	Production (tonnes)	Production potential (tonnes)	Gap (tonnes)
Small	10,507	1198	3152	1954
Medium and large	26,304	343	701	358
Wetlands/beels/ chaurs	240,000	12,000	180,000	168,000
Ox-bow lakes	9500	2090	14,250	12,160
Ponds/tanks	93,296	–	–	–
River basin (including its tributaries)	Ganga, Gandak, Koshi			
	21,300 (km)			
Species/group –wise landing 2021–22 (in lakh Tonnes)				
Major carps (Catla, Rohu, Marigal)	3.42			
Minor carps	1.50			
Exotic carps	0.14			
Murrels	0.47			
Large catfishes (*Wallago attu*)	0.07			
Other fresh water fishes	1.23			
Annual fish production	6.83			
Per capita fish consumption	9.6			
Inland fisher population	60,27,375 (52% male)			
Number of districts	38			
Fishermen population	6,027,375			
Fishermen population per district	158,615			

Sunderbans. West Bengal is second largest producer of inland fisheries after Andhra Pradesh. The state has largest resources of brackish water in country. The state is contribution significantly in fish production and exploited their resources (70% of Inland freshwater and 30% brackish water) and awarded National productivity award for last 17consecutive years.

West Bengal is the second largest fish producer of total fish production 18.43 Lakh MT (11.34%) including marine catch and 16.52 Lakh MT (13.62%) of total inland fish production of country in 2021–22 (Table 12.2). The state is top fish seed producer of India, produced 613,900 Lakh fry in last five years (2017–2022). The state disposes fish in fresh (16.58 Lakh MT), frozen, curing, canning etc. (0.98 Lakh MT). The state has 23 districts and total fishermen population 3,236,261 (2020–21). The engagement of total fishermen population in Inland sector is 2,833,574 of which 58% male population. Inland Fisheries resources of West Bengal River and canals (2526 Km), dominated by small reservoirs or check dams (28,050 ha), Tanks and ponds (2,63,372 ha), Brackish water (2,10,000 ha), Beels/ oxbow lakes/derelict water (42,082 ha) other than parts of river and canals (26,925 ha) (Fig. 12.2). The state contributes to the export market 121,339 MT and 11% of total export value (Handbook of Fisheries Statistics 2022). The state is steadily increasing inland fish

Table 12.2 Fisheries resources and their production potential of West Bengal

Inland resources	West Bengal		
Reservoirs	Area (ha)	Production (tonnes)	Production potential (tonnes)
Small	28,050	593	1561
Medium and large		270	552
Wetlands/beels/chaurs	**42,500**	**29,750**	**63,750**
River basin (including its tributaries)	Ganga, Ichhamati, Padma, Matla and Mahananda		
	2526 (km)		
Species/group –wise landing 2021–22 (in lakh Tonnes)			
Major carps (Catla, Rohu, Marigal)	10.04		
Minor carps	0.64		
Exotic carps	2.02		
Murrels	0.33		
Large catfishes (*Wallago attu*, pangasius)	0.52		
Other fresh water fishes	3.14		
Annual fish production	16.69		
Per capita fish consumption	11.02		
Inland fisher population	32,36,261(50% male)		
Number of districts	23		
Fishermen population	3,236,261		
Fishermen population per district	140,707		

Sources: Sarkar et al. (2021) and Handbook on Fisheries Statistics (2022)

Table 12.3 District wise inland fisheries resources viz., rivers, canals, Beel/Boar and reservoirs (ha) of West Bengal

S/N	District	River	Canal	Beel/boar	Reservoirs	Total area
1	Cooch Bihar	13095.82	1164.58	1168.11	–	15428.51
2	Jalpaiguri	16006.44	562.29	504.38	439.47	17512.58
3	Darjeeling	3395.63	683.94	683.94	14.94	4778.45
4	Uttar Dinajpur	4032.08	1774.30	3548.59	–	9355.69
5	Malda	6497.6	2511.42	4551.55	–	13560.57
6	Murshidabad	36283.23	10013.88	3790.69	–	50087.80
7	Nadia	4191.28	2771.75	4271.61	10.2	11244.66
8	North 24 Parganas	14299.0	8712.00	8861.19	–	31872.19
9	South 24 Parganas	17704.98	13492.47	5749.47	–	36946.92
10	Howrah	107.36	2019.82	118.28	–	2245.46
11	Hooghly	4358.74	3714.94	3884.76	–	11958.44
12	Purba & Paschim Medinipur	12774.12	8766.07	404.01	–	21944.20
13	Bankura	15960.15	11711.04	1973.0	4600	34244.19
14	Purulia	3707.29	–	–	5267.76	8975.05
15	Burdwan	11316.64	5991.36	1939.91	906.428	20154.34
16	Birbhum	795.63	5695.85	632.16	1554.83	8678.47
	Total	164526.70	79585.71	42081.65	12793.45	298987.50

Source: Hand book of Fisheries statistics, Department of Fisheries WB and Das et al. (2017)

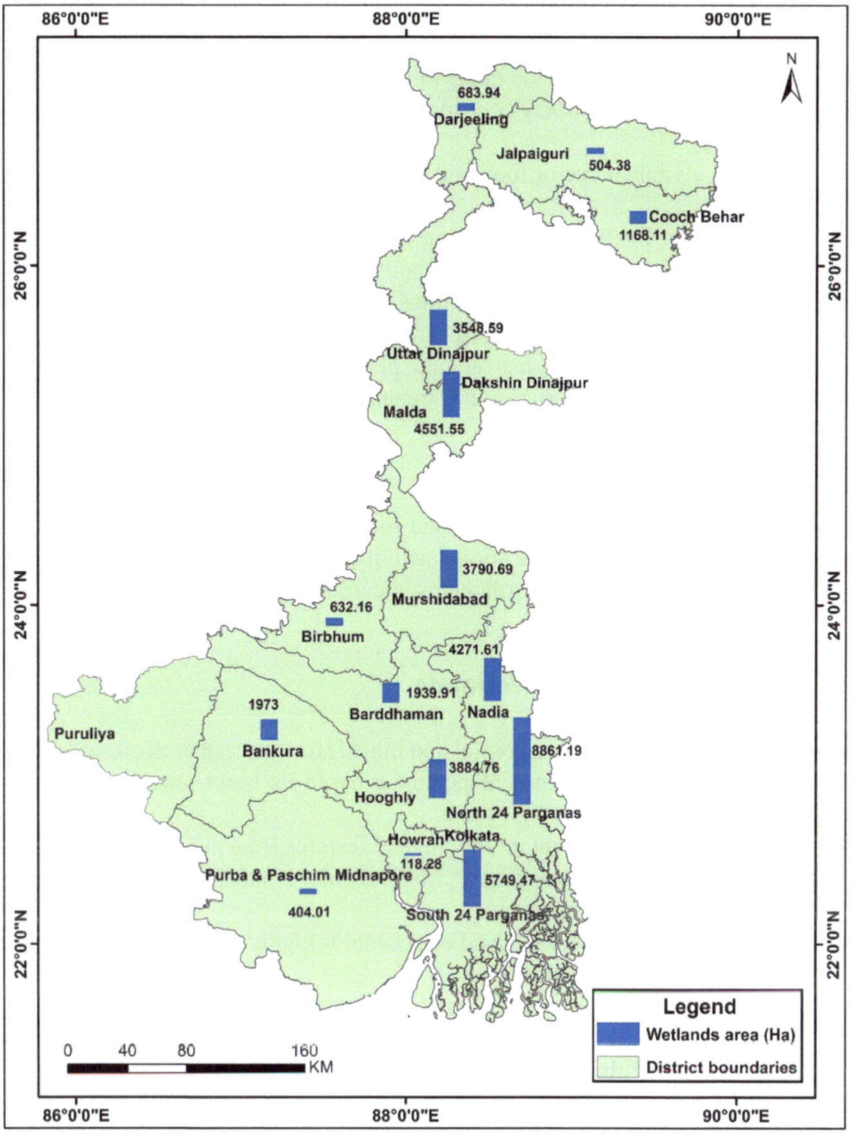

Fig. 12.2 Wetland fisheries resources of Bihar in GIS map

production and contribution to its economy. In order to realize its production potential, need to bridge the gap between production and potential. Fish Farmer's Development Agency of the state has been playing crucial role in sustain fisheries development in ponds and wetlands including seed production. The productivity of pond and tanks has raised from 600–800 kg/ha/ year to 3500–4000 kg/ha/year, wetlands productivity from 750 kg/ha/year to 1500–2000 kg/ha/year.

12.3.3 Importance of Floodplain Wetlands

Floodplain wetlands offer a wide range of valuable services to the communities living along their banks. Their key contributions include:

- Serving as a vital source of food and fodder for local populations as well as for bird species.
- Supporting the livelihoods of fishing communities, while also sustaining the commercial fisheries sector and contributing to nutritional security.
- Providing essential ecological functions such as food supply, habitat, breeding grounds, and shelter for a diverse array of organisms—including mammals, birds, amphibians, reptiles, fish, aquatic plants, insects, and microorganisms.
- Improve water quality in rivers and streams by filtering water and act as kidney for riparian community.
- Remove excess nutrients from water.
- Controlling floods by storing water.
- Sustaining life cycle of 75% of fish and shellfish species.
- Support recreational and educational activities.
- Major carbon reservoirs in earth, helps in carbon sequestration.

12.3.4 Characteristics of Wetlands

- Wetlands are categorized as coastal and inland (freshwater)based on salinity and water course and as seasonal of perennial based its water retention and water holding capacity.
- The wetlands have different morphological features from natural lake with regular contours to horse shoes shaped with irregular contours formed by different course of river or coastal action.
- Ecology and hydrology of wetland ecosystem is highly dynamic and varies with their connectivity with parent river.
- Wetlands of India are moderately infested with aquatic macrophytes (30–50%).
- Natural floodplain wetlands are popularly known as *beels,* in west Bengal and *chaurs/mauns* in Bihar.
- Wetlands are usually shallow, water retention period varies from 8 to 12 months depends on connectivity with parent river channel.
- The growth of fishes in these water bodies is faster compare to reservoirs due to availability of huge reserve of food niche.

12.4 Challenges Within Small Scale Fisheries in India

Small Scale Fisheries basically depend on ecology, hydrology, and self-recruiting fish population of wetland ecosystem. Presently, fish and fisheries of wetlands are declining due to multiple factors anthropogenic and natural cause. There are many challenges in SSF given below.

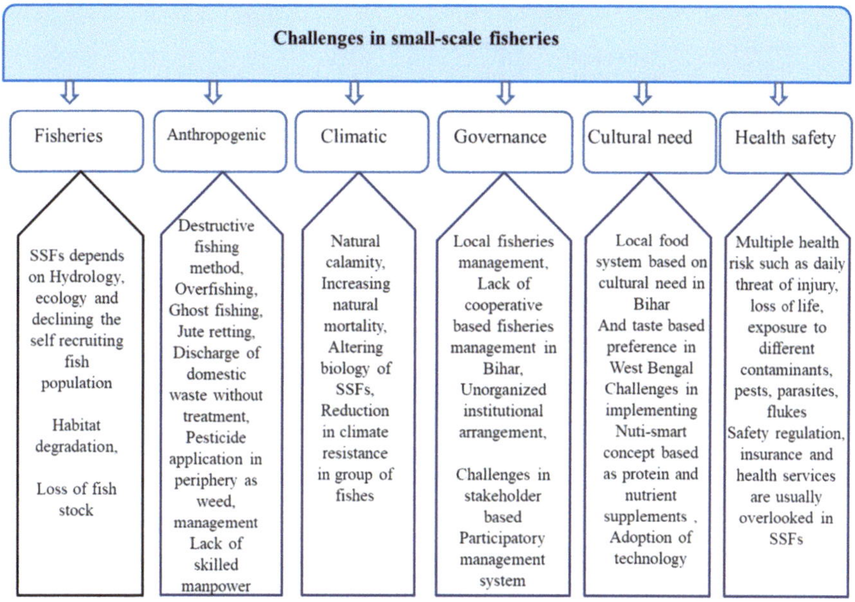

12.5 Opportunity for Vulnerability to Viability

Wetlands in India were used as common resources in concept of open to all, but in last one decade these resources are leased out to individual or a group of fishers or to a cooperative society and exploited unwisely. Small Scale Fisheries of wetlands in tropic regions are highly vulnerable in all three dimensions, economic, social and climatic (Roy and Das 2022). The aim of vulnerability to viability (V2V) to identify the factor affecting the SSF and finding the suitable measures to increase viability in sustainable manner. The investigation was carried out in five wetlands of Bihar: Kararia, Majhari, Sirsa, Rulhi, and Kothia to assess fish diversity and reason for gap in fish production and potential. The estimated fish production potential of wetlands of Bihar is 1000–2000 kg/ha/year but production 20–220 kg/ha/year in year 2017. Major reason for huge gap in production and vulnerability of wetlands were failure in auto-recruitment, shrinkage of link channel, weed infestation, predators, overexploitation of SIFs, lack of coordination among stakeholders, lack of collaborative engagement with institutions and poor capacity building of fishers, unavailability of quantity and quality fish seed, insufficient stocking with inappropriate species composition, lack of seed rearing facility, etc. The above reason made system vulnerable social, ecological, and economical. This comprehensive understanding about the reason for vulnerability faced by SSF and possibility to make system viable with interventions and technological application was also analyzed (Fig. 12.3). In wetland of West Bengal culture-based fisheries practice in almost 50% of wetlands and rest of wetlands are under natural fisheries. The estimated fish production potential of wetland of WB is more than 3000 Kg/ha/yr. but production less than 800 Kg/ha/

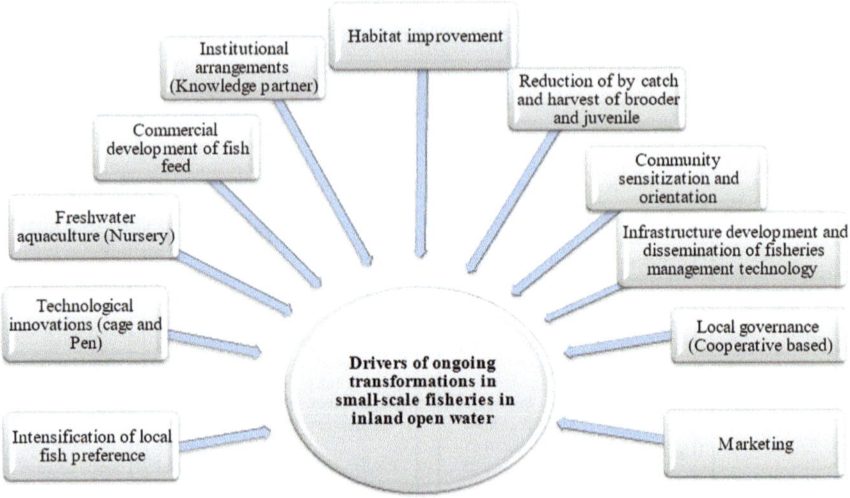

Fig. 12.3 Drivers of ongoing transformation in Small Scale Fisheries towards viability in Bihar and West Bengal

yr. In WB Primary fish cooperative society is active but stocking of IMC and exotic carp is done arbitrarily lead to poor production and underutilization of ecological resources. The aim of fish production is not only for economic viability but also social, political, and ecological viability of wetlands. ICAR-CIFRI intervention for fisheries enhancement was culture based fisheries protocol, nursery infrastructure, scientific enclosure culture practice, training-based building technical capacity of fishers/stakeholders, and institution management support for initial development of wetlands. This technological intention was able to make huge changes in fish production, social, ecological, and knowledge of fishers (NFDB 2022).

12.6 Technological Interventions for Development of Small-Scale Fisheries in Bihar and West Bengal

The challenges associated with Small-Scale Fisheries are many and same time there is tremendous opportunity to transform the Small-Scale Fisheries through technological interventions.

12.7 Fisheries Management Options for Productivity Enhancement

Fish production from floodplain wetlands can be increased, following the scientifically developed management strategies. There are options for fisheries enhancement in wetlands, depending upon the objectivity and prevailing ecological

Table 12.4 Fisheries management options for productivity enhancement

Management options	Practices
Stock enhancement	In open water bodies disconnected from rivers due to embankments and siltation, natural recruitment of commercially important fish has declined due to disrupted auto-stocking and reduced breeding populations. Stocking fingerlings of target species can effectively restore self-recruiting populations and enhance production. Key elements of stock enhancement include species selection, stocking density, and fingerling size. This approach supports either culture-based fisheries focused on recapture or supplements weak natural recruitment in degraded systems.
Species enhancement	Species enhancement involves introducing fast-growing and economically valuable fish species into a water body to occupy its various ecological niches and increase overall fish biomass. This strategy aims to make efficient use of underutilized food resources. However, potential ecological impacts of such introductions must be carefully assessed to avoid adverse effects on the native ecosystem.
Management enhancement	Altering the pattern of management can substantially improve the economic attributes of a water body. This includes adopting culture fishery along with capture, developing sports fishery and opening to managed tourism, declaring protected areas and natural reserves etc., deciding whether private management, management by cooperative societies, co-management, etc. can also alter the economic attributes of the water bodies. *The mauns* and *chaurs*, are ideal for culture-based fisheries development. The *mauns* being the river meanders are perennial, but the *chaurs* are either perennial or seasonal, depending upon the depth of depression and ingress of water during the floods, as such need different management
Culture-based fisheries	The core approach in culture-based fisheries revolves around strategic stocking and recapture. Fish species are stocked according to the availability of natural food niches within the water body. The effectiveness of this practice depends on several key factors, including the size and density of fish at the time of stocking, species selection, fishing intensity, size at harvest, gear selectivity, and appropriate mesh size to ensure sustainable fishery management.

conditions in a particular system. Eco-friendly options before us in augmenting fish production from the water bodies are various enhancement measures (Table 12.4). According to FAO (2004), enhancement refers to the application of technical interventions within existing aquatic resource systems that bring about significant changes in environmental, institutional, and economic characteristics. In India, culture-based fisheries represent the most widely adopted method of enhancement in inland water bodies. This approach involves implementing targeted management practices to achieve both qualitative and quantitative improvements in fish production from these aquatic ecosystems.

12.8 Enclosure Culture Technologies for In-Situ Fish Seed Raising

12.8.1 Pen Culture for Raising Advance Fingerlings

A pen is a fixed enclosure used to confine fish within a section of a beel or wetland for rearing purposes. Pen culture is particularly effective in shallow, neglected, or weed-infested water bodies where conventional aquaculture is not feasible and fishing is challenging. The recommended size for a single pen unit range from 0.1 to 0.2 hectares, and multiple units can be set up as needed based on site conditions and production goals. A production of 400 kg table size Indian major carps (*Labeo catla, Labeo rohita, Cirrhinus mrigala*) can be achieved from a pen unit of 0.1 ha area with stocking of 1000–1500 fingerlings of 10 g/100 mm size; 80 kg freshwater prawn (*Macrobrachium rosenbergii*) can be produced from a pen of same size with stocking of 3000–4500 juveniles of 60–80 mm size, in a grow out period of four months with a survival of 80%. The technology can also be used for raising fingerlings for stocking. A production of 500 kg carp fingerlings of 10 g/100 mm size can be achieved in pens of same size in three months by stocking 50,000 to 75,000 fry of 15–30 mm size with a survival of 80%. The technology is eco-friendly and involves low input cost including supplementary feeding and fish health management. The number of pen units to be put up depends on the size of water body and production requirement.

12.8.2 Cage Culture for Advance Fingerlings and Table Fish Production

Cage culture involves rearing fish in confined enclosures (cages) under managed conditions. This method is especially suitable for deep-water bodies like reservoirs and oxbow lakes with a minimum depth of 5 meters. Cages can be rectangular, square, or circular, typically ranging in volume from 50 to 100 m^3, and multiple units can be deployed depending on site requirements. Ideal locations for cage installation include sheltered zones within reservoirs, deeper areas of floodplain wetlands, and lakes where water depth exceeds 5 meters. Cage culture is effective for both nursery rearing (from fry to advanced fingerlings) and grow-out production of market-size fish. It offers full control over the stock and facilitates high survival and recovery rates.

Stocking densities vary by species and fish size. Major carp fry (40–50 mm) can be stocked at 25–30 fish/m^3. Fingerlings (8–10 cm) of Indian major carps, Pangas, air-breathing fishes, and sea bass can be stocked at 20 fish/m^3 for grow-out production. Fast-growing species like *Pangasianodon hypophthalmus* can tolerate higher densities of 50–70 fish/m^3. With a culture period of 4–5 months, growth increases of 300–500% can be achieved depending on species and management practices. Feeding typically involves extruded or pelleted feeds supplemented with a mix of rice bran and mustard oil cake, enriched with essential vitamins, amino acids, and minerals (at 0.01% concentration). Commercial carp feeds with 25–32% protein are suitable, while catfish diets may require higher protein content for optimal growth.

12.9 Conclusion

Despite the abundance of inland fisheries resources in West Bengal and Bihar, there remains a significant gap in fish production and productivity. In West Bengal, over 90% of total fisheries activity is conducted by small-scale fishers, while in Bihar, the sector is almost entirely dominated by small-scale fisheries (SSF). Effective management of SSF requires empowering communities to actively participate in decision-making processes across social, economic, and ecological domains. Small-scale fisheries hold the potential to tap into growing markets, including exports, by embracing modern technologies. However, pressures from mechanized sectors and declining catches are pushing SSF to adopt innovations. Decentralization and community-based or co-management approaches offer opportunities for fishers to steer their own development.

To strengthen and sustain the SSF sector, the following strategic actions are crucial:

1. Institutional Organization—Formation and strengthening of fisher collectives or cooperatives.
2. Habitat Management—Macrophyte control, desiltation, and restoration of link channels.
3. Fisheries Management—Enforcing mesh size rules, seasonal fishing bans, and preventing destructive methods.
4. Technology Adoption—Promoting culture-based fisheries, pen and cage culture, and nursery rearing systems.
5. Supply Chain Strengthening—Enhancing market access and ensuring efficient input-output linkages.
6. Capacity Building—Training, awareness programs, exposure visits, and provision of fishing tools and gear.
7. Stakeholder Engagement—Building trust between fishers, scientists, and agencies through consultations and workshops.
8. Livelihood Enhancement—Supporting diversified income sources to improve resilience and food security.
9. Adaptive Capacity—Strengthening SSF's ability to cope with climate, economic, and political changes.
10. Data Systems—Developing robust systems to document SSF contributions (including hidden harvests) for policy integration.
11. Policy Focus—Ensuring that policies and guidelines prioritize ecological sustainability and fisher welfare.
12. Knowledge Sharing—Promoting studies and data exchange to increase recognition of SSF's socio-economic role among policymakers.

Points to Remember

- Wetlands in Bihar and West Bengal are ecologically rich and play a vital role in supporting inland small-scale fisheries (SSF) and rural livelihoods.
- Small-scale fisheries (SSF) rely on traditional, low-cost, and environmentally friendly practices, with limited mechanization and high dependence on local knowledge.
- Bihar possesses significant aquatic resources but faces a notable gap between potential and actual fish production due to poor infrastructure, weak institutional support, and ecological challenges.
- West Bengal ranks among the top fish-producing states in India, yet many wetland resources remain underutilized or inefficiently managed.
- The sector faces multiple challenges, including habitat degradation, weed infestation, failure in natural fish recruitment, lack of access to quality fish seed, and weak market linkages.
- The Vulnerability to Viability (V2V) framework offers a holistic approach to understanding and addressing socio-ecological vulnerabilities in SSF systems.
- Technological interventions such as culture-based fisheries, stock and species enhancement, pen and cage culture, and scientific enclosure systems are crucial for enhancing productivity and sustainability.
- Institutional strengthening, including cooperative development, stakeholder coordination, and policy integration, is essential to improve governance and resilience of SSF.
- Capacity building through training, technical support, and knowledge transfer can empower fishers and improve fisheries management.
- A strategic, ecosystem-based approach combining ecological restoration, technological innovation, and inclusive governance is key to unlocking the full potential of SSF in the region.

References

Das BK, Chandra G, Meena DK, Kumari S, Koushlesh SK, Das AK, Ekka A, Mishal P, Karnatak G, Bandopadhyaya MK, Samanta S, Pandit A, Sudheesan D, Manas HM, Mohanty BP, Behera BK, Sahoo AK, Parida PK (2017) Roadmap for development of inland open water fisheries in eastern states of India. ICAR-CIFRI, Barrackpore, p 68. ISSN 0970-616X

FAO (2004) Fisheries country profiles. FAO, Rome. http://www.fao.org/fi/fcp/fcp.asp

Handbook on Fisheries Statistics (2022) Government of India, Ministry of Agriculture and Farmers Welfare, New Delhi

NFDB (2022) SSS: India @75 super success stories from Indian fisheries. Azadi ka Amrut Mahotsav. National Fisheries Development Board, Department of Fisheries, pp 15–17

Ministry of Fisheries, Animal Husbandry & Dairying (2023) Annual Report 2022-23 (English). Government of India

Roy A, Das BK (2022) Resource degradation and conflicts affecting small-scale wetland fishers of West Bengal, India. In: Kerezi V, Chuenpagdee R (eds) Blue justice for small-scale fisheries: a global scan, vol 3. TBTI Global Publication Series, St. John's

Sarkar UK, Mishal P, Borah S, Karnatak G, Chandra G, Kumari S, Meena DK, Debnath D, Yengkokpam S, Das P, DebRoy P, Yadav AK, Aftabuddin M, Gogoi P, Pandit A, Bhattacharjya BK, Tayung T, Lianthuamluaia L, Das BK (2021) Status, potential, prospects, and issues of floodplain wetland fisheries in India: synthesis and review for sustainable management. Rev Fish Sci Aquacult 29(1):1–32

Sugunan VV, Vinci GK, Bhattacharjya BK, Hassan MA (2000) Ecology and fisheries of beels in West Bengal. Bulletin No. 103. CIFRI, Barrackpore, p 53

Small-Scale Fisheries in Small Reservoirs Towards Sustainable Fishery

13

Lianthuamluaia, Suman Kumari, Mishal P., and Gunjan Karnatak

Abstract

Culture-based fisheries (CBF) is a predominant and reliable approach for the management of small reservoirs fisheries in India. Small reservoirs are mostly managed with the involvement of local fishing community. CBF is closely linked with small-scale fisheries (SSF) and plays a vital role in sustaining rural livelihoods, ensuring food security, and maintaining ecological balance. The adoption of CBF in small reservoirs has led to significant increases in fish yields in many small reservoirs across the country. With the involvement of fishing community participation, CBF promotes social and economic development, fish production enhancement, and biodiversity conservation. As a result, CBF serves as a sustainable fisheries management strategy that supports the livelihood of rural communities while maintaining ecological integrity.

Keywords

Culture-based fisheries · Reservoirs · Small-scale fisheries · Sustainable

---

Integrating Culture-Based Fisheries and Small-Scale Fisheries for Sustainable Reservoir Fisheries Management

---

Lianthuamluaia (✉) · S. Kumari · Mishal P. · G. Karnatak
ICAR – Central Inland Fisheries Research Institute (ICAR-CIFRI), Barrackpore, Kolkata, India
e-mail: eltry7@gmail.com

B. K. Das et al. (eds.), *Small-scale Inland Fisheries in India*, Sustainability Sciences in Asia and Africa, https://doi.org/10.1007/978-981-95-3849-2_13

13.1 Introduction

India has vast inland open water fisheries resources comprising of rivers and canals (1.95 lakh km), reservoirs (3.54 million ha), floodplain wetlands and derelict water bodies (0.798 million ha), ponds and tanks (2.4 million ha), mangroves (0.35 million ha), and estuaries (0.28 million ha). Reservoirs refer to man-made water bodies formed by obstructing the natural flow of a river, stream, or watercourse through the construction of a dam. Reservoirs of India have been categorized in 3 categories (small reservoir: <1000 ha, medium reservoir: 1000–5000 ha, and large reservoir: >5000 ha) based on their area for exploiting their maximum potential with suitable management practices (Sugunan 1995). Small reservoirs of India are numerous with 25,251 number covering 1,544,882 ha area spread across 29 states (Fig. 13.1). Small reservoirs are built mainly for irrigation of agriculture crop and water storage; however, these water bodies are also efficiently used for fish production in many places. Small reservoirs are highly productive, shallow, biologically rich system suitable for culture-based fisheries management practice. However, due to several reasons including lack of awareness, poor adoption of scientific technology, lack of proper governance and institutional arrangement, and poor infrastructure facilities (Sarkar et al. 2018), the present fish yield from the small reservoir is far below the estimated fish yield potential (Fig. 13.1). Increased in fish yield from 49.9 to 190 kg/ha/year was achieved through culture-based fisheries (CBF) or stocking and harvesting practices which was further projected to increase its production 343 kg/ha/year by 2026 (Das et al. 2021).

Considering the significant gap between the present fish yield and the estimated potential yield in small reservoirs across India, there exists a huge opportunity to increase fish production. Tapping the reservoir fish potential offers a scope for boosting inland fish production due to its vast resources. To achieve this, it is essential to formulate strategies for sustainable increase of fish production in small

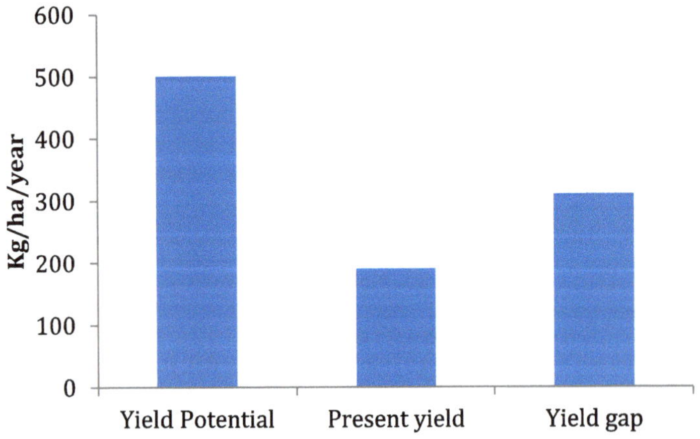

Fig. 13.1 Fish yield gap in small reservoirs of India (Das et al. 2021)

reservoirs. Reservoir fisheries are often community-based which involve local fishers who depend on reservoir fishing as a major source of livelihood and food security. Recognizing the role of fishers and integrating their traditional knowledge and practices into management plans is crucial in most of the inland open waters including reservoirs. Stocking and recapture of fish or CBF is the most reliable option for sustainable enhancement of fish production in reservoirs to prioritize the unique characteristics and needs of small-scale fisheries.

Small-scale fisheries (SSF) and culture-based fisheries in small reservoirs are two interconnected components in fisheries sector, playing a pivotal role in supporting the livelihoods of millions of people worldwide. These fisheries have been instrumental in providing food security, employment opportunities, and income generation for many rural fishing communities (Sarkar et al. 2018). In the context of small reservoirs, culture-based fisheries have emerged as a promising strategy to enhance fish production sustainably, while small-scale fisheries continue to serve as a critical means of subsistence and socio-economic well-being for local communities (Das et al. 2021, 2022). The delicate balance between sustaining fish production, ensuring ecological integrity, and meeting the socio-economic needs of local communities presents a challenging conundrum for fisheries management (FAO 1995). In this regard, culture-based fisheries have garnered considerable attention for their potential to boost fish yields without significantly harming natural ecosystems (Sarkar et al. 2018).

A case study was carried out to assess the impact of CBF on fish production in reservoirs of Chhattisgarh, Odisha, Jharkhand, and Andhra Pradesh. The information of impact of stocking and the data of stocking and production was obtained from various sources Sugunan and Katiha (2004), Sarkar et al. (2020), Lianthuamluaia et al. (2021), and the respective State Fisheries Department (Fig. 13.2).

13.2 Small-Scale Fisheries in Small Reservoirs

Small-scale fisheries are typically characterized by the utilization of labor-intensive techniques for harvesting, processing, and distributing technology to exploit both marine and inland water fisheries resources (FAO 2004). Small-scale fisheries, also known as artisanal fisheries, encompass a diverse range of fishing activities conducted by local fisher folk using traditional methods and relatively small-scale equipment. Small Scale fishery is a widely used terminology for coastal and marine fisheries, presently well discussed for inland fisheries also. A small-scale fishery was highly neglected but according to FAO, contribution of SSF to global fish catch is 90% in developing countries (FAO 2015). SSF includes fish harvest, processing and trading activities using traditional method. The recognition of women's contribution in small-scale fisheries (SSF) has increased significantly. Small-scale fisheries (SSF) play a crucial role in providing nutrient rich food for self, local, national, and international markets. Additionally, they serve as a valuable income for both direct and indirect participants in the sector. As a result, there is a high likelihood of unreliable or incomplete data regarding small-scale fisheries (SSF) available in the

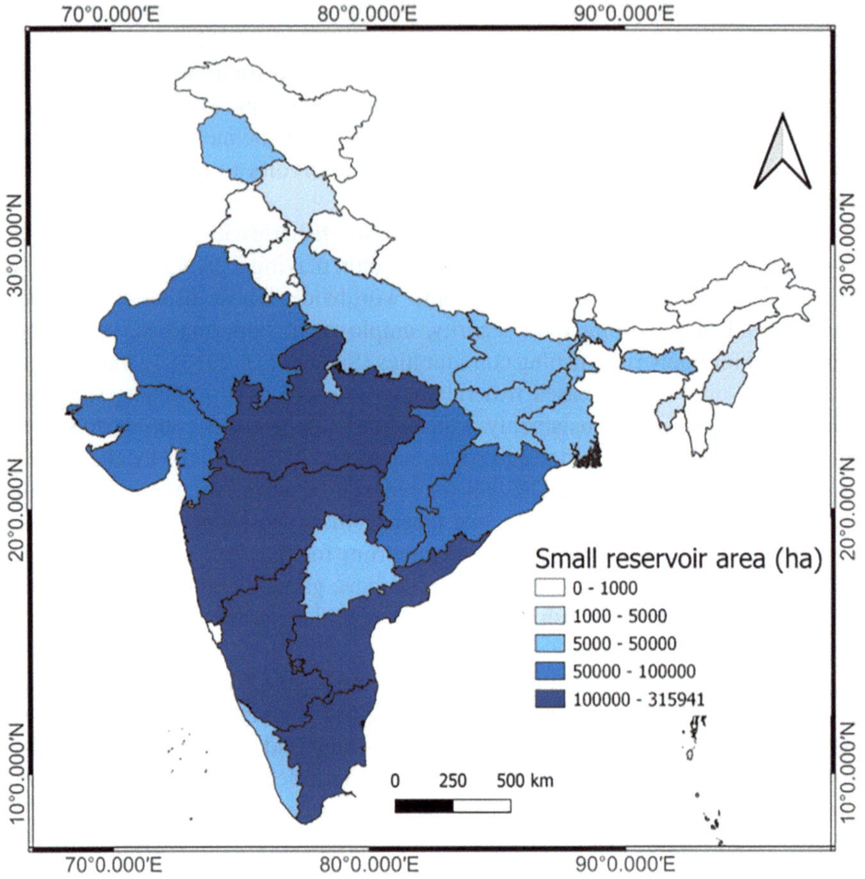

Fig. 13.2 State-wise distribution (area) of small reservoirs in India

public domain. Approximately 81% of total fisheries sector in India is under SSF (Das et al. 2022). Small-scale fisheries (SSF) practices in inland open waters are generally considered to be more environmentally friendly compared to deep-sea marine fishing practices. SSF causes less harm to the ecosystem, generates fewer discards, and results in minimal or no bycatch. These practices are often labor intensive and carried out by members of the same community or family. In SSF, there is a reliance on small-sized vessels or no vessels at all, with fishing activities limited to single-day or few-hour trips. The production costs associated with SSF are typically low. Additionally, SSF commonly employs non-motorized fishing techniques and utilizes non-target fish species for local or household consumption, which contributes significantly to both nutritional and livelihood security.

Small-scale fisheries play a multifaceted role in society. They serve as a vital source of employment and income for millions of people, especially in developing countries, where access to alternative livelihoods is often limited. Additionally, small-scale fisheries contribute significantly to local and national economies by providing fresh and nutritious fish for both domestic consumption and export (FAO 2022). Moreover, these fisheries are vital for food security, ensuring that protein-rich fish products are accessible to communities that heavily depend on inland resources. In small reservoir culture-based fisheries or stocking and harvesting of fish is widely adopted for reservoir fisheries management (Sarkar et al. 2018). In most of the cases harvesting of the stocked fish is done by low-cost gear and craft in small reservoirs. Diverse group of fish is captured with single gear and craft, there is limited gear available for target fisheries. Gill net, drag net, and cast net are the important fishing gear in small reservoir (Sugunan and Sinha 2000). Among all the fishing gear, gill net is the most common fishing gear which is operated by one or two fishers. A small country boat is generally employed for operating the fishing gears.

Here are some key aspects of small-scale fisheries in reservoirs (Sugunan 1997; FAO 2015; Sarkar et al. 2018; Arthur et al. 2021; Das et al. 2022):

- Small-scale fisheries are typically community-based, involving local fisherfolk who have been fishing in the reservoirs for generations. They possess valuable traditional knowledge about fish behavior, fishing techniques, and the local ecosystem.
- Small-scale fisheries are a significant source of income and livelihood for many rural communities. Fishing activities in reservoirs provide employment opportunities and generate income for small-scale fisherfolk and their families.
- Small-scale fisheries contribute to local food security by providing a consistent supply of fresh fish for local consumption. Fish is often a crucial source of protein and essential nutrients in the diets of communities living around the reservoirs.
- Small-scale fisheries tend to use low-impact fishing methods, such as gill net, cast nets, and small boats, which have minimal environmental impact compared to industrial-scale fishing operations.
- Many small-scale fisherfolk have a strong connection to their local environment and are more inclined to engage in sustainable fishing practices. This can include avoiding overfishing, adhering to fishing regulations, and protecting fish spawning and breeding areas.
- Despite their importance, small-scale fisheries in reservoirs face challenges such as declining fish stocks due to overfishing or environmental degradation, lack of access to resources and technology, and limited market opportunities.
- Sustainable management and support for small-scale fisheries are essential to ensure their long-term viability. This can involve providing training in sustainable fishing practices, improving access to credit and resources, and creating market linkages.
- Small-scale fisherfolk are often at the forefront of community-based conservation efforts. Their intimate knowledge of the local ecosystem enables them to actively participate in efforts to protect the reservoir's biodiversity and ensure its sustainability.

13.3 Culture-Based Fisheries (CBF) as a Management Tool for Small Reservoir Fisheries

In small reservoirs, where natural fish populations may be limited by various factors such as overfishing or habitat degradation, culture-based fisheries present an opportunity to restore fish stocks and optimize fish production in a sustainable manner. By selecting appropriate fish species and stocking them at suitable densities, culture-based fisheries can contribute significantly to enhancing the fish production and supporting small-scale fisheries. CBF are generally followed for sustainable enhancement of fish production in small reservoir. In culture-based fisheries management practice, suitable quantity of fish seeds were stocked and allowed them to grow using the available natural food of different niche of water bodies for certain period of time and recaptured the stocked fish. This strategic intervention is designed to complement the existing fish populations, rather than replace them, by striking a balance between fishery enhancement and ecological conservation. Contrary to the practices in aquaculture, where aquatic organisms are cultivated within artificial environments or specific species are commercially produced by managing their life cycle stages, culture-based fisheries (CBF) focus on increasing production within natural environments by controlling certain aspects of the targeted species life history. In CBF, the primary approach involves releasing fingerlings into open waters, enabling them to grow and develop in their natural habitat.

In CBF selected fish species of 80–100 mm size in suitable quantity (estimated based on its productivity) are stocked in reservoir for enhancing fish stock or for propagation of declining self-recruiting fish stock, and after a certain period of time attaining desire size of fish get harvested. This is widely accepted technology used throughout the world and most acceptable in Indian reservoirs/tanks and wetlands. This technology has been accepted by the fishers for increasing production from different water bodies. In CBF, the practice of feeding the fish with artificial feeds or supplements is discouraged, and the fish primarily rely on the natural food available in the water body for their nutrition and growth. The rationale behind this approach is to promote a more sustainable and environmentally friendly method of fish farming. By depending on the natural food present in the reservoir, CBF reduces the need for external inputs like fish feed, which can be costly and may have environmental consequences, such as water pollution and habitat degradation. Monitoring and managing the water quality and ecosystem balance are also crucial to maintaining a healthy and productive fish population in such systems. For achieving the optimum fish yield assessment of potential fish yield, selection of species, estimation of optimum stocking density, size at stocking, and harvesting schedule are the key parameters.

13.3.1 Potential Fish Yield

Estimating the fish yield potential in reservoir is a complex task that requires the consideration of various ecological, environmental, and fishery-related factors. The fish yield potential can vary greatly depending on the size and characteristics of the reservoir, the fish species present, and the management practices implemented. Several methods are available to estimate the fish yield potential in reservoirs. One of the most reliable methods for estimation of fish yield potential is morpho-edaphic index (MEI) which combine the morphometric as well as chemical parameters.

The MEI can be calculated as:

$$\text{MEI} = \text{Specific Conductivity}\left(\mu S/\text{cm}\right)/\text{Mean depth}\left(\text{m}\right).$$

The fish yield potential using MEI can be calculated as

$$\text{Yield}\left(\text{kg}/\text{ha}/\text{year}\right) = 0.9897^{*}\,\text{MEI}^{1.3888}\left(\text{Sugunan et al. 2002}\right).$$

13.3.2 Species Selection

Species selection is very crucial as stocking is the most important part of culture-based fisheries. The success of a reservoir stocking program depends on choosing appropriate fish species that can thrive in the given environmental conditions and utilize the available food resources effectively. Indian Major Carps (IMC) are commonly stocked in many reservoirs in India due to their importance in the local fisheries sector, but consideration of other suitable species is essential for maintaining ecological balance and enhancing overall fish production. The ratio of the species to be stocked is influenced by several factors including the existing food niches. Several aspects or criteria have to be considered while selecting the species for stocking in reservoir (Vass et al. 2009).

- Growth rate
- Food and feeding habits
- Market demand
- Well-suited to the ecological conditions
- Capability to utilize the available natural food
- Availability of fish seed

13.3.3 Stocking Density

Stocking of the right number of fish seed is the key factor for sustainable augmentation of fish production in small reservoir. Stocking rate is calculated to achieve the estimated fish yield potential (Welcomme 1976):

$$S = \left(q \times p / W\right) \times e^{-z(tc - to)}$$

Where

- S is the number of fish to be stocked (in number/ha/year).
- p is the natural annual potential fish yield of the water body (in kg/ha/year).
- q is the proportion of the yield that can be obtained from the species in question
- W is the mean weight at capture (in kg) of the fish species considered for stocking.
- $(t_c – t_o)$ is the period from stocking to harvesting (in years).
- $-z$ is the total mortality rate

13.3.4 Size at Stocking

The success of culture-based fisheries relies on the survival rate of the stocked fish, a factor strongly influenced by their size at the time of stocking. Small-sized fish seeds are more susceptible to predation compared to larger ones. Therefore, it is advisable to stock fingerlings that are larger than 100 mm in reservoirs to enhance their chances of survival.

13.3.5 Harvesting Schedule

Fish harvesting typically occurs when the fish reach a size suitable for the market. Indian Major Carps (IMC) usually attain marketable size within a year after stocking. Therefore, it is recommended to conduct the harvest either after one year or upon reaching marketable size. In situations where fish may escape through the spillway or canals, an annual cropping policy is suggested. Under this approach, the reservoir is stocked around September-October and harvested by the end of June. The specific timing depends on the growth rate of the fish and the overall productivity of the water body.

13.4 Impact of Fish Stocking on Fish Yield: A Case Study

13.4.1 Impact of CBF in Small Reservoirs of Chhattisgarh

Chhattisgarh state has vast resources of reservoirs, a total of 1770 reservoirs covering 0.82 lakh ha area of which 99% are under small reservoir category (Kumar et al. 2021). Data on fish stocking and yield from 2008 to 2016 collected from the State Fisheries Department, Govt. of Chhattisgarh indicated the significant increase in fish yield (from 87.11 to 379.13 kg/ha/year) (t-test, $p < 0.05$) due to stocking in 24 small reservoirs of Chhattisgarh (Lianthuamluaia et al. 2021). The findings clearly indicated that stocking or culture-based fisheries program implemented had a significant positive effect on fish yield (Fig. 13.3).

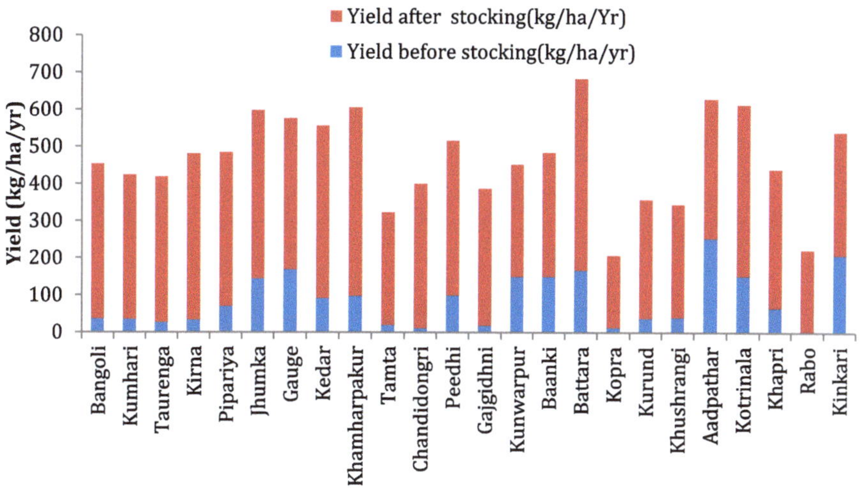

Fig. 13.3 Impact of fish seed stocking in small reservoir of Chhattissgarh (Lianthuamluaia et al. 2021)

13.4.2 Impact of CBF in Small Reservoirs of Odisha

Odisha is rich in small reservoir resource with 1433 (<1000 ha each) of covering 65,047 ha area. Fish stocking and yield data were collected from 58 small reservoirs from the State Fisheries Department of Odisha to assess the impact of stocking in fish yield. The results revealed a significant increase in fish yield ($p < 0.05$, t-test) from 204 kg/ha/year in 2012–13 to 323 kg/ha/year in 2013–14 following the adoption of CBF (Sarkar et al. 2020) (Fig. 13.4). This increase in fish yield led to a corresponding rise in per capita fish production from 398 kg/fisher/year to 702 kg/fisher/year, benefiting over 8000 fisher households. Among the different size groups of reservoirs, those ranging from 100 to 500 ha showed the most substantial response in fish yield due to fingerling stocking. The size of the stocked fingerlings and the reservoir area were found to have a significant impact on enhancing fish yield (Fig. 13.4).

13.4.3 Impact of CBF in Reservoirs of Jharkhand

Jharkhand state is located in the Eastern part of India and is blessed with abundant water resources, including rivers, canals, wetlands, and reservoirs. Among the major rivers that flow through the state are Ganga, Damodar, Son, Subarnrekha, Punpun, Koel, and Mayurakshi, among others. To harness these water resources for power generation and irrigation, the state has constructed several dams along the Ganga, Damodar, Son, and Subarnrekha rivers. In total, there are 258 reservoirs in the state, covering an extensive area of approximately 115,000 ha. These reservoirs play a vital role in supporting the livelihoods of around 1.35 lakh fishers in the region. The

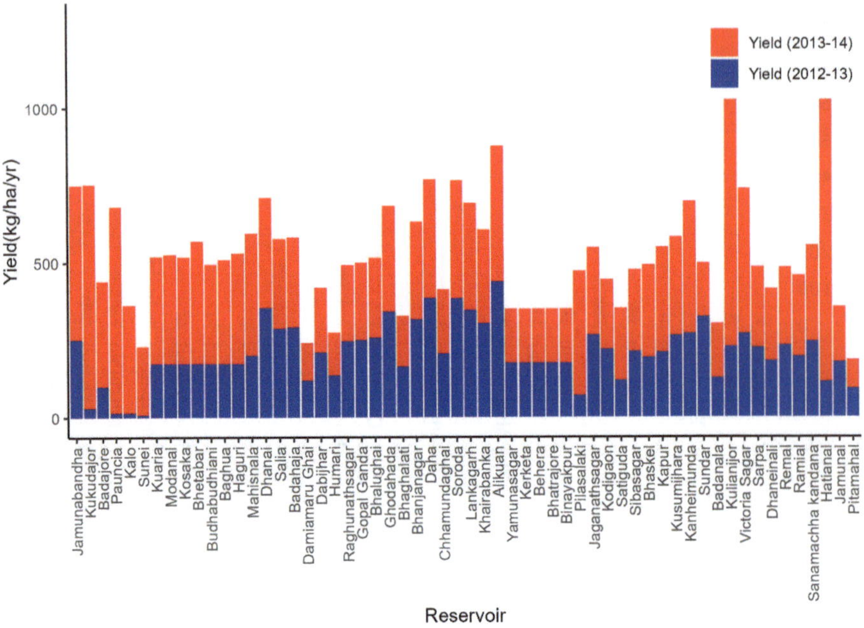

Fig. 13.4 Impact of CBF on fish yield in 58 small reservoirs of Odisha. (Source: Sarkar et al. 2020)

reservoirs in Jharkhand are categorized based on their size, with small reservoirs occupying an area of 30,089 ha, medium-sized reservoirs covering 14,200 ha, and large reservoirs sprawling across 71,711 ha.

Since 2009, there has been a notable surge in the improvement of fisheries and fish production in reservoirs due to the intervention of ICAR-CIFRI. As a result, the implementation of CBF and enclosure culture practices has been initiated. To facilitate the development of fisheries in Jharkhand, a comprehensive roadmap was created based on in-depth studies of the ecological conditions and fisheries potential. The stocking density of fingerlings per hectare was a crucial factor considered in this roadmap, taking into account the edaphic factors. For small, medium, and large reservoirs, the standardized fingerling stocking densities have been set at 1000, 500, and 300 fingerlings per hectare, respectively (Das et al. 2017). These standardized stocking densities aim to optimize fish growth and enhance sustainable fish production in the reservoirs of Jharkhand.

A stocking program was initiated in the reservoirs of Jharkhand, and its impact was studied after ten years of its implementation. The assessment of the impact of stocking, combined with enclosure culture practices, revealed a significant increase in fish production in the state. Over the last decade (from 2010–11 to 2021–22), fish production has risen from 0.7 lakh metric tonnes to 2.65 lakh metric tonnes, representing a remarkable four-fold increase.

13.4.4 Impact of Stocking in Small Reservoirs of Andhra Pradesh

To understand impact of stocking in reservoirs of Andhra Pradesh, Sugunan and Katiha (2004) analyzed production of 40 small reservoirs of Andhra Pradesh to understand the relationships between surface area, stocking density, and fish yield. They observed significant positive correlation ($r = 0.6$) between stocking density and fish yield. However, the size of the reservoirs (surface area) showed a negative correlation with both stocking density ($r = -0.57$) and fish yield ($r = -0.62$).

It is clear from the case study that stocking of fish seed plays an important role for increasing fish production in reservoirs in different geographical region of India. In all the cases, the fish production increased after the stocking of fish fingerlings implemented through various programmes. Artificial feed was not provided; instead, the natural food availability of the reservoirs was efficiently utilized making the process eco-friendly and cost effective. The increase of fish productions from these small reservoirs will enhance the livelihood of the fishers depending on the fisheries of these reservoirs.

13.5 Conclusions

The synergy between small-scale fisheries and culture-based fisheries in small reservoirs holds the potential to promote sustainable fishery management. Integrating these two approaches is instrumental in achieving a balance between conserving natural ecosystems and meeting the growing demands for fish as a vital food resource. CBF in small reservoirs requires a comprehensive understanding of local ecological dynamics, fish species preferences, and socio-economic context. It necessitates engaging local communities as key stakeholders in decision-making processes, as they possess the knowledge and vested interest in ensuring the long-term sustainability of their fishing resources. CBF is an effective strategy that optimize fish yields, safeguard traditional fishing practices, and conserve aquatic biodiversity for future generation. In summary, culture-based fisheries in small reservoirs of India with a special focus on small-scale fisheries aim to combine ecological sustainability, social inclusiveness, and economic benefits. By involving local communities in the process and promoting responsible fish farming practices, these initiatives have the potential to improve fish production, enhance livelihoods, and contribute to food security in rural areas.

Points to Remember

- Small-scale fisheries and culture-based fisheries in is an integral part for sustainable fishery management.
- Involvement of fishing community is the one of the key factors for sustainable management of small reservoir.

References

Arthur RI, Skerritt DJ, Schuhbauer A, Ebrahim N, Friend RM, Sumaila UR (2021) Small-scale fisheries and local food systems: transformations, threats and opportunities. Fish Fish 23(1):109–124

Das BK, Sarkar UK, Samanta S, Das AK, Hassan MA (2021) Strategic plan for inland open water fisheries development under PMMSY. ICAR-Central Inland Fisheries Research Institute, Barrackpore, Kolkata

Das BK, Parida PK, DebRoy P, Roy A (2022) A situational analysis of small-scale inland open water fisheries in India: from vulnerability to viability, V2V Working Paper 2022-8. V2V Global Partnership. University of Waterloo, Canada

FAO (2004) Paper COFI/2003/9 presented to the Committee on Fisheries, 25th Session – Rome, Italy 24–28 February 2003, FAO Fisheries Report No. 735. FAO, Rome, p 198

FAO (2015) Voluntary guidelines for securing sustainable small-scale fisheries in the context of food security and poverty eradication. Food and Agriculture Organization of the United Nations, Rome

FAO (2022) The state of world fisheries and aquaculture 2022 – towards blue transformation. Food and Agriculture Organization of the United Nations, Rome

Kumar M, Kujur R, Honnananda BR (2021) Cage culture practices in Kabirdham district of Chhattisgarh: a case study. J Entomol Zool Stud 9(3):170–174

Lianthuamluaia L, Sarkar UK, Mishal P, Zama S, Singh KK, Chakraborty SK, Ghosh A (2021) Assessment of the impact of fish seed stocking on fish yield in small reservoirs of Central India: towards sustainable management. Environ Sci Pollut Res 28:36464–36471

Sarkar UK, Sandhya KM, Mishal P, Karnatak G, Kumari L, Lianthuamluaia S, Panikkar P, Palaniswamy R, Karthikeyan M, Mol SS, Paul TT, Ramya VL, Rao DSK, Khan MF, Panda D, Das BK (2018) Status, prospects, threats, and the way forward for sustainable management and enhancement of the tropical Indian reservoir fisheries: an overview. Rev Fish Sci Aquac 26(2):155–175

Sarkar UK, Lianthuamluaia L, Panda D, Kumari S, Parida PK, Karnatak G, Mishal P (2020) Evaluation and impact assessment of culture-based fisheries to enhance fish yield in small reservoirs in Odisha State, India. Fish Manag Ecol 27:481–489

Sugunan VV (1995) Reservoir fisheries of India. Food and Agriculture Organization of the United Nations, Rome, p 243

Sugunan VV (1997) Fisheries management of small water bodies in seven countries in Africa, Asia and Latin America, FAO Fisheries Circular No. 933 FIRI/C933. FAO, Rome

Sugunan VV, Katiha PK (2004) Impact of stocking on yield of small reservoirs in Andhra Pradesh, India. Fish Manag Ecol 11:65–69

Sugunan VV, Sinha M (2000) Guidelines for small reservoirs fisheries management in India, CIFRI Bulletin No. 93. Central Inland Fisheries Research Institute, Barrackpore

Sugunan VV, Mandal SK, Rao DSK (2002) Fish yield prediction through morpho-edaphic index and estimation of stocking density for Indian reservoirs. Indian J Fish 49(4):369–378

Vass KK, Shrivastava NP, Katiha PK, Das AK (2009) Enhancing fishery productivity in small reservoirs in India: a technical manual, WorldFish Center Technical Manual No. 1949. WorldFish, Penang, p 19

Welcomme RL (1976) Some general and theoretical considerations on the fish yield of African rivers. J Fish Biol 8(3):51–64

Indian Rivers in the Context of Small-Scale Fisheries: Now and Then

14

Dibakar Bhakta, Basanta Kumar Das, and Piyashi DebRoy

Abstract

Indian rivers have historically served as a crucial foundation for various small-scale fisheries, providing livelihoods, nutrition, and a sense of cultural identity to millions. Over time, these riverine fisheries have undergone a transformation-from traditional, sustainable practices to a complex interaction of ecological degradation, policy oversight, and conflicting water-use priorities. This chapter delves into the historical evolution, current challenges, and future prospects of SSF in Indian rivers. The text highlights the profound changes in river ecosystems resulting from dam construction, pollution, overfishing, and climate change, which have adversely impacted fish populations and the communities dependent on fishing. Given these challenges, SSFs continue to demonstrate resilience by leveraging local knowledge and community institutions and adapting to evolving policy frameworks effectively. The manuscript promotes a revolutionary strategy for river governance that emphasises the perspectives of small-scale fishers, restores ecological flows, and ensures equitable access, underscoring the significance of sustainability, justice, and cooperative management.

Keywords

Small scale fisheries (SSF) · Riverine ecosystems · Ecological degradation · Sustainable livelihoods · Fisheries governance

Small-scale riverine fisheries provide a means for rural river-associated fishermen to rise out of poverty

D. Bhakta · B. K. Das (✉) · P. DebRoy
ICAR – Central Inland Fisheries Research Institute (ICAR-CIFRI), Barrackpore, Kolkata, India

B. K. Das et al. (eds.), *Small-scale Inland Fisheries in India*, Sustainability Sciences in Asia and Africa, https://doi.org/10.1007/978-981-95-3849-2_14

14.1 Introduction

India is bestowed with very rich inland open water resources consisting of backwaters, canals, estuaries, reservoirs, rivers, and wetlands (floodplain and coastal). Most of the inland fisheries of India are small-scale, and fishers mainly earn their livelihood and nutritional security from fishing activities (Jena and George 2018). Small-scale fisheries usually differ from large-scale fisheries in several economical, technical, and social attributes (Chaves 2022). SSF trends to optimize human work and reduce fuel consumption, as such fishing mainly uses passive gears such as gillnets, hand lines, fish traps, etc. (FAO 2016). Small-scale fisheries (SSF), or "artisanal" fisheries, may be defined as traditional fisheries involving fishing households, using relatively small amount of capital and energy, relatively small fishing vessels (if any), making short fishing trips, close to shore in the marine sector, and mainly for the local consumption (FAO 2015). The Food and Agricultural Organization (FAO) reports that approximately 90% of individuals, out of nearly 40 million registered globally, are classified as small-scale fisheries, with 97% of this demographic originating from developing countries (FAO 2021). Data concerning the livelihoods of small-scale fishing communities in developing nations is notably scarce (Bailey and Jentoft 1990; Cinner et al. 2010).

The importance of SSFs in improving food security in developing countries has been recognised, particularly for those whose livelihoods depend on them (Kawarazukan 2010; Maldonado et al. 2022). In addition to providing a source of income for fishermen and other directly engaged parties, SSF is essential for guaranteeing nutritional security in local communities and has an impact on both domestic and foreign markets. Several studies have demonstrated that inland industries related to SSF may contribute to local nutrition and food security and create more job possibilities than their marine equivalents (56 million against 52 million) (World Bank 2012; Bennett et al. 2018; Funge-Smith 2018). Among the components of the SSF workforce, women contribute nearly half of global SSF activities (World Bank 2012; Bennett et al. 2018; Das et al. 2022). In the past, SSF was generally represented as a type of fishery with low production and low yield rate (Smith 1979; FAO 2003). Presently, SSF is considered a family-based enterprise, private, or self-owned property, regardless of the farm size, species reared, and volume produced (Shrestha and Pant 2012).

14.1.1 Background

Historically, small-scale fisheries in Indian rivers relied on traditional knowledge, seasonal fishing methods, and community-based governance structures. The contributions of these fisheries extend beyond household nutrition and local economies; they are also essential for maintaining ecological balance. In recent decades, there has been a notable shift in river ecology and fishery practices. Contemporary infrastructure initiatives, alterations in water systems, and the introduction of non-native species have significantly impacted indigenous fish populations and sidelined traditional fishing communities (Gozlan et al. 2010).

14.1.2 Examination of Issues

Small-scale fishers face intensified challenges due to the decline in fish catches, the erosion of biodiversity, increased competition with commercial entities, and the reduction of traditional fishing rights (Arthur et al. 2022). Despite their ecological and socio-economic significance, small-scale riverine fisheries have not received adequate attention in policy discussions. Furthermore, there is a lack of comprehensive records that document their historical importance and the contemporary challenges they encounter.

14.1.3 Purpose and Objectives

This chapter aims to examine the evolution of small-scale fisheries in Indian rivers by contrasting historical perspectives with present-day realities. The aims of this chapter are to:

- Investigate the historical significance and customary methods of small-scale fisheries within the prominent river systems of India.
- Examine the ecological, institutional, and economic transformations impacting these fisheries.
- Examine the primary obstacles encountered by small-scale fishers in the present scenario.
- Examine case studies that demonstrate the decline and resilience of small-scale fisheries.
- Propose sustainable strategies and policy suggestions for the rejuvenation of small-scale fisheries within riverine ecosystems.

This chapter is organized into the following sections to meet the outlined objectives:

- *Historical overview*—Examining the development of small-scale fisheries and their cultural and economic importance.
- *Ecological changes in river systems*—Effects of human activities on river biodiversity and fish habitats.
- *Current issues*—Examination of risks such as habitat degradation, competitive pressures, governance challenges, and climate-related effects.
- *Case studies*—A detailed examination of localised studies from significant Indian rivers, including the Ganga, Brahmaputra, Godavari, Narmada, etc.
- *Examination of policy and management frameworks*—Assessment of current regulations, identification of gaps, and evaluation of community-based management practices.
- *Future perspectives and recommendations*—Proposals for sustainable development, collaborative management frameworks, and river restoration aimed at enhancing fisheries.

This thorough analysis aims to provide a detailed insight into the transformation of small-scale fisheries in Indian rivers and explores strategies for their preservation and promotion amidst a shifting socio-environmental context.

14.2 Riverine Fisheries Resources

India has a vast network of river systems that spans over 45,000 km and covers a catchment area of 3.12 million km^2, comprising 113 river basins (Ayyappan et al. 2019). The Indian mainland comprises 15 major rivers with drainage basins exceeding 20,000 km^2, 45 medium rivers ranging from 2000 to 20,000 km^2, and over 102 minor rivers with basins smaller than 2000 km^2, alongside numerous ephemeral streams in the western arid region (Fig. 14.1). The Indian River systems are classified based on their origins, which include Himalayan rivers and peninsular rivers, and their flow direction, which differentiates between east-flowing and west-flowing rivers (Table 14.1). The Ganga-Brahmaputra-Meghna river system

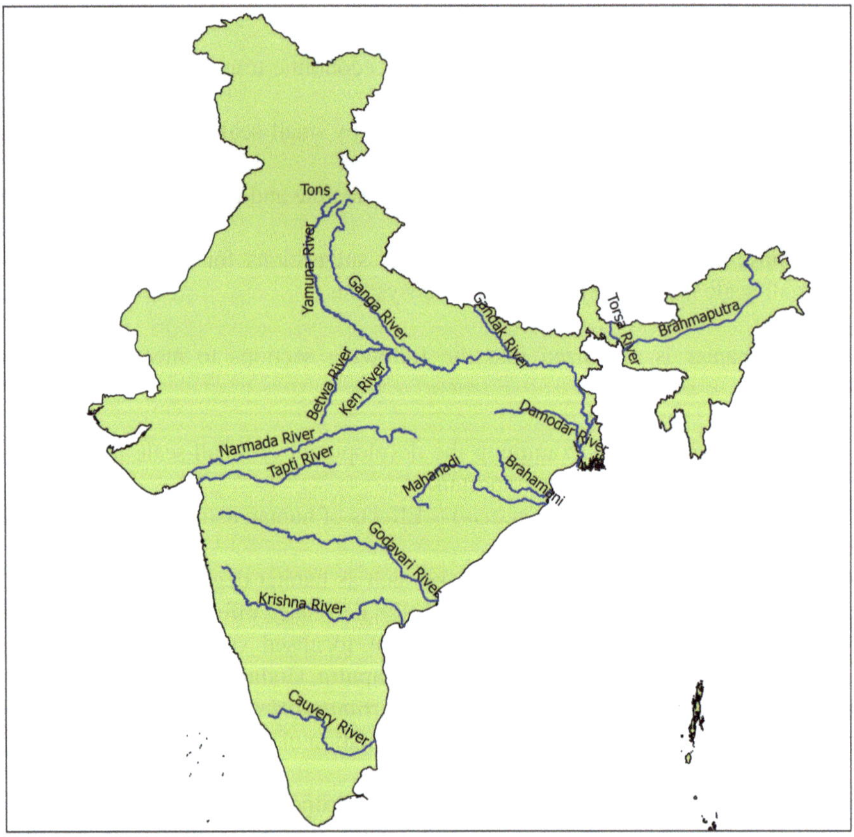

Fig. 14.1 Map showing the important river systems in India

Table 14.1 Details profile of important river systems of India

River system	Name of the main river	Approx. length (in km)	State(s) covered
Himalayan Ganga	Ganga	2525	Uttarakhand, Uttar Pradesh, Madhya Pradesh, Jharkhand, Bihar, West Bengal
	Ramganga	569	Uttar Pradesh
	Gomti	940	Uttar Pradesh
	Ghagra	1080	Uttar Pradesh, Bihar
	Gandak	300	Bihar
	Kosi	492	Bihar
	Subernarekha	395	Bihar, Odisha, West Bengal
	Yamuna	1376	Uttarakhand, Haryana, Delhi, Uttar Pradesh
	Chambal	1080	Madhya Pradesh, Uttar Pradesh, Rajasthan
	Tons	264	Uttarakhand
	Sone	784	Uttar Pradesh
	Ken	360	Madhya Pradesh, Uttar Pradesh
	Damodar	592	Jharkhand, West Bengal
Brahmaputra	Brahmaputra	4000	Arunachal Pradesh, Assam
	Dibang, Siang		Assam
	Lohit, Manas		Nagaland
	Duri Dihang		Sikkim
	Dhansiri, Koppili		Manipur
Indus	Jhelum	400	Jammu and Kashmir
	Chenab	330	Jammu and Kashmir, Himachal Pradesh
	Beas	460	Himachal Pradesh, Punjab
	Sutlej	1450	Himachal Pradesh, Punjab
	Ravi	725	Jammu and Kashmir, Himachal Pradesh, Punjab
Peninsular Rivers			
East Coast	Mahanadi	857	Chhattisgarh, Madhya Pradesh, Odisha
	Brahmani	799	Odisha, Bihar
	Baitarani	360	Odisha
	Godavari	1465	Maharashtra, Andhra Pradesh
	Krishna	1401	Maharashtra, Andhra Pradesh, Karnataka
	Cauvery	800	Tamil Nadu, Karnataka
	Pennar	597	Karnataka, Andhra Pradesh
	Bhima	861	Karnataka
West Coast	Narmada	1312	Madhya Pradesh, Maharashtra, Gujarat
	Tapti	720	Madhya Pradesh, Maharashtra, Gujarat
	Mahi	583	Gujarat
	Sabarmati	371	Rajasthan, Gujarat

consists of the main river, the Ganga, and its tributaries, such as the Ramganga, Gomti, Gandak, Koshi, Sone, and Yamuna, which flow into the Bay of Bengal through Bangladesh. The Indus system includes the Indus River and its tributaries: the Jhelum, Chenab, Beas, Sutlej, and Ravi. These rivers originate in the north and primarily flow westward or southwestward into the Arabian Sea through Pakistan. The rivers of the east coast, such as the Mahanadi, Brahmani, Baitarni, Subarnarekha, Damodar, Krishna, Godavari, and Cauvery, flow into the Bay of Bengal along the eastern shoreline. Narmada and Tapti rivers originate in the central peninsular and flow in a western direction into the Arabian sea at Gujarat. Other west-flowing rivers usually carry relatively little flow such as Mahi, Sabarmati, and Luni are flowing through Rajasthan and Gujarat flow into the Arabian sea at Gujarat and some are lost through internal drainage.

Indian rivers host around 250–300 species of freshwater fish, showcasing a rich variety of carps including *Labeo rohita*, *Labeo catla*, and *Cirrhinus mrigala*, as well as catfishes, featherbacks (*Notopterus* spp.), murrels, and many small indigenous species (SIS). Diverse rivers showcase distinct biodiversity profiles. The Ganga River system is home to a rich diversity of more than 200 fish species, including prominent species like hilsa (*Tenualosa ilisha*), Mahseer (*Tor* spp.), and the Giant river catfish (*Bagarius bagarius*) (Das et al. 2021). The Brahmaputra River system is noted for its considerable endemism, supporting over 160 species, including several SIS like *Esomus danricus* and *Puntius* spp. (Vass et al. 2009a). The Godavari and Krishna Rivers support rich fisheries, especially of carps and catfishes, with productivity significantly increasing during the monsoon season as a result of improved floodplain connectivity (Sugunan 1995).

Riverine fisheries yield around 0.2–0.25 million tonnes each year, a figure that may appear modest in comparison to aquaculture, yet carries significant local importance (FAO 2022). The fisheries in question primarily rely on capture methods, exhibiting limited stocking or management practices, and play a crucial role in ensuring nutritional security for communities residing along rivers and riparian zones. The composition of catches has evolved over the decades, with traditional catches that were once dominated by Indian major carps and Hilsa now being increasingly supplanted by exotic or low-value species. This shift can be attributed to factors such as ecological degradation, overfishing, and modifications to water flow.

A notable decrease in the presence of large-bodied fish species in Indian rivers has been observed, attributed to factors such as damming, pollution, and encroachment (Pinder et al. 2019). More than four million individuals engaged in inland fishing, a significant number of whom operate on a small scale, rely directly or indirectly on resources from rivers. These fishers are generally structured within traditional caste-based systems or cooperatives. Fishing typically follows seasonal patterns and is often combined with other livelihood strategies such as agriculture or daily wage labour. The contributions of women in processing, marketing, and SIS collection are vital, though frequently overlooked (Béné et al. 2007). Although the primary emphasis of institutions has transitioned towards aquaculture, organizations such as ICAR-CIFRI and NMCG (Namami Gange) have embarked on projects aimed at restoration and stock enhancement (CIFRI 2020). Initiatives are underway to protect unique and at-risk species, enhance community-driven fisheries management, and advance data gathering and support programmes for fisher (Table 14.1).

14.3 Riverine Fishing Communities

Riverine fisheries are also a part of small-scale fisheries and directly cater to the livelihood and nutritional securities of the millions of riverine fisheries communities. The Indian rivers are subjected to excessive stress and the consequent effects are poor fish landing and loss of livelihood of fishing communities. Being a community-based development process, fisheries have a direct holding on the rural economy. In developing countries, fisheries play a vital role in creating employment, and income generation, helping in poverty alleviation in society, as well as to provide nutritional security (Thilsted et al. 1997). Riverine fisherfolk are the poorest of the lot, and they have been denied a fair deal so far (Fig. 14.2). Disruption of the environment and the creation of man-made water bodies on rivers have displaced fisherfolk to some extent and effected adversely their livelihood (IFSI 2006). Traditionally, in general, the rivers are managed as a common property resource, and these resources have multiple uses for the riparian area population. Members of the fishing communities mainly depend on riverine resources for their livelihood security.

Fig. 14.2 Community participation in riverine fish harvest

Table 14.2 Key factors influencing fish yield from Indian rivers and their implications

Factor	Explanation
Unstable catch composition	Shifts in species availability due to overfishing, habitat degradation, and ecological disturbances, exemplified by the substitution of key species such as major carps or hilsa with lower-value species.
Conflicts between multiple users of the riverine waters	Competition among fishers, irrigation, industrial usage, hydropower, and navigation often limits access and degrades fish habitats.
Cultural beliefs lead to nutritional loading and pollution	Ritual offerings, idol immersion, and festivals lead to both organic and inorganic pollution, resulting in decreased water quality and negative impacts on fisheries.
Inadequate infrastructure and supporting services	Inadequate landing facilities, insufficient cold storage, weak market linkages, and transportation challenges hinder the optimal utilisation of the catch and reduce fisher incomes.
Lack of understanding of the fluvial systems	A limited understanding of river ecology, fish migration patterns, and seasonal dynamics affects management decisions and conservation strategies.
Poor data management in riverine fish catch	The lack of reliable, real-time data on fish catch, effort, and species composition impedes effective monitoring and policy development.
Socio-economic and socio-cultural determinants, etc.	The interaction of poverty, caste dynamics, gender inequality, and inadequate institutional support makes small-scale fishers vulnerable, limiting their ability to adapt and innovate.

14.3.1 Factors Influencing Fish Yield from the Indian Rivers

Key factors influencing fish yield from Indian rivers and their implications are provided in the Table 14.2.

14.4 Institutional Frameworks

In the Indian fisheries context, SSF has intrinsic advantages, in comparison with other fishing methods, as SSF has fewer negative impacts on the ecosystem. In SSF, due to the use of more selective fishing gears, the overall fishing activities are less destructive and henceforth produced less by-catch. In most cases, SSF is involved with the traditional way of fishing, and very often protects, and preserves cultural heritage and environmental knowledge. Due to the lower running cost, and minimal fuel consumption, SSFs are involved with more fisheries. There is a need to improve participation and transparency by ensuring the participation of stakeholders for small-scale fishers during policy development, legislation, and management decision while developing the institutional framework (Mariette Correa 2017). For the last few years, ICAR-CIFRI initiated river ranching program after successful breeding of the wild stock population and reared them up to fry/ fingerling stages under the NMCG flagship program. This could be a classic example of the institutional framework for the regeneration of depleted stock in the river systems for sustainable fisheries.

Table 14.3 Key institutions and governance bodies

Institution/framework	Role in riverine fisheries management
The Department of Fisheries (DoF)	Holds the responsibility for crafting fisheries policies, executing development initiatives, and granting fishing licenses at both the central and state levels in designated states.
ICAR institutes, such as CIFRI	ICAR institutes are involved in scientific inquiry, perform stock evaluations, rehabilitate fish habitats, and offer technical assistance.
Inland Fisheries Co-operatives / Fisher Associations	Cooperatives focused on inland fisheries and associations of fishers support the coordination of communities, credit systems, marketing strategies, and, at times, the self-regulation of fishing practices.
Water Resources Departments	Water Resources Departments frequently prioritise water usage for irrigation and hydropower, often overlooking the requirements for fisheries.
Panchayati Raj Institutions (PRIs)	Panchayati Raj Institutions (PRIs) embody decentralised governance structures in rural India, designed to enhance local self-government and promote community engagement in decision-making processes.
Legal frameworks (e.g., Indian Fisheries Act 1897, state fisheries acts)	Local governance structures can shape the allocation of fishing rights and the management of conflicts within communities. The legal frameworks include the Indian Fisheries Act of 1897 along with various state fisheries acts. The current legal framework governing fisheries is frequently antiquated or poorly enforced, especially concerning river fisheries.

Small-scale fisheries in riverine environments in India are shaped by a multifaceted interplay of institutions operating at central, state, and local levels, in conjunction with customary and community-based systems. The governance of riverine fisheries frequently experiences challenges due to disjointed institutional frameworks, resulting in inadequate collaboration among various sectors. Table 14.3 provides a list of key institutions and governance bodies, along with their respective roles in riverine fisheries management in India.

14.5 Small-Scale Fisheries in Indian River System: Few Cases Studies

The majority of riverine fisheries operate on a small scale, providing livelihoods and ensuring nutritional security for those who fish in local water resources. The ICAR-Central Inland Fisheries Research Institute has regularly studied the social and economic lives of small-scale fishers from inland waters, such as estuaries, floodplain wetlands, reservoirs, and backwaters, especially looking at river areas (Fig. 14.3). In recent times research has been done, and is ongoing in the rivers the Brahmaputra, Cauvery, Chaliyar, Ganga, Krishna, Narmada, Siang, Tapti, Torsa, and associated estuaries such as Hooghly-Matlah, Mahanadi, Narmada, etc. The study indicated that fishing is the main occupation of most of the fisheries, and that too contributes around >60% of their household income. A few case studies of river systems with special reference to SSFs are provided below:

Fig. 14.3 Small-scale riverine and estuarine fisheries components

14.5.1 River Ganga, and SSFs

River Ganga is a very rich ecosystem and harbors 265 species of fish out of which 181 fish species are reported from freshwater, and around 10–13 million riverine fishers are involved with the systems (SANDRP 2014; Ayyappan et al. 2019). The system supports the livelihood and nutritional security of millions of people, as a source of irrigation, water supply, fishing activities, etc. A recent study (2017–2019) revealed that the average monthly income of the fishers from systems was Rs. 4345/–, Rs. 5866/–, and Rs. 7283/– from the upper, middle, and lower stretches, respectively. While day-wise income from fishing activities was estimated at Rs. 145, Rs. 196, and Rs. 243 per day during the study period of 2017–2019, in contrast to Rs. 40.03 per day was reported by Tyagi (2009) from the Bihar and Uttar Pradesh stretch of river Ganga. While Pandit et al. (2019) could find the monthly income of fishers' households around Rs. 9391/– during the study period of 2016 in the Bhagirathi-Hooghly stretch of the Ganga River. Fishers are involved with fishing activities on small scales mainly by gill nets, seine net, bag net, meen jal, hook and line, lift net, etc.

Small-scale fisheries predominantly depend on capture methods that employ traditional gear, including gillnets, cast nets, and traps. Fishing activity shows notable increases during the monsoon and post-monsoon months in the rive Ganga (Vass et al. 2009b; Montana et al. 2011). The river Ganga is essential for sustaining the livelihoods of many small-scale fishing families, particularly those from communities like Mallah, Bind, Nishad, and Kaibarta. Fishing often overlaps with agricultural activities or paid labor. Women are involved in fish processing, marketing, and the gathering of SIS. Small-scale fishing is vital for improving local nutrition and generating income; however, fishermen often face socio-economic marginalisation (Roy et al. 2023). Fishing practices are deeply intertwined with cultural rituals, festivals like Chhath Puja, and the beliefs that highlight the sacred essence of the Ganga. The significance of fish is evident in both the dietary habits and spiritual ceremonies of the local community. The river holds significant spiritual value, influencing how communities perceive and make use of its resources.

The Ganga exemplifies a multifaceted river system distinguished by a historically important diversity of life forms. However, the ecological integrity of the area has been undermined by various factors, including flow regulation via dams and barrages, pollution, habitat fragmentation, and the presence of invasive species. The influence of these stressors on fish breeding, recruitment, and overall productivity is considerable (CIFRI 2020).

State departments of fisheries primarily manage fisheries governance, overseeing licensing, leasing, and certain conservation initiatives. National initiatives like Namami Gange integrate aspects of biodiversity conservation. Co-operative societies exist; nonetheless, they often exhibit a lack of robustness. The incorporation of small-scale fisheries into river basin management is still restricted, as fishers play a negligible part in water governance decisions (Vass et al. 2009b; NMCG 2021).

14.5.2 River Cauvery, and SSFs

A very recent study by ICAR-CIFRI (2017–2020) at river Cauvery showed that an average year of experience by the riverine fisheries is around 17 years and an average income of Rs. 10,500/– (CIFRI 2020). Previously fisheries were involved in fishing mainly by bamboo-made coracles, which were the main fishing gear, and gradually it has been replaced by fiber-made coracles. Gill, or entangling gears, falling nets, hook and lines, traps, and encircle gears are mainly employed to catch fish. Along with small-scale fisheries fish-based eco-tourism is suggested at all the dam/barrage sites, which might give fishers to have a better livelihood option through the promotion of fish-based catering systems within the tourism.

14.5.3 River Tapti, and SSFs

The Tapti River basin covers almost 2% of the entire territory of India and provides livelihood security to the fishers' community of the states of Madhya Pradesh, Maharashtra, and Gujarat. A recent study (2017–2020) indicated that the average income of the riverine fisheries is about 5500/– per month, and fisheries are engaged with fishing activities 6–7 months in a year (being the seasonal river), and the rest of the period worked as agricultural labor, stone mining, and own business (CIFRI

2020). Fishers are mainly involved with fishing activities using gill net, cast net, dragnet, hook and line, pole and line traps, etc. by small-scale fishing crafts such as inflated motor vehicle tube, boat, dugout canoe, catamaran, thermocol platform, etc. Survey also revealed that the decline in fish catch and less availability of different types of fish species are one of the major problems of the riverine fisheries, followed by erratic rainfall, destructive fishing methods, siltation, less access to the market, and interference of the middleman. River Tapti is a diverse habitat type, and its flow mainly depends on rainfall. So judicious release of water from the number of dams and barrages built on the river is required to maintain the minimum flow in the river for sustaining the biodiversity and livelihood of the local fishers.

14.5.4 River Chaliyar, and SSFs

River Chaliyar is the fourth largest river in the state of Kerala and contributes significantly to the local small-scale fisheries. A recent study by the ICAR-CIFRI indicated that the average income of the fisheries from the riverine fisheries is around Rs. 8000/– per month and fisheries are involved with fishing activities around 6–7 months a year (CIFRI 2020). Fisheries mainly catch their fish using gill net, followed by hook and line, and cast net.

14.5.5 River Siang, and SSFs

The Brahmaputra known as Siang/Dihang in Arunachal Pradesh is a major trans-boundary river of South Asia. Though the recent study revealed that, fishing by the riverside population is not the main livelihood source. The main occupations of the sampled population were found to be agriculture (66.0%), followed by livestock farming (18.0%), and fishing from the river (16.0%) (CIFRI 2020). The average income from the river fisheries was found to be Rs. 5500/– per month and for the migratory fishermen around Rs. 8000/– per month. Fishing is done mainly traditional way, and traditional bamboo traps locally known as Porang were the most common gear in the upper and middle stretch of the river, along with cast nets. In the lower stretch drag nets were generally used for commercial fishing. Subsistence fishing along the lower stretch of the river is carried out mainly by gill nets, traditional bamboo traps, cast nets, lift nets, etc.

14.6 Spawn Collections and Small-Scale Fisheries

The collection of finfish and shellfish spawns from the lower stretches of river systems (estuarine zones) is a classic example of the small-scale fishery from the riverine ecosystems. The demand for fin and shellfish spawns is increasing significantly,

Fig. 14.4 Collection of shellfish post-larvae from the lower stretch of Hooghly-Matlah estuarine systems Left panel: A woman in a red outfit operating a scope net for the collection of shrimp juveniles. Right panel: A group of people, including children, are near the water's edge. One lady is segregating shrimp juveniles, while others are watching it. The sites are surrounded by mangroves, which are ideal for collecting shrimp seeds

attributed to the expansion of brackish water aquaculture in coastal areas (Bhakta et al. 2021). Approximately 420,000 individuals engage in the collection of prawn seeds from estuarine and coastal areas of India (World Bank 2012). Estuaries function as essential seed collection centers for coastal aquaculture, and numerous species use these habitats as spawning grounds (Goswami and Goswami 1993; Brinda et al. 2010). Fishermen, operating individually or collectively, primarily employed specific traditional gear, with some modifications, to collect finfish seeds and shellfish post-larvae from estuarine and coastal environments. Examples of these devices include shooting nets, stake nets, scoop nets, drag nets, bundles of brushes, bag nets, and cradle traps. Fish spawn also collected from the Indian major river systems such as the Ganga and Brahmaputra in the form of eggs, spawn (<8 mm in length), fry (>8 mm to <40 mm in length), and fingerlings (>40 mm to >100 mm). In river Ganga spawns are mainly collected from the state of Bihar, West Bengal, and Uttar Pradesh from May to September. For the first time, ICAR-CIFRI initiated the study on spawn prospectus in Indian rivers with a special focus on the Ganga in 1964 and reported that private fishers used to collect spawns on a very large scale for private trade. A recent study also indicated that overall major carps spawn availability at the previously potential sites of the river Ganga has drastically reduced, mainly due to the destruction of breeding sites, and other anthropogenic factors (Fig. 14.4).

14.7 Threats to Riverine Fisheries

Indian river ecosystems are experiencing significant human intervention, leading to habitat loss and degradation. Consequently, several fish species in these riverine habitats have become vulnerable. The construction of dams, the introduction of invasive exotic species, pollution, overfishing, and climate change are significant factors contributing to the decline of fish diversity and fisheries (Rosenberg et al. 1997; Nilsson et al. 2005; Bhakta et al. 2018). Several factors are responsible for which Indian riverine fisheries are in great threat, and some of them are as:

14.7.1 Dams and Water Abstraction with Flow Regulation

The construction of dams in the rivers divides the main steam into several huge reservoirs leading to fish habitat fragmentation (Kang et al. 2009). To meet the energy consumption of the ever-increasing population, mainly in the Asian countries many hydropower dams are planned and contracted, accordingly. There are huge impacts on the hydrology and biotic components downstream of the dams. Furthermore, dams will change the physical characteristics of natural waterways by eliminating shallow banks, changing the flow regime, and destroying various river habitats, such as rapids and pools. Such construction would ultimately result in the extinction of fish species and a slow decline in natural fish recruitment, which mostly relies on unimpeded flows. The construction of dams and barrages on the rivers are the main ecohydrological alterations that are responsible for the loss of wetlands and floodplain habitats including water diversions.

14.7.2 Pollution

Riverine pollution can originate from various sources, including factory or mining discharges, diffuse runoff from agricultural areas, livestock waste, sewage (which may contain pharmaceuticals), factory effluents, landfall seepage, oily runoff from roads and impermeable surfaces, agrochemicals (such as fertilizers and pesticides), and both macro and microplastics (Fig. 14.5) (Burns et al. 2018). The impact of pollutants can manifest in various forms, including direct, indirect, lethal, or sub-lethal effects. Additionally, the interactions among these pollutants may lead to unforeseen consequences for both inland open water fisheries and riverine fisheries. Anticipated increases in pollution burdens are likely due to heightened wastewater discharge resulting from the intensification of livestock farming, industrialisation, and urbanisation, among other factors (Wen et al. 2017).

Fig. 14.5 Macro-plastic an immerging pollutant in Indian river/estuarine systems

Fig. 14.6 Overfishing of fishes using small-meshed gill nets

14.7.3 Overexploitation/Overfishing

Overexploitation is considered is one of the main reasons for the decline of fisheries in Indian river systems. Increasing fishing pressure can lead to attribute more numbers of fisheries with more efforts of readily available fishing gear (Fig. 14.6). The increasing fishing pressure leads to the initial targeting of larger fish, subsequently affecting smaller species, which ultimately contributes to a decline in fisheries and poses a risk of collapsing aquatic ecosystems (Pauly et al. 1998; Jackson et al. 2001).

14.7.4 Alien Invasive Species

Many exotics species have entered the river systems, probably from the reservoirs, flooded aquaculture ponds, etc. Introductions of aquatic species have both advantages and disadvantages. Few of the negative impacts are genetic contaminations, the introduction of diseases, ecological interactions leading to loss of native fish fauna, etc. Some of the exotic species which are abundant in the Indian river systems viz. *Clarias gariepinus*, *Pangasius sutchi*, *Oreochromis mossambicus*, *Oreochromis niloticus*, *Puntius javonicus*, *Cyprinus* carpio, *Pterygoplichthys disjunctivus*, *Amphilophus trimaculatus*, *Atractosteus spatula*, etc. (Das et al. 2020; Kumar et al. 2021; Manna et al. 2021). The entry of such exotic species into the Indian river systems might have increased the biological stress of existing native species.

14.7.5 Climate Change

Climate changes include rising air and water temperatures, shifts in regional monsoons, an increase in drought occurrences, and a significant rise in severe storm incidents in coastal areas, among other factors. The impacts of climate change are becoming more apparent in inland aquatic resources and their related fisheries. Examination of time series data over a 30-year period from existing literature

Table 14.4 Primary threats to riverine fisheries in India

Category of threats	Description/primary elements
Dams and water abstraction with flow regulation	The erection of dams, barrages, embankments, and water diversion structures alters flow regimes, impedes fish migration, and degrades breeding and nursery habitats.
Pollution and nutrient enrichment	Industrial effluents, sewage discharges, agricultural runoff (comprising pesticides and fertilisers), and cultural or religious practices increase toxic loads, reduce oxygen levels, and negatively impact fish health and biodiversity.
Habitat degradation and riverbed disturbance	Habitat degradation and riverbed disturbance sand mining, encroachment, deforestation in watershed regions, and riverbank erosion severely impact ecosystems critical for fish life cycles.
Overfishing and detrimental fishing	Practices the use of illegal equipment (fine mesh nets, poisons, explosives) leads to unsustainable exploitation, while insufficient regulation of fishing activities exacerbates the problem.
Alien invasive species	The introduction and proliferation of species such as *Clarias gariepinus* (African catfish), *Pangasiodon hypophthalmus*, and Tilapia compete with native species, alter food webs, and upset ecosystem equilibrium.
Climate change and extreme weather	Alterations in precipitation patterns, increased temperatures, and erratic flood-drought cycles affect fish reproduction, recruitment, and productivity, hence exacerbating the susceptibility of small-scale fisheries.

indicated that in the river Ganga, there has been a rise in minimum water temperature of 1.5 °C in the colder regions and between 0.2 and 1.6 °C in the aquaculture farms of West Bengal situated in the Gangetic plains. The impacts of climate change are evident in the reproductive failures of Indian major carps, which have led to a significant decrease in the availability of fish spawn in the river Ganga. Fish production in the middle stretches of the river Ganga has steadily declined over the last twenty years, with the share of Indian major carps (IMC) dropping from 41.4% to 8.3%. In contrast, the proportions of miscellaneous fish and catfish have risen (Ayyappan et al. 2019). As a result, specific adaptation strategies can be put into action to tackle rising water temperatures, extreme occurrences like floods and droughts, storms, and water scarcity, along with various other challenges.

Below is a tabular arrangement (Table 14.4) that lists the main threats to river fisheries, summarizing the category of threats and respective key points.

14.8 Conservation Measures

The construction of fish passages in respective dams, so that fish could move freely up to the spawning grounds would be an enormous benefit from a conservation point of view (Moffitt et al. 1982). The creation of specially targeted protected areas is an important step to protect the natural habitat of fishes and provide suitable spawning conditions. In addition to such important spawning habitats, it is also very important to conserve the habitats of endangered species. Further, for the protection of aquatic resources and associated flora and fauna of the rivers and respective tributaries, there is much needed to remove various stress on the aquatic systems (Sarkar et al. 2010). Habitat restoration and rehabilitation need to be carried out to improve the natural stocks of riparian stocks. Stocking of indigenous fish species (from the wild population) through ranching in rivers will be very much helpful for the restoration of threatened and disappeared fish species. For successful conservation of fishes, there is an urgent need to carry out research on life-history traits and biology of fishes along with habitat parameters and to establish proper facilities.

14.9 Conclusions

The gradual demands for economic development, human activities, overfishing, and especially the construction of dams have structurally and temporally affected the fish species richness and composition, both up and below the dams of river systems. Such types of changes in species composition gradually impact the decline of biodiversity and fisheries of the ecosystems. The biodiversity losses also will be accompanied by the invasion of invasive exotic species. Some basic knowledge of the biology of some commercially important and key species with their respective artificial propagation will help to restock the systems and ultimately recover and improve the native fish species in the basin. Riverine fishing communities mainly depend on small-scale fisheries for their livelihoods, and these are poorly understood. To properly address, their vulnerability issues require understanding the socio-economic context in which SSFs operate for designing policies but unfortunately, such information is usually incomplete in developing countries. To guarantee the sustainability of the SSF, it may be necessary to implement stringent and sufficient conservation strategies. This may be accomplished through the implementation of alternative income sources, including compensation schemes, social protection, or policies that encourage alternative livelihoods.

Points to Remember

- Indian riverine small-scale fisheries serve as more than a mere livelihood; they represent essential socio-ecological systems intricately integrated into the essence of rural life.
- Despite ecological degradation, policy neglect, and marginalization, these fisheries demonstrate remarkable resilience.
- To revive and sustain these systems, it is essential to implement inclusive governance for river resources that honors traditional knowledge, restores ecological flows, and guarantees equitable access for all stakeholders.
- Centering small-scale fishers in policy and conservation initiatives is crucial for achieving environmental sustainability and promoting social justice.
- Small-scale fisheries (SSF) are traditional fisheries that mainly produce food for local consumption while using relatively little energy and capital.
- India is endowed with a vast number of river systems that span 45,000 km and have a combined catchment area of 3.12 million km^2, made up of 113 river basins.

- The millions of communities that depend on riverine fisheries for their livelihood and nutritional security are directly supported by these small-scale, riverine fisheries.
- Due to significant human involvement that is destroying Indian river ecosystems and resulting in habitat loss, several fish species that depend on riverine environments are currently in danger.
- Among the strategies to safeguard small-scale riverine fisheries are the construction of fish passages, the establishment of protected areas, the reduction of anthropogenic stressors, habitat restoration and rehabilitation, etc.

References

Arthur RI, Skerritt DJ, Schuhbauer A, Ebrahim N, Friend RM, Sumaila UR (2022) Small-scale fisheries and local food systems: Transformations, threats and opportunities. Fish Fish 23(1):109–124

Ayyappan S, Moza U, Gopalakrishnan A, Meenakumari B, Jena JK, Pandey AK (2019) Handbook of fisheries and aquaculture. Indian Council of Agricultura Research, New Delhi, p 1116

Bailey C, Jentoft S (1990) Hard choices in fisheries development. Mar Policy 14(4):333–344

Béné C, Macfadyen G, Allison EH (2007) Increasing the contribution of small-scale fisheries to poverty alleviation and food security. FAO Fisheries Technical Paper No. 481

Bennett A, Patil P, Kleisner K, Rader D, Virdin J, Basurto X (2018) Contribution of fisheries to food and nutrition security: current knowledge, policy, and research. NI Report

Bhakta D, Meetei WA, Vaisakh G, Das SK, Manna RK (2018) Impacts of water regulation on *Tenualosa ilisha* in the Narmada Estuary, Gujarat, India. J Fisheries 6(1):563–568

Bhakta D, Manna RK, Samanta S, Das BK (2021) Spawn collection from estuarine systems for coastal aqua-farming: sustainability issues and way forward. In: Mandal, U. K., Lama, T.D., Burman, D., Mandal, S. and Raut, S. (Eds.). Souvenir, international symposium on coastal agriculture: transforming coastal zone for sustainable food and income security, 16–19 March, 2021. Indian Society of Coastal Agricultural Research, ICAR-Central Soil Salinity Research Institute, Regional Research Station, West Bengal:105

Brinda S, Srinivasan M, Balakrishnan S (2010) Studies on Diversity of fin fish larvae in Vellar Estuary, South east coast of India. World J Fisheries Marine Sci 2:44–50

Burns EE, Carter LJ, Kolpin DW, Thomas-Oates J, Boxall ABA (2018) Temporal and spatial variation in pharmaceutical concentrations in an urban river system. Water Res 137:72–85

Chaves PTC (2022) Juveniles and undersized fish in small-scale fisheries: gillnets are not less implied than trawling. Marine Fishery Sci 35(2)

CIFRI (2020) ICAR-CIFRI Annual Report 2020:299

Cinner JE, Mcclanahan TR, Wamukota A (2010) Differences in livelihoods, socio-economic characteristics, and knowledge about the sea between fishers and non-fishers living near and far from marine parks on the Kenyan coast. Mar Policy 34(1):22–28

Das BK, Ray A, Manna RK, Roshith CM, Baitha R, Karna SK, Bhor M (2020) Occurrence of exotic vermiculated sailfin catfish *Pterygoplichthys disjunctivus* from the lower stretch of River Ganga, West Bengal, India. Curr Sci 119(12):2006–2009

Das BK, Bhakta D, Kumar L, Nair SM, Koushlesh SK, Sajina AM, Roshith CM, Samanta S (2021) Ichthyofaunal diversity of the major Indian rivers: a review. J Inland Fish Soc India 53(1&2):22–35

Das BK, Roy A, Som S, Chandra G, Kumari S, Sarkar UK, Pandit A (2022) Impact of COVID-19 lockdown on small-scale fishers (SSF) engaged in floodplain wetland fisheries: Evidences from three states in India. Environ Sci Pollut Res 29(6):8452–8463

FAO (2003) Advisory committee on fisheries research (ACFR). Report on the second session of the working party on small-scale fisheries, FAO Report No. 735. Bangkok, Thailand
FAO (2015) Voluntary guidelines for securing sustainable small-scale fisheries in the context of food security and poverty eradication. Food and Agriculture Organization of the United Nations, Rome
FAO (2016) Technical and socio-economic characteristics of small-scale coastal fishing communities, and opportunities for poverty alleviation and empowerment, by Uwe Tietze. FAO Fisheries and Aquaculture Circular 1111, p 136
FAO (2021) Small-scale fisheries and the human right to adequate food-Making the connection: exploring synergies in the implementation of the SSF Guidelines and the Right to Food Guidelines. Rome. https://doi.org/10.4060/cb4939en
FAO (2022) The state of world fisheries and aquaculture 2022: towards blue transformation. FAO, Rome. https://doi.org/10.4060/cc0461en
Funge-Smith S (2018) Review of the state of the world fishery resources inland fisheries. In: FAO fisheries and aquaculture circular. Food and Agriculture Organization of the United Nations, Rome
Goswami SC, Goswami U (1993) Seasonal variability in Penaeid Prawn larval Abundance in the Mandovi and Zuari Estuaries, Goa. J Indian Fish Assoc 23:45–54
Gozlan RE, Britton JR, Cowx I, Copp GH (2010) Current knowledge on non-native freshwater fish introductions. J Fish Biol 76(4):751–786
ICAR–Central Inland Fisheries Research Institute (2017–2020). Annual reports (2016–17, 2017–18, 2018–19, 2019–20). ICAR–CIFRI
IFSI (2006) River fisheries in India—issues and current status. In: Vass KK, Mitra K, Suresh VR, Katiha PK, Shrivastava NP (eds) Proceedings of the national seminar on Management Challenges in Fisheries of Rivers and Associated Ecosystems-Issues and Strategies, Barrckpore, India. 16–17 April 2005, p 319
Jackson JB, Kirby MX, Berger WH, Bjorndal KA, Botsford LW, Bourque BJ, Warner RR (2001) Historical overfishing and the recent collapse of coastal ecosystems. Science 293(5530):629–637
Jena J, George G (2018) Small-scale fisheries in India: an appraisal. In: Giri S (ed) Small-scale Fisheries in South Asia. South Asian Association for Regional Cooperation Agriculture Centre, Bangladesh, pp 15–41
Kang B, He D, Perrett L, Wang H, Hu W, Deng W, Wu Y (2009) Fish and fisheries in the Upper Mekong: current assessment of the fish community, threats and conservation. Rev Fish Biol Fish 19(4):465–480
Kawarazukan B (2010) Linking small-scale fisheries and aquaculture to household nutritional security: an overview. Food Secur 2:343–357
Kumar L, Kumari K, Gogoi P, Manna RK, Madayi RC, Salim SM, Das BK (2021) Risk analysis of non-native three-spot cichlid, *Amphilophus trimaculatus*, in the River Cauvery (India). Fish Manag Ecol 28(2):158–166
Maldonado JH, del Pilar M-SR, Vargas-Morales ME, Leguízamo E (2022) Livelihoods characterization of a small scale fishing community in the Colombian Caribbean. Marine Fish Sci 35(2):181–207
Manna RK, Ray A, Bayen S, Bera T, Palui D, Das BK (2021) First record of exotic alligator gar, *Atractosteus spatula* (Actinopterygii: Lepisosteiformes: Lepisosteidae), from Ganga River system, India: a possible threat to indigenous riverine fish diversity. Acta Ichthyol Piscator 51:385
Mariette Correa ICSF (2017) Report on workshop on enhancing capacities of women fish workers in India for the implementation of the SSF guidelines. International Collective in Support of Fishworkers (ICSF), Chennai, p 46
Moffitt CM, Kynard B, Rideout SG (1982) Fish passage facilities and anadromous fish restoration in the Connecticut River basin. Fisheries 7:2–11
Montana CG, Choudhary SK, Dey S, Winemiller KO (2011) Compositional trends of fisheries in the River Ganges, India. Fish Manag Ecol 18(4):282–296
Nilsson C, Reidy CA, Dynesius M, Revenga C (2005) Fragmentation and flow regulation of the world's large river systems. Science 308(5720):405–408

NMCG (2021) Annual report. National Mission for Clean Ganga, MoWR, GoI

Pandit A, Ekka A, Das BK, Samanta S, Chakraborty L, Raman RK (2019) Fishers' livelihood diversification in Bhagirathi–Hooghly stretch of Ganga River in India. Curr Sci 116(10):1748

Pauly D, Christensen V, Dalsgaard J, Froese R, Torres F Jr (1998) Fishing down marine food webs. Science 279(5352):860–863

Pinder AC, Britton JR, Harrison AJ, Nautiyal P, Bower SD, Cooke SJ, Raghavan R (2019) Mahseer (Tor spp.) fishes of the world: status, challenges and opportunities for conservation. Rev Fish Biol Fish 29:417–452

Rosenberg DM, Berkes F, Bodaly RA, Hecky RE, Kelly CA, Rudd JW (1997) Large-scale impacts of hydroelectric development. Environ Rev 5(1):27–54

Roy A, Manna RK, Ghosh S, Sahu SK, Das BK (2023) Assessment of women's contribution in small-scale fisheries in Indian Sundarbans: Issues, strategies, and way forward for sustainability. Fish Manag Ecol 30(4):378–391

SANDRP (2014). https://sandrp.wordpress.com/2014/08/30/dams-fish-andfishing-communities-of-the-ganga-glimpses-of-the-gangetic-fisheries-primer/

Sarkar UK, Gupta BK, Lakra WS (2010) Biodiversity, ecohydrology, threat status and conservation priority of the freshwater fishes of river Gomti, a tributary of river Ganga (India). Environmentalist 30(1):3–17

Shrestha MK, Pant J (2012) Small-scale aquaculture for rural livelihoods: In Proceedings of the symposium on small-scale aquaculture for increasing resilience of rural livelihoods in Nepal. WorldFish, Kathmandu, Nepal

Smith IR (1979) A research framework for traditional fisheries, ICLARM Studies and Reviews, No. 2. International Center for Living Aquatic Resources Management, Manila

Sugunan VV (1995) Reservoir Fisheries of India. FAO Fisheries Technical Paper No. 345. FAO, Rome

Thilsted SH, Roos N, Hassan N (1997) The role of small indigenous fish species in food and nutrition security in Bangladesh. Naga-The ICLARM Quarterly, pp 13–15

Tyagi RK (2009) Socio-economic status of fishers of River Ganga. Bulletin No. 160. ICAR-Central Inland Fisheries Research Institute, p 55

Vass KK, Das MK, Srivastava PK, Dey S (2009a) Assessing the impact of climate change on inland fisheries in River Ganga and its plains in India. Aquat Ecosyst Health Manage 12(2):138–151

Vass KK, Tyagi RK, Pathak V (2009b) Status and strategies for the conservation and management of small indigenous fish species in riverine ecosystems. J Inland Fish Soc India 41(2):18–25

Wen Y, Schoups G, van de Giesen N (2017) Organic pollution of rivers: combined threats of urbanization, livestock farming and global climate change. Sci Rep 7:43289

World Bank (2012) Hidden harvest: the global contribution of capture fisheries. WorldBank; WorldFish; FAO, Washington, DC

Does Small Scale Fishery Contribute to the Artha-Ganga Concept?

15

Basanta Kumar Das, Piyashi DebRoy,
Bandana Das Ghosh, and Gitashree Thengal

Abstract

The Ganga River holds immense ecological, cultural, and economic significance in India, supporting the livelihoods of millions through its ecosystem goods and services. Among these, small-scale fisheries play a crucial role, especially for riparian communities, contributing significantly to household incomes and food security. It explores the interconnections between riverine ecosystems and community livelihoods within the framework of the Artha-Ganga initiative economic pillar added to the Namami Gange Programme to strengthen the river-people relationship through sustainable resource use. Despite the river critical contributions, it faces multifaceted threats, including pollution, over-exploitation, habitat destruction, climate change, and the introduction of exotic species. These pressures have led to declining fish diversity and productivity, directly impacting the socio-economic well-being of fishing communities. Artha-Ganga proposes a self-sustaining model centered on promoting natural farming, reuse of treated wastewater, ecotourism, and inclusive growth. Recognizing the part of small-scale fisheries in the blue economy, this study advocates for their integration into national river management policies. The approach aligns with sustainable development goals, aiming to restore the ecological health of the Ganga while enhancing economic resilience and social inclusion among riparian populations.

The Ganga's flow carries not just water, but the lifeblood of millions-where every fish caught is a thread in the fabric of rural resilience.

B. K. Das (✉) · P. DebRoy · B. D. Ghosh · G. Thengal
ICAR – Central Inland Fisheries Research Institute (ICAR-CIFRI), Barrackpore, Kolkata, India

B. K. Das et al. (eds.), *Small-scale Inland Fisheries in India*, Sustainability Sciences in Asia and Africa, https://doi.org/10.1007/978-981-95-3849-2_15

Keywords

Artha-Ganga · Namami Gange · Ganga basin · Water economy

15.1 Introduction

Rivers play an important role in the lives of the Indian people. The contribution of the riverine system is recognized on a large scale in agriculture, transportation, drinking water, electricity, and also providing livelihoods for a large number of people in riparian communities. This is the probable reason for the location of the major cities of India near the banks of river. It is considered sacred in Hindu mythology in India. Eighty-three percent of India's drainage basin area is covered by the 14 main rivers (Ganga, Brahmaputra, Brahmani, Cauvery, Godavari, Indus, Krishna, Mahanadi, Mahi, Narmada, Periyar, Sabarmati, Tapti, and Subarnarekha) and their numerous tributaries (Das et al. 2007). The major river systems are the Ganga River System, the Indus River System, the Brahmaputra River System, the East Coast River System, and the West Coast River System. The *Namami Gange* (National Mission for Clean Ganga) is the first integrated river conservation mission in India on the sustainable management of waters of river Ganga launched by the Government of India in 2014 with the involvement of different ministries/departments and other stakeholders, institution and community organization (Balkrishna et al. 2024). *Artha-Ganga* was launched in 14 December 2019 at the Meeting of the National Ganga Council (NGC) in Kanpur to establish a sustainable and feasible model for economic development and to strengthen the river-people connection through an economic bridge. It was added as a vertical of the Namami Gange programme in addition to the four other verticals, namely *Nirmal Ganga* (unpolluted river), *Aviral Ganga* (flowing river), *Jan Ganga* (people's participation) and *Gyan Ganga* (research and knowledge management). The concept of Artha-Ganga centers around restoration and conservation of river Ganga to strengthen the economic development of communities dwelling around the Ganga basin, and also on fostering India's economy from the ecosystem goods and services generated by the river Ganga system on a macro-scale (Fig. 15.1). Ecological and environmental parameters play a vital role to maintain the sustainability in the aquatic ecosystem. The two main components that have been historically important in the economy of river Ganga are material resources and spiritual resources that are connected with the Ganga. Ecosystem goods from the Ganga that promote livelihood generation include fish, medicinal plants, and local handicrafts and ecosystem services include river water tourism, navigation, power generation, and others. The fish catch of river Ganga is estimated to be 56,387.24 tonnes per annum (Das et al. 2024). The self-sustaining nature of the Artha-Ganga concept implies that the financial and other resources required for restoration and management of the river system should be generated from the river itself. This would be achieved through strengthening the local economy, enhancing resource use efficiency, water conservation, boosting capacity and public awareness.

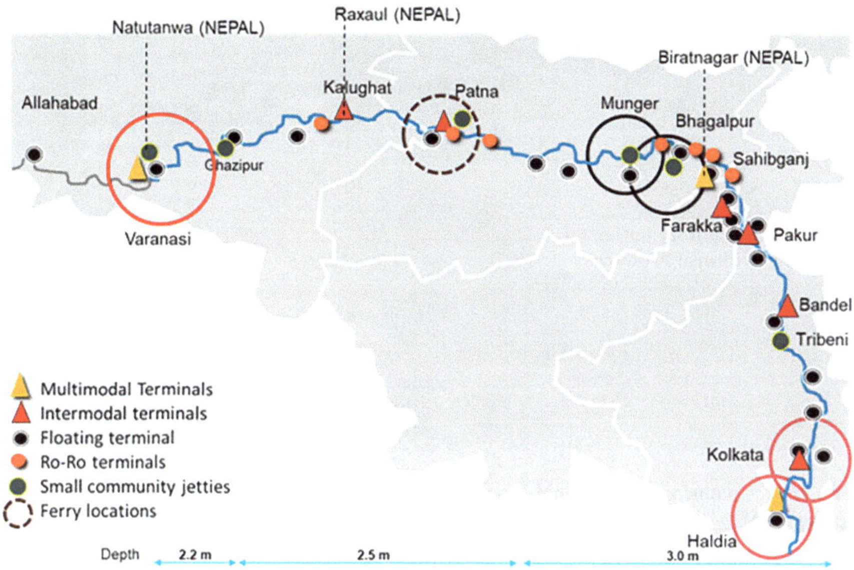

Fig. 15.1 Master Plan for Artha-Ganga. (Source: https://www.pib.gov.in/Pressreleaseshare.aspx?PRID=1624562, 15 May 2020)

The six pillars of Artha-Ganga are (i) promotion of natural farming on 5 km band on the either side of river Ganga, (ii) Monetization and reuse of treated water and sludge from STPs, (iii) Improvement of livelihood opportunities, particularly for women, (iv) Promotion of tourism and cultural heritage, (v) People's participation, and (vi) building of capacities and institutions. With the Mission of "Reform, Perform and Transform," the cargo volume in the Ganga is poised to increase by four times with special focus on developing inland waterways through boosting inclusive growth for farmers, small traders, and villagers.

15.2 Ecosystem Goods and Services of River Ganga Supporting Its Economy

The riverine resources provide essential ecological services, support ecosystem health, and influence fish production for fisheries. According to the Millennium Ecosystems assessment, ecosystem services are the benefits people derive from ecosystems. These services are categorized into provisioning services (food and water), regulating services (flood and disease control), cultural services (spiritual, recreational, and cultural benefits), and supporting services (nutrient cycling). The benefits obtained from the cultural services of the Ganga River are significant (Sharma and Kaushik 2018). The Ganga offers immense benefits across all categories of ecosystem services, given its high socio-cultural, economic, and ecological significance (Fig. 15.2) (Singh and Singh 2020). The riverine ecosystem provides water

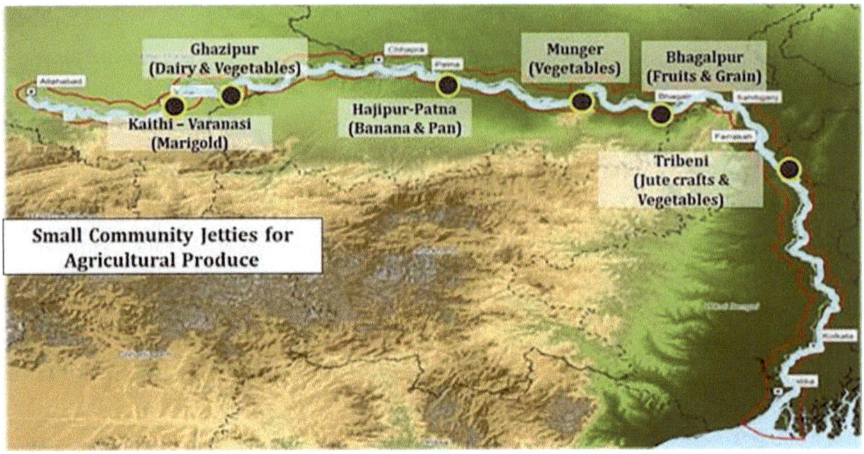

Fig. 15.2 Commercial agricultural goods in river Ganga system of India. (Source: https://sagarmala.gov.in)

resources, fish, and other aquatic food, assists in flood protection, and offers space for recreation, all of which are crucial for human well-being. The Indian riverine fishery resources are very productive, yielding abundant finfish and shellfish that have not yet been integrated into the Artha-Ganga framework. The Ganga supports a diverse array of fish species upon which thousands of people depend for their daily livelihoods. Provisioning services such as fisheries, agriculture, and employment generation for women and men in fisheries, along with contributions to food and nutrition security; cultural services; regulating services like food web dynamics, bioremediation, and carbon sequestration; and supporting services that maintain biodiversity are provided by the wetlands in the Gangetic plains (Ekka et al. 2024).

15.3 Contribution of Small-Scale Fisheries in Artha-Ganga

15.3.1 Livelihood of Riverine Fishing Communities

Riverine fisheries are a complex system that includes artisanal, subsistence, and traditional fisheries. Fisheries resource management of the Ganga River is conducted by dividing the river into three stretches: (i) the upper stretch in the Northern Himalayan area (Uttarakhand), (ii) the middle stretch (Uttar Pradesh and Bihar), and (iii) the lower stretch (Bihar, Jharkhand, and West Bengal). The highly scattered and isolated nature of fishing, extensive range of landing areas, diversification in fishing gear operations, migratory fishers, and an unorganized marketing system contribute to the complexities of aquatic resource management. The targeted fisheries consist of a multi-gear, multi-species fishery that is open access to all. The ICAR-Central Inland Fisheries Research Institute has been continuously monitoring and assessing the ecology and fisheries, including fish diversity, fish catch, fishermen livelihoods, and river habitat along with tributaries, since the establishment of the Institute, creating a repository of biological and ecological data on the river system. Several studies and research efforts have documented different inland fisheries resources, their utilization and actual growth rates, the scope for development of riverine resources, improvement of resource abundance levels, strategies for sustainable resource exploitation, ensuring food security, economically viable exploitation processes, and conservation of riverine resources (Balkrishna et al. 2024).

Fisheries engage a large number of fishermen and provide livelihoods to millions of rural individuals (Katiha and Sarkar 2010). The fisheries sector supports the livelihoods of approximately 14.5 million fishers in India. Riverine fisheries is one of the oldest occupations for fishers and involves artisanal and small-scale fisheries. Riverine fishing is generally a community-based occupation, has a direct impact on the rural economy, and helps maintain daily livelihoods for riverine fishing communities. The traditional riverine fishing communities in India are widely scattered and cover a vast area. A major part of this community engages in artisanal, small-scale fishing as well as subsistence fishing, where fish is consumed for household purposes.

Riverine fishermen's economic situation is very poor to moderate (Katiha et al. 2017), and the main reason is the decline in the number and quality of fish produced from inland open waters over the past few decades, particularly economically important fish species. Several studies revealed their unsatisfactory education, lack of alternative occupation, lack of facilities and other infrastructures as the main causes of deprivation of fishermen. Another facet is that capture-based riverine fishery resources are declining due to resource degradation and over-exploitation, which has a direct impact on the livelihood of fishers. Therefore, the communities are being economically, socially, and educationally deprived, with many of them below the poverty line. The health status of riverine fishing communities hints at serious issues as a lack of awareness regarding nutrition, hygiene, and other health-related aspects. Comprehensive studies on the socio-economic status of riverine fishermen communities are very rare and inadequate. Inadequate information regarding social, cultural, and economic aspects leads to a serious issue related to socio-economic conditions for the riverine fishers.

15.3.2 Livelihood and Income of Fishing Communities

The income of fishers in the lower stretch of river Ganga was ₹ 11,273.70, that in the middle stretch was ₹ 8430, and that in the upper stretch was ₹ 5950.39 (Das et al. 2024). The income from fisheries in the lower stretch was significantly higher than the income from fisheries in the upper and middle stretches of river Ganga. ICAR—CIFRI studies have found that fisheries constitute 46.27% of the total household income of fishers. Fisheries as a livelihood is more important in the lower stretches (61.27% of household income) than in the upper and middle stretches. Livelihood diversification is more in the upper stretch with a Simpson Diversification Index of

0.46. The abundant fishery resources in the lower stretch reduce the necessity for livelihood diversification among the fishers. The research and developmental efforts especially ranching activities carried out as a part of *Namami Gange* activities have helped in shifting the scale of economies of the small-scale fishers of river Ganga as evident from the increases in their incomes (Das et al. 2024). Increase in income of fishers observed in the upper stretch was 36.94%, in the middle stretch was 43.7% and that in the lower stretch was 54.8% from 2017–19 to 2022–23 (Das et al. 2024).

15.4 Threats to Riverine Fisheries

The major threats to the river Ganga fisheries are declining fish population, dams and barrages not taking into account fish movement and migration, sand mining, aquatic pollution, encroachments along river banks, and juvenile fishing. In the recent past, the river has undergone drastic changes due to rapid ecological degradation caused by man-made interruptions in the river basin which causes a noticeable decline in both fish production and productivity. The Ganga basin is facing serious threats from accelerated deterioration of irrigated fields, severe soil erosion, and nutrient leaching into the river basin. River systems of India are home to a variety of wild fish germplasm. River system like Ganga alone harbours about 190 fish species (Das et al. 2023). As a result, the Gangetic ecosystem must be preserved and restored in a sustainable manner, which will ultimately include the biotic components as a whole.

15.4.1 Impact of Climate Change

Climate change in India is increasingly evident through rising air temperatures, shifting monsoon patterns, more frequent droughts, and intensified storm activity, all of which are having profound effects on inland aquatic ecosystems and fisheries, especially in the Ganga River basin. Research indicates that minimum water temperatures have risen by 0.99 °C in the upper stretches of the Ganga and from 0.5 to 1.4 °C in aquaculture fish farms across West Bengal (Das et al. 2013). These thermal changes, along with increased post-monsoon rainfall, have disrupted the breeding cycles of Indian Major Carps (IMC), leading to a noticeable decrease in the availability of fish spawn. Warm-water fish species such as *Glossogobius gurius* and *Xenentodon cancila* are now migrating to cooler upstream regions, and the predator-prey balance in the river's middle stretch has declined sharply over the past three decades. Over the last 20 years, fish production has also shifted, with the IMC share dropping from 41.4% to just 8.3%, while the presence of miscellaneous and catfish species has grown (Vass et al. 2009). These changes, coupled with increased vulnerability to extreme weather events and water stress, pose serious threats to the livelihoods of millions dependent on the Ganga's fisheries and highlight the urgent need for adaptive and climate-resilient fishery management strategies.

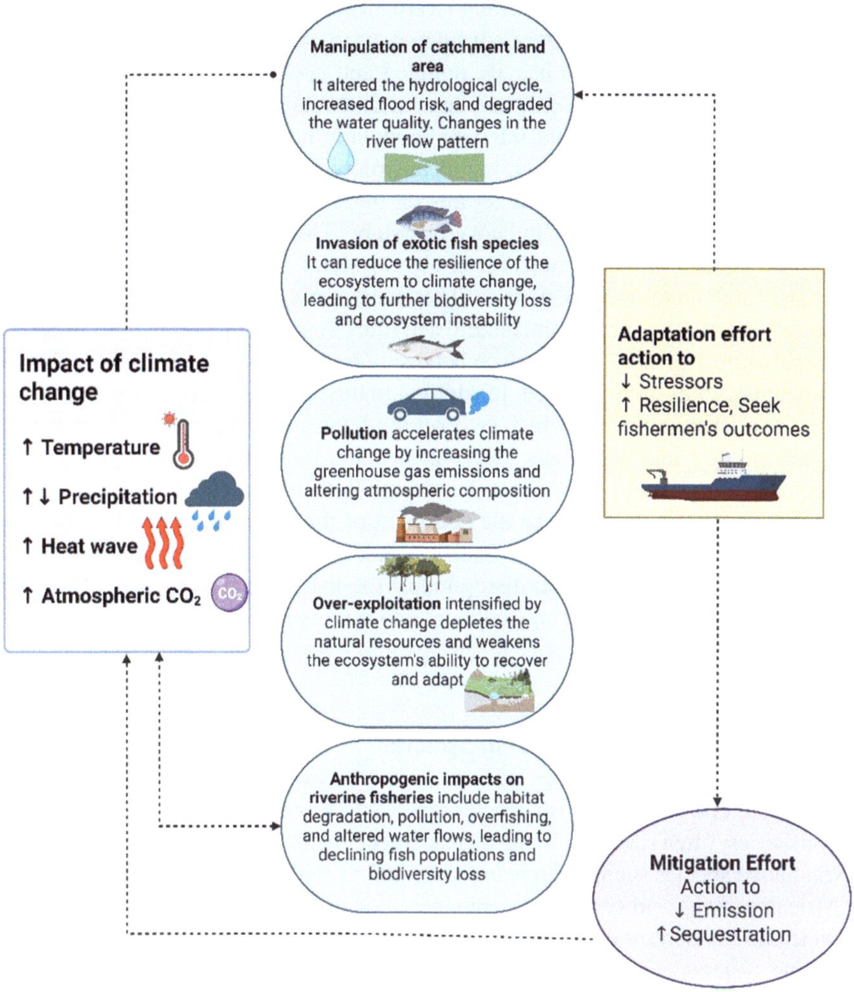

Fig. 15.3 Schematic diagram showing the climate change impact on the River Ganga

Climate change has a complex and wide-ranging impact on the River Ganga, influencing its hydrological dynamics, ecological balance, and socio-economic systems. These effects vary significantly across different regions and stretches of the river, depending on local climatic, geographic, and anthropogenic factors (Fig. 15.3).

15.4.2 Manipulation of Catchment Land Area

River encroachment and taming of the riverine catchment area has emerged as a rising concern, particularly in urban and suburban regions, where rivers serve as vital lifelines for communities. However, rapid and unregulated urban expansion, coupled with industrial growth along riverbanks, has led to severe degradation of these critical water bodies. Once considered clean and life-sustaining, many urban rivers have now become so heavily polluted and contaminated that they are now very much unsafe even for basic human contact. Linking to the construction of barrages and human impediments, the river floodplains has been impacted severely due to reduced water flow. Moreover, rapid construction and development of unauthorized residential and commercial set ups on these ecologically sensitive areas has created havoc menace and regulatory oversight. This encroachment severely disrupts the natural flow and function of river systems like preventing groundwater recharge and increased flood risks. The rivers natural ecology has now greatly affected due to construction and landfill activities. A suburban populations grow and infrastructure expands, the interface between human activity and the natural environment becomes more strained. In addition to this, a particularly harmful intervention is river channelization, which involves dredging and realigning the river courses to increase flow and floodwater carrying capacity. While this method may provide temporary flood control, it eliminates essential riverine habitats and accelerates water flow unnaturally, often to the detriment of the river's ecological health. The impact can be directly seen upon the fish populations, causing degradation of water quality, variation in depth and disruption of the fish spawning grounds. Increased runoff often results in carrying sediment and pollutants into the river, resulting in habitat degradation.

15.4.3 Invasion of Exotic Fish Species

The introduction of exotic fish species represents a major ecological threat to riverine fisheries, often resulting in the displacement and decline of native ichthyofauna. Non-native species such as *Oreochromis niloticus* (Nile tilapia), *Clarias gariepinus* (African catfish), and *Cyprinus carpio* (common carp) are frequently introduced for aquaculture enhancement or commercial exploitation, but their presence in natural riverine systems can lead to intense interspecific competition for food, space, and spawning grounds. These species often exhibit high environmental tolerance, rapid growth, and aggressive foraging strategies, which enable them to establish and proliferate quickly. Their introduction can also facilitate the transmission of novel pathogens and parasites, increasing disease susceptibility among native fish populations. Furthermore, hybridization between exotic and closely related indigenous species may result in genetic introgression, leading to the loss of genetic purity and adaptive traits in native stocks. Cumulatively, these impacts can cause significant alterations in trophic structure, biodiversity loss, and ecosystem function, thereby undermining the ecological integrity and socio-economic sustainability of riverine fisheries.

15.4.4 Pollution

It is a critical environmental concern, driven by a combination of industrial, agricultural, and domestic activities along its course. The river receives large volumes of untreated sewage, industrial effluents, solid waste, and agricultural runoff containing pesticides and fertilizers, leading to severe degradation of water quality. The high levels of heavy metals, pathogenic microbes, chemical oxygen demand (COD), and biological oxygen demand (BOD) render portions of the river unfit for aquatic life, drinking, and bathing. Urban expansion and inadequate waste management systems in towns along the Ganga exacerbate the problem. Domestic and industrial pollutants have severely contaminated the Ganga River downstream of Haridwar. Below Kannauj, particularly around the meeting point of the Ramganga and Kali rivers, the pollution reaches a worrying level and persists until at least Varanasi. The high level of pollutants in a river is lethal to its biota (Singh et al. 2022). Pollution not only affects aquatic biodiversity but also poses serious health risks to millions who rely on the river for daily use. Despite various initiatives like the *Namami Gange Programme*, effective pollution control remains a major challenge, requiring stronger enforcement, community participation, and investment in sustainable wastewater treatment infrastructure.

15.4.5 Over-Exploitation

Over-exploitation poses a critical threat to the sustainability of riverine fisheries, leading to the depletion of fish stocks, disruption of aquatic food webs, and long-term ecological imbalance. Intensive and unregulated fishing practices—such as the use of fine-mesh nets, destructive gear, and indiscriminate harvesting during breeding seasons—target both mature and juvenile fish, reducing recruitment and altering population structure. In many river systems, commercially important species, including Indian Major Carps, have shown significant population declines due to excessive fishing pressure. The situation is further exacerbated by increased demand for fish protein, lack of effective enforcement of fishing regulations, and limited awareness among fishers about sustainable harvest practices. Over time, such pressure not only reduces biodiversity but also threatens the livelihoods of fishing communities that depend on these resources. Sustainable fisheries management, including regulated fishing seasons, gear restrictions, and community-based conservation initiatives, is essential to mitigate the impacts of overexploitation and ensure the long-term viability of riverine fisheries.

15.4.6 Anthropogenic Impacts on Riverine Fisheries

Anthropogenic activities have significantly impacted on riverine fisheries, altering aquatic ecosystems and threatening the sustainability of fish populations. Infrastructure such as dams and embankments disrupt natural river flow and block migratory routes critical for spawning, while activities like sand mining, deforestation, and urbanization degrade habitats and increase sedimentation. Pollution from industrial discharge, agricultural runoff, and untreated sewage leads to water quality deterioration, eutrophication, and fish mortality. Overfishing and harmful fishing practices further stress fish stocks, reducing biodiversity and resilience. Nutrient enrichment, a term used to describe anthropogenic influences on river nutrient concentrations, raises the levels of nitrogen, phosphorus, and other nutrients present in wastewater from homes, businesses, and agriculture. On the other hand, nutritional deficits are frequently overlooked (Bowes et al. 2020). Climate change adds to these pressures by altering hydrological cycles and water temperatures, disrupting fish life cycles. Collectively, these impacts compromise the ecological integrity of river systems and the livelihoods of communities dependent on inland fisheries, underscoring the urgent need for sustainable management and conservation efforts (Fig. 15.3).

15.5 Conclusion

The Artha-Ganga framework offers a transformative opportunity to enhance fisher livelihoods and promote sustainable use of riverine fisheries in the Ganga basin. By recognizing the ecosystem goods and services such as fish production, employment, and cultural value emanating from the river, Artha-Ganga can strengthen the role of small-scale fisheries in India inland economy. Prioritizing these services within river management policies would support food and nutritional security, generate income for riparian communities, and contribute to a more inclusive blue economy. Integrating riverine fisheries into Artha-Ganga aligns with sustainable development goals and ensures ecological conservation alongside economic development. This approach can bridge ecological health with socio-economic growth, making the Ganga not only a sacred river but also a pillar of sustainable inland fisheries and rural livelihoods.

Things to Remember

- Fishing is a mainstay occupation of riverine communities residing around river Ganga with fishery-related income contributing to more than half of the total household income.
- Community participation and inter-sectoral co-operation are essential for boosting the circular water economy of river Ganga.
- To improve and make it easier to preserve and promote the Ganga cultural heritage and tourism in the surrounding areas.

References

Balkrishna A, Singh SK, Ghosh S, Banerjee S, Verma S, Arya V (2024) An analytical review on the integrated management of river resources through Namami Gange. Water Policy 26(5):462

Bowes MJ, Read DS, Joshi H, Sinha R, Ansari A, Hazra M, Simon M, Vishwakarma R, Armstrong LK, Nicholls DJE, Wickham HD, Ward J, Carvalho LR, Rees HG (2020) Nutrient and microbial water quality of the upper Ganga River, India: identification of pollution sources. Environ Monit Assess 192(8):533

Das MK, Samanta S, Saha PK (2007) Riverine health and impact on fisheries in India. Policy Paper No. 01, Central Inland Fisheries Research Institute, Barrackpore, Kolkata, p 120

Das MK, Sharma AP, Sahu SK, Srivastava PK, Rej A (2013) Impact and vulnerability of climate change in the Ganga River system in India. Aquat Ecosyst Health Manag 16(4):415–424

Das BK, Ray A, Johnson C, Verma SK, Alam A, Baitha R, Manna RK, Roy S, Sarkar UK (2023) The present status of Ichthyofaunal diversity of river ganga India: synthesis of present v/s past. Acta Ecol Sin 43(2):307–332

Das BK, Thengal G, Johnson C (2024) Fish catch and fishers socio-economics of River Ganga. ICAR–Central Inland Fisheries Research Institute, p 95

Ekka A, Das BK, Roy A, Pandit A, Swain PR, Lianthuamluaia L, Chanu TN, Parida PK, Debroy P, Bhattacharya S, Chakraborty S (2024) Ecosystem services assessment of Beledanga oxbow Lake in the Gangetic Plains: pathways to sustainable conservation. Front Freshw Sci 17(2):1491720

Katiha PK, Sarkar S (2010) Sustainability in Indian fisheries. In: Sen RK (ed) Environment and sustainable economic development. Deep and Deep Publications Pvt Ltd, New Delhi, pp 243–264

Katiha PK, Pandit A, Ekka A, Sharma AP (2017) Socioeconomic status of riverine fisher communities in India. Aquat Ecosyst Health Manag 20(1–2):188–197

Sharma P, Kaushik A (2018) Drivers of ecosystem change: a case study of River Ganga. East West J Sci Technol 13(28):167–176

Singh R, Singh GS (2020) Integrated Management of the Ganga River: an Ecohydrological approach. Ecohydrol Hydrobiol 20:153–174

Singh D, Shukla AK, Yadav S, Pandey G, Dutta V (2022) The Ganga River water pollution status in India characterized with river Gomti. Ecol Environ Conserv 28:1636–1643

Vass KK, Das MK, Srivastava PK, Dey S (2009) Assessing the impact of climate change on inland fisheries in river Ganges and its plains in India. Aquat Ecosyst Health Manag 12(2):138–151

Small Indigenous Fishes vis-a vis Small Scale Fisheries

16

Basanta Kumar Das, Mishal P., and Gunjan Karnatak

Abstract

Small Indigenous Fishes (SIFs) are vital for food security, nutrition, ecosystem sustainability, and rural livelihoods, particularly within India's small-scale inland fisheries (SSIF). These species, generally under 30 cm in length, are nutrient-dense and culturally important, yet they remain overlooked in mainstream fisheries policy and data collection. SIFs are rich sources of essential amino acids, omega-3 fatty acids, vitamins (A, D, B12), and minerals like calcium, zinc, and iron, making them crucial for combating malnutrition and hidden hunger, especially among vulnerable groups. Predominantly found in rivers, wetlands, ponds, and reservoirs, SIFs contribute substantially to local diets and household income, often consumed directly due to their affordability and availability. Despite their importance, SIF populations are under threat due to overfishing, pollution, habitat loss, and invasive species. Climate change adds further stress, altering habitats and reducing population viability. The chapter highlights the role of SIFs in achieving Sustainable Development Goals (SDGs), particularly those targeting hunger (SDG 2), poverty (SDG 1), gender equality (SDG 5), and biodiversity conservation (SDG 14 & 15). Conservation strategies should include habitat restoration, regulation of fishing practices, pollution control, and promotion of SIF culture through pens, backyard ponds, and polyculture systems. Community participation and capacity-building are essential for sustainable management. Integrating SIFs into nutritional and conservation policies can significantly benefit human health and environmental sustainability.

From nutrient-rich flesh to ecosystem harmony, small indigenous fish are nature's invaluable treasures

B. K. Das (✉) · Mishal P. · G. Karnatak
ICAR – Central Inland Fisheries Research Institute (ICAR-CIFRI), Barrackpore, Kolkata, India

B. K. Das et al. (eds.), *Small-scale Inland Fisheries in India*, Sustainability Sciences in Asia and Africa, https://doi.org/10.1007/978-981-95-3849-2_16

Keywords

Small Indigenous Fish (SIF) · Rivers · Wetlands · Floodplains · Nutrition · Vulnerabilities

16.1 Introduction

Fisheries provide livelihood to over 120 million individuals globally, with approximately 97 percent residing in impoverished nations, and over 90 percent are engaged in small-scale fisheries (World Bank 2012). Small-Scale Inland Fisheries (SSIF) represent artisanal fisheries where fishers or fish farmers typically engage in fishing or fish culture within inland water bodies or small artificial ponds, employing traditional methods that require relatively modest capital and energy, often using small traditional boats. In India, the total fish production from inland capture stands at 1.46 million tonnes (FAO 2022), with nearly major chunk (>90%) originating from the small-scale sector. However, the country's inland small-scale fishing sector remains largely unorganized, lacking comprehensive data. Small-scale fisheries hold significant potential in fostering economic, social, traditional, and cultural empowerment among marginalized fishers at the individual, local, regional, and national levels besides playing significant role in meeting the nutritional needs. However, policymakers have consistently relegated SSF to a low-importance category both at the national and global levels (Smith and Basurto 2019). Despite being neglected for extended periods, recent decades have witnessed a notable surge in the popularity of Small-Scale Inland Fisheries (SSIF), accompanied by much-needed scientific attention. In India, the hidden harvest in SSF is mostly contributed by SIFs that play important part in addressing the hidden hunger in fisher community.

The global demand for protein-rich food is rising, particularly in underdeveloped and developing countries. Fish, being a cost-effective source of high-quality dietary protein, essential fatty acids, and micronutrients, plays a crucial role in addressing this need. Food and Agriculture Organization (FAO) of the United Nations underscores the crucial role of fish as a vital food source, constituting approximately 16% of the world's animal protein intake. Fish plays a crucial role as a primary source of dietary protein, ranking just after cereals and milk. For a considerable portion of the global population, especially in regions where access to other animal protein sources is limited, fish serves as the sole source of animal protein. Small-scale fisheries provide more than half of the world's catch, with 90–95 percent of the small-scale catch consumed locally (FAO 2022) in which SIFs forms a major chunk. SIFs, which typically grow to a maximum length of 25–30 cm (Felts et al. 1996), are significant contributors to both nutrition and livelihood security, particularly in rural areas. Direct consumption of fish contributes significantly to nutritional intake, while Small-Scale Fisheries (SSF) play a crucial role in generating income, thereby enhancing purchasing power, which directly impacts household food and nutritional security (Kawarazuka and Béné 2010).

India, known for its remarkable biodiversity, boasts a rich array of fish species especially SIFs. However, this diversity is under significant threat due to rapid decline, posing a serious challenge to conservation efforts (Lakra et al. 2010). The capture and cultivation of fish, spanning inland waters, are vital for ensuring the nutritional security of rural populace. SIFs are known for their resilience and adaptability, thriving in various habitats such as ponds, tanks, irrigation canals, and rivers, often with minimal management requirements. A multitude of anthropogenic influences, such as the use of fine-meshed gears, intense fishing activity, insecticide application, aquatic pollution, alterations in habitat and the introduction of non-native species, play a part in declining SIF stocks in natural water bodies besides the impact of climate change. Despite their nutrient-rich and naturally occurring status, the significance of these small fish is often underestimated and not factored into commercial fisheries management. Local knowledge about the nutritional and medicinal value of inland fish in India has been underutilized (Daniels 2010). Conserving small indigenous fishes is vital for maintaining the health and resilience of aquatic ecosystems and supporting the nutrition and livelihoods of millions of people worldwide. By implementing holistic conservation strategies that address the various threats facing these species, we can ensure their long-term survival and contribute to the overall sustainability of our planet's freshwater and coastal environments.

16.2 Contribution of SIFs

According to NBFGR (2016), India harbors 2936 native fish species, comprising 936 freshwater, 113 brackish water, and 1887 marine species. Within the freshwater category, 450 are classified as SIF species (Sarkar and Lakra 2010). The highest diversity of SIFs has been documented in the North East region (216 species), followed by the Western Ghats (196 species) and Central India (120 species). Out of the total 450 SIFs identified, 62 are classified as food fish, while 42 are considered ornamental fishes. In the Ganga basin and associated floodplains, the dominant SIF groups include minnows (*Chela* sp., *Rasbora* sp., *Amblypharyngodon* sp., and *Salmophasia* sp.), barbs (*Puntius* sp.), scheilbids (*A. coila* and *Eutropichthys* sp.), clupeids (*Goniolosa manmina* and *G. chapra*), and bagrids (*H. menoda* and *Mystus* sp.) according to Sarkar and Lakra (2010). In Bangladesh, Felts et al. (1996) identified 45 fish species as SIFs, encompassing 18 species of carps and minnows, 9 species of catfishes, 9 species of perches, and others. Notably, SIFs are not accounted for in catch statistics that heavily rely on catches from organized and defined landing centers, as the SIFs are primarily sold and consumed locally and contribute to hidden harvest and household nutrition. This observation underscores the significance of local consumption and demand, as highlighted by Halwart and Settle (2008) and Roos et al. (2007). Additionally, Sarkar and Lakra (2010) have identified specific SIF species with high demand in the market. The utilization SIF holds the potential to enhance the income of fishing communities, leading to the empowerment of these communities with improved access to education, health, nutrition, and better living standards (Roy et al. 2020) (Fig. 16.1).

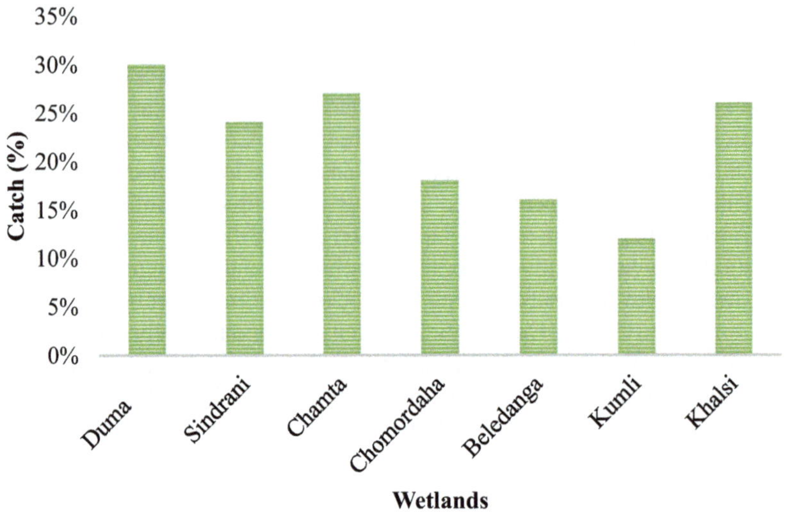

Fig. 16.1 Contribution of SIFs to total catch in selected wetlands of West Bengal

16.3 SIF in Rivers and Wetlands

SIFs are predominantly harvested by local villagers from rivers, canals, beels, channels, and various other water bodies, serving as a crucial source of sustenance in rural diets. Affordability plays a significant role in their consumption as they are readily accessible for daily meals. From the extensive array of fish species documented by ICAR-CIFRI in the river Ganga (Table 16.1), comprising a total of 190, 63 are recognized as small indigenous fish species, spanning 33 families and 14 orders. Among these, four species are classified as Near Threatened (NT), one as Near Endangered (NE), and one as Endangered (EN) according to the IUCN Red List. Cypriniformes emerged as the predominant order, representing 36% of the SIF fish group, followed by Siluriformes (16%) and Clupeiformes (11%) based on findings from the present study. Notably, 43 fish species inhabit freshwater environments, while 11 thrive in brackish water, and 10 are adaptable to both freshwater and brackish water zones (Fig. 16.2). Previous studies have emphasized the significant contribution of SIFs to the overall abundance of indigenous fish species in the river Ganga. Upon analyzing the species data, it was observed that SIFs, notably including *Cabdio morar* (31.46%), *Salmostoma bacaila* (18.08%), *Puntius sophore* (6.95%), *Securicula gora* (3.96%), *Parambassis ranga* (2.82%), *Osteobrama cotio* (2.07%), *Barilius barila* (1.68%), *Gudusia chapra* (1.29%), and *Tariqilabeo latius* (1.05%), made the most significant contribution to the relative abundance of indigenous fish species in the river Ganga.

Table 16.1 SIF reported from floodplain wetlands of West Bengal

Family	Scientific name	Common name
Cyprinidae	*Puntius terio*	One spot barb
	Puntius chola	Swamp barb
	Pethia gelius	Golden dwarf barb
	Pethia phutunio	Spottedsail barb
	Pethia conchonius	Rosy barb
	Pethia ticto	Twospot barb
	Systomus sarana	Olive barb
	Puntius sophore	Pool barb
	Cirrhinus reba	Reba carp
	Esomus danricus	Flying barb
	Rasbora daniconius	Slender rasbora
	Labeo bata	Bata
	Amblypharyngodon mola	Mola carplet
	Salmostoma bacaila	Large razorbelly minnow
	Salmostoma phulo	Finescale razorbelly minnow
Ambassidae	*Parambassis ranga*	Indian glassy fish
	Parambassis lala	Highfin glassy perchlet
	Chanda nama	Elongate glass-perchlet
Nandidae	*Nandus nandus*	Asian leaf fish
Mastacembelidae	*Macrognathus pancalus*	Indian spiny eel
	Mastacembelus armatus	Zig-zag eel
	Macrognathus aral	One-stripe spiny eel
Synbranchidae	*Monopterus cuchia*	Cuchia
Belonidae	*Xenontodon cancila*	Freshwater garfish
Botiidae	*Botia dario*	Bengal loach
Gobiidae	*Glossogobius giuris*	Tank goby
Badidae	*Badis badis*	Blue perch
Anabantidae	*Anabas testudineus*	Climbing perch
Cobitidae	*Lepidocephalichthys guntea*	Guntea loach
Channidae	*Channa punctata*	Spotted snakehead
	Channa gachua	Red seam snakehead
Osphronemidae	*Trichogaster chuna*	Honey gourami
	Trichogaster lalius	Dwarf gourami
	Trichogaster fasciata	Banded gourami
Clariidae	*Clarias batrachus*	Walking catfish
Heteropneustidae	*Heteropneustes fossilis*	Stinging catfish
Aplocheilidae	*Aplocheilus panchax*	Blue panchax
Bagridae	*Mystus cavasius*	Gangetic Mystus
	Mystus tengara	Tengara catfish
	Mystus gulio	Long whiskers catfish
Clupeidae	*Gudusia chapra*	Indian river shad
Ocellated pufferfish	*Leiodon cutcutia*	Ocellated pufferfish

Source: ICAR-CIFRI

Fig. 16.2 SIF caught from floodplain wetlands

In a recent study by ICAR-CIFRI, out of the total 121 species reported from River Godavari, around 60 species fall under the category of SIFs. The major group in these SIFs are barbs and minor carps (Family: Cyprinidae) with representation of 61 species. The prominent among them are *Amblypharyngodon* spp., *Osteobrama* spp., *Puntius* spp., *Pethia* spp., *Salmostoma* spp. A rich fishery of *Amblypharyngodon* sp. was observed in the Sriramsagar reservoir, a large reservoir in the middle stretch of Godavari in Telangana. As the local demand for fish was less, most of the catch went for drying and dried fish were sold for human as well as poultry consumption. Around 10 species of catfishes contributed to the SIF fishery of the river, the major ones being Bagrids like *Mystus* spp., and *Rita gogra*, *Clarias magur* and *Pachypterus atherinoides*. The Asiatic glassfishes (Family Ambassidae) including *Chanda nama*, *Parambassis lala,* and *P. ranga* also formed an important component in the SIFs. Some species of Gobies (Family Gobeidae) are also significant SIFs and they are dominant in the lower Godavari, flowing through the East and West Godavari districts of Andhra Pradesh and also the Yanam.

SIFs play a vital role in the ecological dynamics of wetlands, serving as essential components of aquatic food webs and contributing significantly to the biodiversity and productivity of these ecosystems. Wetlands provide crucial habitats for the breeding, spawning, and feeding of SIFs, offering a diverse array of microhabitats that support their life cycles. The presence of SIFs in wetlands contributes to the sustenance of local communities, as these fish species are often integral to the diets and livelihoods of riparian populations through subsistence fishing and small-scale aquaculture practices. More than 40 SIF species have been reported from floodplain wetlands of West Bengal which is close to 50% of the total fish diversity reported from these water bodies (Figs. 16.3 and 16.4).

Fig. 16.3 Rich fishery of *A. mola* in the Sriramsagar reservoir

Fig. 16.4 Some of the common SIFs found in floodplain wetlands (**a**. *Puntius sophore*, **b**. *Rasbora daniconius*, **c**. *Amblypharyngodon mola*, **d**. *Parambassis ranga*, **e**. *Parambassis lala*, **f**. *Parambassis nama*, **g**. *Colisa fasciata*, **h**. *Anabas testudineous*, **i**. *Nandus nandus*, **j**. *Osteobrama cotio*, **k**. *Macrognathus pancalus*, **l**. *Heteropneustes fossilis*, **m**. *Channa punctata*, **n**. *Channa gachua*, **o**. *Mystus tengara*)

16.4 SIF and SDGs

Small indigenous fish species play a crucial role in advancing several United Nations Sustainable Development Goals (SDGs), including food security, poverty alleviation, and environmental conservation. Rich in essential nutrients, they address hunger and malnutrition, supporting SDG 2 (Zero Hunger). Additionally, they serve as a vital income source for communities, contributing to poverty reduction (SDG 1) and promoting sustainable livelihoods. The nutritional benefits of small indigenous fish can contribute to combating malnutrition, particularly in children and pregnant women. Access to diverse diets, including small fish, supports overall health and well-being, reducing the incidence of diet-related diseases and hence contribute to SDG 3 (Good Health and Well-being). In many societies, women play key roles in small-scale fisheries, from harvesting to processing and marketing of small indigenous fish. Empowering women within these fisheries can enhance gender equality (SDG 5) by providing economic opportunities and decision-making roles. Sustainable management of small indigenous fish stocks contributes to the conservation of aquatic ecosystems and biodiversity. By promoting responsible fishing practices and safeguarding habitats, SDG 14 goals of conserving and sustainably using marine resources can be advanced.

Sustainable harvesting practices associated with these fish align with SDG 12 (Responsible Consumption and Production), preserving marine ecosystems. Moreover, they drive economic growth (SDG 8) by creating employment opportunities, particularly in rural and coastal areas. Small indigenous fish are often linked to ecosystems such as rivers, lakes, and wetlands, which are interconnected with terrestrial habitats. By ensuring the sustainability of these fish stocks and their habitats, SDG 15 objectives related to the conservation, restoration, and sustainable use of terrestrial ecosystems can be achieved. Recognizing the significance of SIF in the context of sustainable development aligns with several SDGs, including those related to poverty alleviation, food security, nutrition, gender equality, and environmental sustainability. Prioritizing the conservation and sustainable management of these fish resources can contribute to achieving multiple development goals and promoting inclusive and sustainable growth.

16.5 SIF and Household Nutrition

Small indigenous fishes hold immense potential as a source of household nutrition, particularly in regions where access to diverse and nutritious foods is limited. SIFs are typically found in local water bodies such as rivers, lakes, and ponds. They encompass a wide variety of species depending on the geographical location. Unlike larger commercial species that require extensive cultivation and management practices, these fishes can be sustainably harvested from natural water bodies. This characteristic makes them an attractive option for households looking to improve their nutrition without placing undue strain on the environment or requiring significant financial investment. Despite their small size, these fishes are nutritional rich and quality source of macro and micronutrients. They are abundant sources of high-quality protein, containing all the essential amino acids necessary for human health. Additionally, they are rich in omega-3 fatty acids, vitamins (such as vitamin D, B12, and A), and minerals (such as calcium, iron, and zinc), all of which are critical for optimal growth, development, and overall well-being. Incorporating these fishes into daily diets can significantly enhance the nutritional intake of households, especially for vulnerable populations such as children and pregnant women. Moreover, small indigenous fishes are often more resilient to environmental changes and require minimal resources for cultivation, making them a sustainable and cost-effective source of nutrition even in challenging conditions. Additionally, their small size and rapid reproductive rates enable quick replenishment of stocks, ensuring long-term sustainability. By promoting the consumption of these nutrient-dense fishes, communities can improve food security, combat malnutrition, and foster sustainable practices that benefit both human health and the environment (Thilsted 2010). Increased consumption of small fish could help alleviate micronutrient deficiency, reduce poverty and infant mortality, and improve maternal health. Efforts should be made to motivate people to diversify their diets, strengthen micronutrient intake, and develop sustainable, low-cost technologies for the management, conservation, production, and accessibility of small fish (Thilsted and Wahab 2014).

16.6 Nutritional Value of SIFs

Fish is often regarded as "nature's superfood," capable of fulfilling protein, micronutrient, and vitamin requirements. Natural freshwater ecosystems harbor rich fish biodiversity, including SIFs, also known as "Nutrifish," which are rich in minerals and vitamins (Mohanty et al. 2016, 2019). Research conducted in the Indian Sundarbans by Roy et al. (2020) illustrates that incorporating SIS into daily dietary patterns can alleviate micro-nutrient deficiencies in targeted populations. Researchers have explored the benefits of consuming nutrient-rich SIFs, highlighting the importance of consuming them whole, including the head, bones, and eyes, to maximize nutritional benefits by utilizing all available nutrients, including micronutrients. These fishes are rich source of trace elements such as magnesium, selenium, zinc, and iron, along with vitamins A, D, E, and B-group vitamins. Additionally, SIFs offer high-quality protein, making them a valuable addition to diets high in carbohydrates, especially in economically disadvantaged populations. The protein content in small freshwater fish species typically ranges from 14 to 22% of their live body weight (Mahanty et al. 2014). Consequently, even small quantities of SIFs can significantly enhance the nutritional value of diets and contribute to improved nutritional security. Research suggests that the vitamins and minerals present in one kilogram of SIFs are equivalent to those found in larger size fishes like Indian Major Carps (Mohanty et al. 2010).

The significance of mola (*A. mola*) has been emphasized as an excellent source of vitamin A (Alam 1985). Consuming these fish could provide a simple and effective solution to addressing vitamin A deficiency. Small fish, in comparison to other animal sources, are particularly rich in zinc. Hansen et al. (1998) found that calcium absorption from small fish is comparable to that from skimmed milk. SIFs can serve as natural supplements to address and prevent both micronutrient deficiencies and protein-calorie malnutrition (Mahanty et al. 2014). Certain small fish species, such as *Osteobrama* sp., *Gudusia* sp., *Esomus* sp., *Salmophasia* sp., and *Mystus* sp., contain high iron content. Similarly, species like *Mystus tengra*, *Mystus vittatus*, *Pseudeutropius atherinoides*, *Parambassis ranga*, *Chela bacaila*, and *Osteobrama cotio cotio* are notable for their high zinc content (Mahanty et al. 2014). SIFs are rich sources of vitamins, particularly vitamins A, D, E, and various B-group vitamins (B1, B2, B3). Among commonly consumed SIFs such as *Amblypharyngodon mola, Osteobrama cotio, Esomus danricus*, and *Parambassis ranga*, significant amounts of vitamin A are present in the form of retinol and 3,4-dehydroretinol isomers. Additionally, SIFs provide high-quality protein, including essential amino acids like lysine, as well as sulphur-containing amino acids such as methionine and cysteine.

Variation in the proximate composition of fish flesh was observed by Salam et al. (1995), and Jacquot (1961), with factors such as species, season, age, and feeding habits influencing these variations. In Bangladesh, SIFs are noted as the primary sources of vitamin A and calcium for rural communities (Roos et al. 2003). The role of SIFs extends to the prevention of malnutrition, particularly protein and micronutrient deficiencies, thereby ensuring the nutritional and livelihood security of rural populations. The successful integration of human nutrition and fisheries to tackle micronutrient deficiencies has great relevance for underdeveloped and developed countries. ICAR-CIFRI has undertaken comprehensive investigations into the nutrient profiles of selected SIFs from inland waters. The data repository can be accessible at http://www.cifri.ernet.in/nutrifishin/index.php, which encompasses exhaustive data pertaining to the gross chemical composition, amino acid composition, fatty acid composition, and micronutrient content of various food fish species (Fig. 16.5).

Fig. 16.5 SIF catch from river Ganga

16.7 Vulnerabilities of SIFs

Small indigenous fishes, despite their nutritional benefits and ecological significance, are vulnerable to various threats (Fig. 16.6) that can endanger their populations and the ecosystems they inhabit. These vulnerabilities stem from a combination of environmental, socio-economic, and anthropogenic factors. SIFs are often subject to overexploitation due to high demand for their nutritional value and economic importance in local communities. Unsustainable fishing practices, such as the use of fine mesh nets and dynamite fishing, can lead to depleted stocks and disruption of natural breeding cycles. The use of small meshed lift-nets exerts high pressure on small indigenous fish species (SIFs) (Suresh et al. 2010). Destruction of aquatic habitats, including rivers, lakes, and wetlands, through deforestation, urbanization, and agricultural expansion, poses a significant threat to small indigenous fishes. Loss of breeding grounds, pollution from agricultural runoff and industrial waste, and habitat fragmentation can severely impact fish populations and reduce their ability to thrive (Daniels 2010). Rapid changes in climate patterns, including rising temperatures, altered precipitation regimes, and extreme weather events, can disrupt the delicate balance of aquatic ecosystems. SIFs may struggle to adapt to these changes, leading to shifts in distribution patterns, changes in reproductive cycles, and increased susceptibility to diseases. The majority of SIFs are captured before

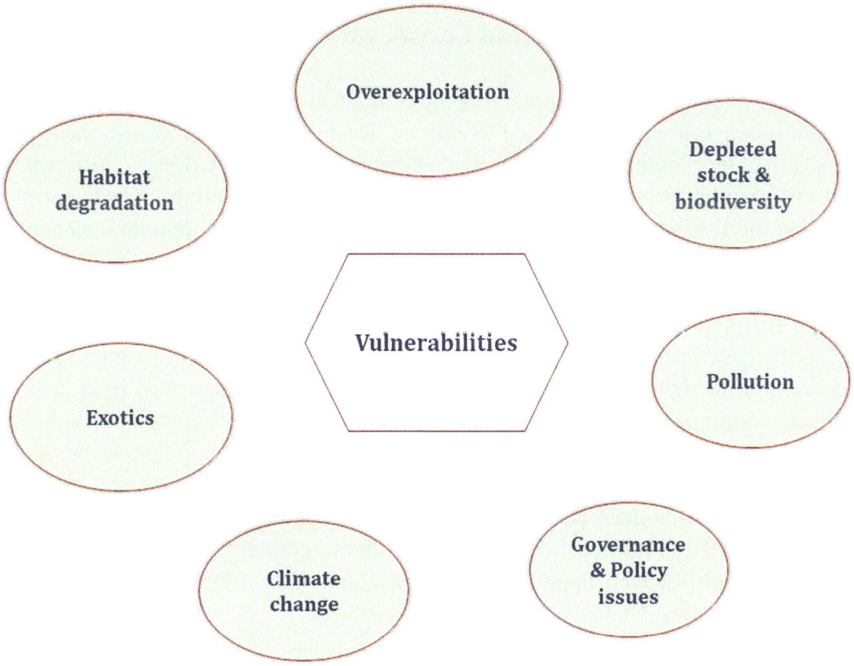

Fig. 16.6 Major threats to SIFs in inland open waters

reaching maturity, leading to concerns about their sustainable population levels. Sinha and Roy (2016) reported a noteworthy decline in the availability of SIF like *Amblypharyngodon mola, S. bacalia, M. vittatus, P. sarana*, and *N. nandus* in the water channels of the Sundarbans.

Introduction of non-native species, can have detrimental effects on indigenous fish populations. Invasive predators may outcompete small indigenous fishes for food and habitat resources or prey upon them directly, leading to population declines and ecosystem imbalances. Illicit introduction of carnivorous and voracious feeders like *Aristichthys nobilis*, *Clarias gariepinus*, and *Serrasalmus nattereri* to India poses a serious threat to SIF if these exotic fish establish themselves in natural water bodies (Frey 1961). Water pollution from industrial discharge, agricultural runoff, and improper waste disposal can have severe consequences for small indigenous fishes. Toxic chemicals, heavy metals, and excess nutrients can accumulate in their tissues, leading to bioaccumulation and biomagnification along the food chain, ultimately posing risks to human health. In many regions, there is limited awareness about the importance of small indigenous fishes and the need for their conservation. Weak governance, inadequate enforcement of fishing regulations, and a lack of community involvement in resource management exacerbate the vulnerabilities faced by these species. Addressing the vulnerabilities of small indigenous fishes requires a holistic approach that integrates ecosystem-based management, sustainable fishing practices, habitat restoration, and community engagement.

16.8 SIF Management and Conservation

SIFs play a crucial role in maintaining the ecological balance of aquatic ecosystems besides their role in nutrition and livelihood. Despite their significance, these species often face numerous threats that endanger their populations. Conservation efforts aimed at protecting small indigenous fishes are essential not only for preserving biodiversity but also for sustaining the livelihoods of communities dependent on these aquatic resources. SIFs contribute to ecosystem functioning through various ecological roles such as controlling insect populations, serving as prey for larger fish and birds, and participating in nutrient cycling. Moreover, they often form the base of the food web, supporting the survival of numerous other species. In addition to their ecological significance, small indigenous fishes hold cultural and economic importance in many regions, providing a vital source of protein and income for local communities. Despite their importance, SIF populations are under severe threat from various anthropogenic activities and environmental stressors. Climate change-induced alterations in temperature and precipitation patterns also impact the SIFs. Therefore, conservation efforts targeting small indigenous fishes require a multi-faceted approach involving collaboration between governments, local communities, NGOs, and scientific institutions.

16.9 Key Strategies

Some key strategies include (Fig. 16.7) habitat restoration, effective fisheries management, pollution control, research and monitoring, community mobilization and sensitization. Initiatives aimed at restoring and protecting critical habitats such as wetlands, rivers, and mangroves are essential for supporting the breeding and feeding grounds of small indigenous fishes. Implementing sustainable fishing practices, regulating intensive fishing, mesh size regulations, closed seasons and establishing no-take zones can help prevent overexploitation of SIF populations. Effective conservation strategies require the enforcement of laws and regulations to protect SIFs, especially during breeding. Strict enforcement of pollution control measures and promoting eco-friendly practices can mitigate the adverse effects of pollution on aquatic ecosystems and small indigenous fishes. Protection of SIFS in their natural habitats is crucial. Conducting research on the ecology, distribution, and population dynamics of small indigenous fishes is crucial for informing conservation strategies and monitoring the effectiveness of interventions.

The research endeavour must incorporate an integrated ecosystem-centric approach, incorporating both in-situ and ex-situ interventions. This encompasses initiatives such as conservation aquaculture, live gene banking, tissue banking, genetic resource valuation and registration, and monitoring critical parameters. Moreover, the strategy should encompass the replenishment of natural stocks through ranching methods, community engagement, as well as the enhancement of

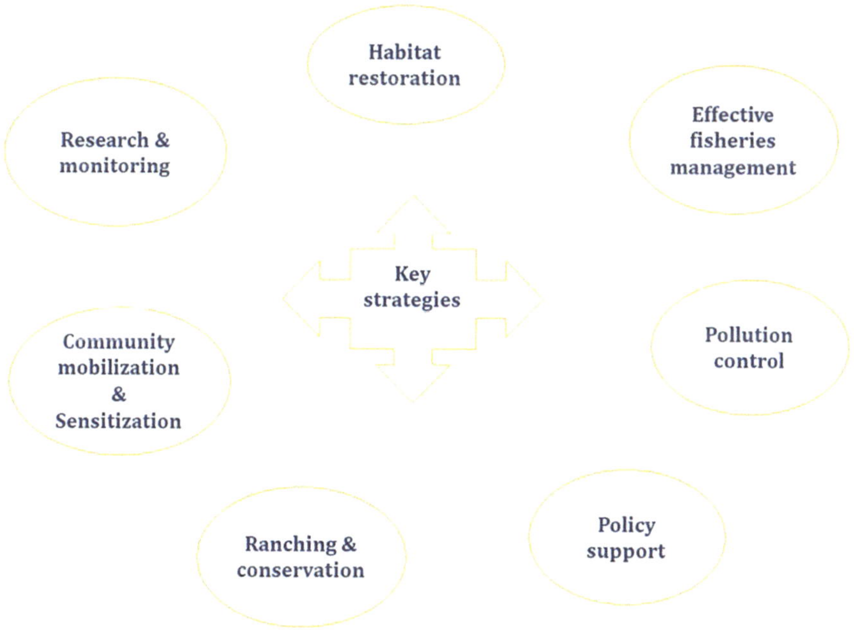

Fig. 16.7 Key strategies for SIF management and conservation

human resource capabilities through structured capacity building programs (Lakra et al. 2010). Planning and research should focus on prioritization, refinement of culture and technology, market development, and the formulation of policies and support systems (Poonniah and Sundaray 2010). Culture of SIF need to be promoted in monoculture or poly culture with carps. Backyard ponds or hapas can be used for culture of these fishes. Enclosure such as cages and pens can also be engaged for SIF farming. Pens have been found to support self-recruiting small indigenous fishes in wetlands. Dubey et al. (2014) proposed that *A. mola* and *P. ticto* exhibit potential suitability as candidate species for aquaculture diversification within the marginally saline environments of the Sundarbans.

Involving local communities in conservation efforts through education, awareness campaigns, and capacity-building initiatives will play a greater role in conservation efforts. Implementing awareness programs targeting both local communities and the general public is essential. These programs should highlight the importance of SIFs in maintaining biodiversity and ecosystem health, as well as the significance of conserving them for future generations. Sinha et al. (2017) attempted to raise awareness within the community about incorporating small indigenous fish (SIFs) into their diet, aiming to encourage small-scale entrepreneurs to cultivate SIFs for both nutritional security and livelihood improvement. Capacity-building programs should be developed to empower local communities, including women, with the knowledge and skills necessary for sustainable management of SIFs and their habitats. This can include training in sustainable fishing practices, habitat conservation techniques, and entrepreneurship skills for those involved in fish culture systems.

Local communities should be actively engaged in conservation efforts, as they often hold valuable traditional knowledge about the local environment and species. Involving them in decision-making processes and incentivizing their participation through financial benefits can lead to more effective conservation outcomes. A trial in the Sundarbans aimed at initiating in situ conservation through community participation yielded encouraging results (Sinha and Roy 2016). Eco-restoration efforts can include the ranching of SIFs with the active involvement of local communities in freshwater canals or water channels. Traditional conservation ethics, as highlighted by Deb and Malhotra (2001), can play a significant role in safeguarding biodiversity. These ethics should be promoted and integrated into conservation policies and practices, with a focus on fostering a sense of stewardship and responsibility among local communities.

Utilizing mass and social media can help raise public awareness about the importance of SIF conservation and the need for sustainable management practices. These campaigns should be tailored to reach diverse audiences and convey key messages effectively. Providing financial incentives for communities engaged in conservation efforts can be a powerful motivator. This can include revenue-sharing schemes from ecotourism activities, sustainable fishing practices, or direct payments for ecosystem services provided by healthy SIF populations. By implementing these measures, policymakers can work towards the conservation and promotion of SIFs while ensuring access for disadvantaged groups, particularly women. This holistic approach recognizes the interconnectedness of environmental conservation, socioeconomic development, and cultural preservation.

16.10 Conclusion

Promoting the consumption of small indigenous fishes can have a transformative impact on household health and well-being. Their resilience, affordability, accessibility, and nutritional density make them an ideal choice for improving diets, particularly in areas where access to other protein-rich foods is limited. However, despite their potential benefits, small indigenous fishes often face numerous challenges, including overfishing, habitat degradation, pollution, and competition from introduced species. Addressing these challenges requires a multifaceted approach that involves sustainable management practices, community engagement, and policy interventions to ensure the long-term viability of these valuable resources. In conclusion, small indigenous fishes represent a valuable yet often overlooked resource for improving household nutrition and food security. By harnessing the nutritional potential of these species and implementing sustainable management practices, we can empower communities to combat malnutrition, promote healthier lifestyles, and safeguard the biodiversity of our aquatic ecosystems for generations to come. Efforts to conserve these valuable resources must prioritize the protection of their habitats, promotion of sustainable livelihoods for fishing communities, and implementation of effective governance mechanisms to ensure their long-term viability.

Points to Remember

- *Promote Climate-Resilient SIF Species*: Identify hardy and resilient SIFs that can tolerate temperature fluctuations, low oxygen, and habitat variability.
- *Habitat Restoration and Protection*: Restore riverine and wetland ecosystems, ensure connectivity, and reduce anthropogenic pressures like pollution, siltation, and encroachment to maintain viable habitats for SIFs.
- *Adopt Low-Input Culture Systems*: Utilize cost-effective and eco-friendly systems such as backyard ponds, pens, and integrated polyculture with major carps to enhance SIF availability and income under resource-limited, climate-stressed conditions.
- *Community-Based Management*: Engage local fishers in monitoring, conserving, and sustainably harvesting SIFs with traditional knowledge and adaptive practices to build resilience against climate risks.
- *Integrate SIFs into Climate Adaptation Strategies*: Recognize the nutritional, economic, and ecological value of SIFs in national food security and climate-smart fisheries policies.

References

Alam AKMA (1985) Mini Pond. CARITAS Bangladesh, Dhaka

Daniels RJ (2010) The miscellaneous fish in rural food baskets: knowledge and rights (abstract). In: Workshop on small indigenous freshwater fish species: their role in poverty alleviation, food security and conservation of biodiversity. Central Inland Fisheries Research Institute, Barrackpore, Kolkata

Deb D, Malhotra KC (2001) Conservation ethos in local traditions: the West Bengal Heritage. Soc Nat Resour 14:711–724

Dubey SK, Trivedi RK, Rout SK, Chand BK, Choudhury A (2014) Of two small indigenous fish species Amblypharyngodon mola and Pethia ticto from Indian Sundarban. J Aquac Res Dev 5(249):2

FAO (2022) The state of world fisheries and aquaculture 2022: towards blue transformation. FAO, Rome. https://doi.org/10.4060/cc0461en

Felts RA, Fajts F, Akteruzzaman M (1996) Small indigenous fish species culture in Bangladesh (technical brief), IFADEP Sub Project 2. Development of Inland Fisheries

Frey H (1961) Illustrated dictionary of tropical fish. TFH Publ Inc, New Jersey

Halwart M, Settle W (2008) Participatory training and curriculum development for farmer field schools in Guyana and Suriname, a field guide on integrated pest management and aquaculture in rice, 116

Hansen M, Thilsted SH, Sandström B, Kongsbak K, Larsen T, Jensen M, Sørensen SS (1998) Calcium absorption from small soft-boned fish. J Trace Ele in Med Biol 12(3):148–154

Jacquot R (1961) Organic constituents of fish and other aquatic animals: fish as food. In: Borgstrom Academic Press, New York/London, pp 145–209

Kawarazuka N, Béné C (2010) Linking small-scale fisheries and aquaculture to household nutritional security: an overview. Food Sec 2(4):343–357

Lakra WS, Sarkar UK, Gopalakrishnan A, Kathirvel, Pandian A (2010) Threatened freshwater fishes of India. NBFGR Publ. ISBN: 978-81-905540-5-3

Mahanty A, Ganguly S, Verma A, Sahoo S, Mitra P, Paria P, Sharma AP, Singh BK, Mohanty BP (2014) Nutrient profile of small indigenous fish Puntius sophore: proximate composition, amino acid, fatty acid and micronutrient profiles. Natl Acad Sci Lett 37(1):39–44

Mohanty BP, Behera BK, Sharma AP (2010) Nutritional significance of small indigenous fishes in human health. Central Inland Fisheries Research Institute (ICAR), Barrackpore, Bulletin No. 162, ISSN: 0970-616X, 63

Mohanty BP, Sankar TV, Ganguly S, Mahanty A, Anandan R, Chakraborty K, Sridhar N (2016) Micronutrient composition of 35 food fishes from India and their significance in human nutrition. Biol Trace Elem Res 174(2):448–458

Mohanty BP, Mahanty A, Ganguly S, Mitra T, Karunakaran D, Anandan R (2019) Nutritional composition of food fishes and their importance in providing food and nutritional security. Food Chem 293:561–570

NBFGR (2016) Annual report of the ICAR-National Bureau of fish genetic resources 2015–2016. ICAR-NBFGR, Lucknow, India, p 103

Poonniah AG, Sundaray JK (2010) The strategy for promoting culture of small native fish species. In: Workshop on small indigenous freshwater fish species: their role in poverty alleviation, food security and conservation of biodiversity. Central Inland Fisheries Research Institute, Barrackpore, Kolkata

Roos N, Islam M, Thilsted SH (2003) Small fish is an important dietary source of vitamin a and calcium in rural Bangladesh. Int J Food Sci Nutr 54(5):329–339

Roos N, Thorseng H, Hamnan C, Larsen T, Gondolfu B, Thilsted SH (2007) Iron content in common Cambodian species: perspective for dietary iron intake in poor rural households. Food Chem 104(3):12226–11235

Roy A, Aftabuddin M, Parida PK, Sinha A, Das BK, Adhiguru P (2020) Small indigenous fishes to boost nutritional security: a roadmap for Nutri-Smart Village in deltaic Sunderbans. In: Agricultural extension: socio economic imperatives, pp 63–67

Salam MA, Alam N, Nasiruddin M, Nabi R, Howlader MZH (1995) Biochemical composition of body muscles and its caloric contents of tawes (Puntius gonionotius, Bleeder). J Bangladesh Sci Res 13(2):205–211

Sarkar UK, Lakra WS (2010) Small indigenous freshwater fish species of India: significance, conservation and utilization. Aquac Asia 15(3):34–35

Sinha A, Roy A (2016) Sunderban Me Ajibika Abam Poshakiya Suraksha Hetu Chhoti Deshi Machhlika Utpadan Abam Sangrakhan (in Hindi). Central Inland Fisheries Research Institute, Barrackpore, Bulletin No 195, p 24

Sinha A, Roy A, Das BK, Mohanty BP (2017) Small Indigenous Fish (SIF) for livelihood and nutritional security: awareness and sensitization initiative by ICAR-Central Inland Fisheries Research Institute, Barrackpore, Report No. ISSN: 0970-616X

Smith H, Basurto X (2019) Defining small-scale fisheries and examining the role of science in shaping perceptions of who and what counts: a systematic review. Front Mar Sci 6:236

Suresh VR, Vinci GK, Mukherjee A, Jha BC, Das AK, Sahu SK (2010) Fishery of wetlands in West Bengal and contribution of wild fish stocks to production and fisher's income. In: Workshop on small indigenous freshwater fish species: their role in poverty alleviation, food security and conservation of biodiversity. Central Inland Fisheries Research Institute, Barrackpore, Kolkata

Thilsted SH (2010) The role of SIFFS in improving nutrition in rural populations (abstract). In: Workshop on small indigenous freshwater fish species: their role in poverty alleviation, food security and conservation of biodiversity. Central Inland Fisheries Research Institute, Barrackpore, Kolkata

Thilsted SH, Wahab MA (2014) Production and conservation of nutrient-rich small fish (SIS) in ponds and wetlands for nutrition security and livelihoods in South Asia. WorldFish

World Bank (2012) Hidden harvest: the global contribution of capture fisheries

17 Women in Small-Scale Fisheries of in Sundarbans: Bargaining with Risk and Uncertainty of Life Under a Changing Climate

Aparna Roy, R. K. Manna, and Basanta Kumar Das

Abstract

Fisherwomen engaged in small-scale fisheries (SSF) in the Sundarbans are at the frontline of an increasingly uncertain and precarious existence shaped by the intensifying impacts of climate change. Their lives are deeply intertwined with the fragile estuarine ecosystems, where rising sea levels, erratic weather patterns, frequent cyclones, and salinity intrusion have not only disrupted traditional fishing practices but also intensified the risks to their livelihoods and well-being. This study explores how women in the Sundarbans negotiate these multifaceted vulnerabilities through adaptive strategies, informal networks, and community-based resilience. It sheds light on their silent yet significant contributions to household income, food security, and ecosystem stewardship, often underrecognized in mainstream fisheries policy. The findings underscore the urgent need for gender-sensitive climate adaptation frameworks and inclusive fisheries governance that account for the lived realities of these fisherwomen who continually bargain with risk and uncertainty under a changing climate.

Keywords

Small scale fisheries · Climate change · Women · Sundarbans

Resilience in the Tides: Women Fishers of Sundarbans Navigating Livelihoods Amidst Climate Change

A. Roy · R. K. Manna · B. K. Das (✉)
ICAR – Central Inland Fisheries Research Institute (ICAR-CIFRI), Barrackpore, Kolkata, India

B. K. Das et al. (eds.), *Small-scale Inland Fisheries in India*, Sustainability Sciences in Asia and Africa, https://doi.org/10.1007/978-981-95-3849-2_17

17.1 Introduction

Sundarbans, a Ramsar wetland and a UNESCO World Heritage site, is the largest mangrove forest of the world; is basically a cluster of islands spreading over 26,000 km^2 across India and Bangladesh. One third of the Sundarbans (9629 km^2) lies in India. The Indian Sundarbans possesses an intricate coastline blessed with mangrove vegetation, cluster of deltas, interlinked with channels, creeks and estuaries, are a treasure of fin and shell-fish seeds. Sundarbans is ecologically very active, being highly susceptible to continuous change induced by higher tidal amplitude and frequent devastating cyclonic storms and other extreme environments.

The Sundarbans biosphere is very rich in floral and faunal diversity, hence has enormous ecological and socio-economic significance. During 1775, human settlements were recorded in some parts of Sundarbans by reclaiming forest lands. Afterwards, the British settled people there to reclaim forests in that area. The growth in population was quite rapid in the Sundarbans after Indian Independence. The decennial population growth was 21.47%, 29.55% and 17.4% during 1971–81, 1981–1991 and 1991–2001 (Katiha et al. 2012). Out of 102 islands of Indian Sundarbans, 54 islands are inhibited by 4.5 million people as per the 2011 Indian census, whereas the population was 3.75 million in 2001 (Census 2001, 2011). Though the population density of the Sundarbans is high, the inhabited areas are hard access due to numerous tidal rivers and creeks.

The people in the Sundarbans region have to depend on ferry services for transport, which is also dependent on ebb and flow of tides. About, 45,000 sq. km is human settled but only 280 kms are connected with roads and 42 km by rail (Jha 2012). People are directly and indirectly dependent on mangrove forests for their livelihoods as it is the home of 481 vertebrate species, 1104 invertebrate species, 1 hemicordate, 106 protozoan, and 84 Mangrove species (Raha et al. 2013) including Royal Bengal Tiger. It is also the natural breeding and nursery ground of many important fish species which support the estuarine and coastal fisheries along the Bay of Bengal. In the vast deltaic region of Indian Sundarbans, the faunal assemblage includes a list of 267 fish species (Mandal et al. 2013); Recently a comprehensive checklist mentioned existence of 350 fish species from Indian Sundarbans (Mishra and Gopi 2017). The variation in abundance of fish species distribution in Indian Sundarbans is dependent on the location, salinity gradient, freshwater influx, and season (Manna et al. 2014). The deltaic Sundarbans contribute 95% of total catch of Hooghly estuary and it has reached a figure over 65,000 t during 2012 (Sharma 2012). A large number of people are dependent on inland fisheries in Sundarbans for earning their livelihood but it is often threatened by various level of risks like cyclonic storms, embankment failure due to tidal surge, etc. (Rajshekar 2011). After the extreme cyclonic storm *Aila* in 2009, the availability of fishes has been reduced considerably as per the local fishers (Sinha et al. 2014). Thus, issues like salinity, erosion, tidal waves, natural hazards etc. have added to the problems of the island dwellers and also threaten the ecosystem of the Sundarbans. In face of growing human population and reducing carrying capacity of natural habitat, the future of the island dwellers is confronted with increased uncertainty. The situation of the fishermen community is more vulnerable as they are marginalized and underprivileged belonging to the lower strata of the society. Remoteness of human habitation, especially that of the fishermen community, is one of the major reasons for their poor development. The people of fishermen community have strong belief on local deities. The fishers do not go for fishing without paying homage to local deities like *Bon bibi*, *Dakhina Rai*, *Gazi*, *Kasthadevi*, *Sha Jangali*, *Danga* etc.

The relationship between ecology and environment is very delicate. It is now well evidenced that anthropogenic activities largely in the name of development have resulted in loss of green cover, excessive pollution due to use of fossil fuel, etc. which have resulted in damaging that delicate balance subsequently leading to what is now recognized as climatic change, which is a change in climatic patterns in variance from past patterns. This has affected the Sundarbans also. Studies reveal that from 1986 to 2012, an area of 124.418 sq. km. of mangrove forest of the Sundarbans has been lost due to sea level rise (Hazra et al. 2002; Hazra et al. 2004; Singh 2017). The island dwellers are the worst hit, with life and livelihoods directly affected due to this. In the last decade, extreme climatic events (severe cyclonic storms) like *Aila*, *Cider*, and *Hudhud* have caused devastation along the coast line of Sundarbans. Many people lost their lives, home, land, animals, and also their livelihood options. Basati, Kultali, and Patharpratima were worst affected blocks in Aila. A survey in a village in Kultali Block in which 60 families were interviewed, 36 of these families had lost their houses in *Aila* and rest reported that their houses were partially damaged (Nandi 2010). People were compelled to migrate to peri-urban areas of the state and even to other States of India for earning their livelihood. The major livelihood options of the populace of Sundarbans are agriculture, fisheries, aquaculture, coastal fishing etc. However, due to climate change, uncertainty in crop and fish yields are causing livelihood insecurity among the inhabitants. About 93% of people (of 2188 households surveyed in a study) are unwilling to depend on forests and on nature for their livelihoods any longer (Priyadarshini 2015).

Like other parts of India (and the world), in the Indian Sundarbans gender discrimination is the norm. A patriarchal, patrilineal society shapes women's lives. Women frequently face numerous disadvantages, such as having to work hard to obtain food and a living, having less control over their income and possessions, being intimidated and subjected to violence, having a lower social status, and being underrepresented in policy and decision-making.

. Women of Sundarbans bear the additional burden of tide, tiger, tragedy, and tears. Due to destruction of forests, and natural habitats, tigers have become more ferocious and are seen to be attacking human beings and the cost is mainly paid by the women as their male counterparts who go fishing, collecting honey and other non-timber forest products into the deep forest areas and often become targets of tigers. The widows of these men have to take load of all additional responsibilities of their families including the social stigma of becoming "tiger widow." In addition, due to climate change related adverse situation, most of the men are now migrating in search of employment and women are handling the social and economic affairs of the households which is slowly leading to a major change in social structure in Indian Sundarbans. Roy and Bhaumik (2012) observed that more than 50% of fisher women in Hooghly estuary belong to fishermen families living in abject poverty. Women often bear the brunt by eating least amount and by seeking alternate livelihood to meet the basic need of households (Hossain 2009). The roles of women involved in fishing activities in the changing scenario of climatic conditions need greater understanding. No systematic studies have been carried out so far on the impact of extreme environments on the livelihood of women community of Sundarbans and how they have adapted themselves to sustain their livelihood in this fragile eco-region.

In this chapter, we attempt to describe how women are facing the challenges in the Sundarbans especially in view of changing climatic condition and synthesizes the available information from the region. The study also discusses issues and challenges and suggest strategies to minimize the risk.

17.2 Livelihood Options of Fishers of Sundarbans

Fishing activities in the Sundarbans are predominantly undertaken by small-scale fisheries (Islam and Chuenpagdee 2013). A large population in Sundarbans is dependent on capture fisheries and fishery related activities and is treated as the backbone of Sundarbans economy (Chandra and Sagar 2003). Rural livelihoods are integrally connected with seasonality (Mukherjee 1993). Each season has its problems and opportunity (Roy et al. 2017) and the inhabitants set strategies for their livelihoods based on them. The livelihoods of the fishers of Sundarbans are mainly dependent on capture fishery and it may be divided into the following categories:

17.2.1 Fishing for Fry (Collection of Juveniles)

The mangrove ecosystem of Indian Sundarbans acts as nursery ground of various commercially important fishes and prawn species and harvesting fry is one of the major livelihood activities of fishers in this region. Collected juveniles are sold in live condition for aquaculture in coastal wetlands (bheri). Fishers, especially women, collect post-larvae and juveniles of fishes and prawns using shooting net or shore seine. Post-larvae and juveniles of *Bagda chingri (Penaeus monodon)*, Chapra chingri (*P. indicus*), Chingri/Yellow Shrimp (*Metapenaeus brevicornis)*, Chingri/Pink shrimp (*M. monoceros*), Narir chingri/Haity river prawn (*Machrobrachium rude*), Bhegatara chingri (*Parapenaeopsis sculptilis*), Kiddi chingri (*Expalaeomon stylifera*), Bhangad (*Liza tade*), Parse (*Liza parsia*), Gurjali (*Eleutheronema tetradactylum*), Pairatali (*Scatophagus argus*), Bhetki (*Lates calcarifer*), Banded pearl spot (*Etroplus suratensis*), *etc* are often caught in shooting nets or shore seine but, in most occasions, the process is causing huge destruction of untargeted species (Bhaumik et al. 1993).

17.2.2 Capture Fisheries

The fishery in Sundarbans may be categorized as artisanal fishery using traditional fishing gears like small and large mesh gill nets, seine nets, bag nets, lift nest, push nets, set barrier net, various types of fishing traps like box trap (*Ghuni, Atal*), triangular trap (*Banki*), etc. *Anodontostoma chacunda* (locally known as *Lona Khoira*) generally dominates in gill net catch in most of the rivers in Sundarbans but, in westernmost Hooghly zone of Sundarbans, Ilish *(Tenualosa ilisha)* dominated the catches. Other dominating fish species in rivers of Sundarbans are Gang Mouti *(Escualosa thoracata),* Loitha (*Harpadon nehereus*), Phasa (*Setipinna taty*), Phasa (*Setipinna phasa*), Double spotted queen fish/ Rani machh (*Scomberoides lysan*), Ilish (*Tenualosa ilisha*), Bhola (*Otolithoides pama*), Pomphret (*Pampus argenteus*), Ribbon Fish (*Lepturacanthus savala*), Amodi (*Coilia dussumieri*), Amude (*Coilia ramcarati*), Gule (*Odontamblyopus rubicundus*), Nona tangra (*Mystus gulio*), Bele (*Sillago sihama*), etc.

17.2.3 Winter Migratory Bag Net Fishery

Winter migratory bag net fishery contributes more than 65% of total estuarine catch in West Bengal (Sharma 2012). Species like Bhola (*Otolithoides pama*), *Loitha/ Bombay duck* (*Harpadon nehereus*), *Amude/Amodi* (*Coilia* spp.), ribbon fishes, *etc* and prawns generally dominate the catch.

17.2.4 Bheri Fishery

The tidal wetlands and the embanked waterbodies in Sundarbans are called *Bheri*. *Bheri* fisheries has a major contribution to the livelihood security of the local people. The wetlands situated in low-lying areas adjacent to the creeks, tidal rivers, and estuaries are used for aquaculture. The traditionally cultivated *bheries* are auto-stocked during tidal ingress but some are managed commercially with stocking of huge amount of fish / prawn juveniles collected from rivers. In such *bheries*, commercially important fishes and prawns like Bagda /Tiger Prawn (*Penaeus monodon*), Yellow Shrimp (*Metapenaeus brevicornis*), Bhetki (*Lates calcarifer*), *Parse* (*Liza parsia*), etc. are produced.

17.3 Identified Roles of Fisher Women of Sundarbans

Women of fisherfolk community may be divided into two groups (i) those who are directly attached with the fishing and fishing related activity and (ii) those whose husbands or sons are fishers and she herself is engaged in some other activity mostly related to agriculture or animal husbandry (Mukherjee 1993). Based on focus group discussion and by observation, it has been observed that fisherwomen of Sundarbans are engaged in a wide range of activities in fisheries and others related activities which are discussed below.

In Sundarbans, women are active in both artisanal and commercial fisheries, besides taking care of their families or working in other sectors. Due to social taboo, women do not go for fishing in the sea but they perform other fishery related activities generally to meet their household needs. The fisheries activities include:

- *Fish Seed collection:* Harvesting fin-fish and shellfish seeds from the rivers, creeks, and estuaries is an important livelihood activity of the womenfolk of coastal fisherman community of Indian Sundarbans. Fish seed/fry collection is a family occupation, with even children taking part in seed collection. The peak season lasts from March through May. A woman can collect 50–70 pieces of Triger prawn/Bagda (*Penaeus monodon*) seed in a day during the peak season and earn about Rs. 150–200/–. Seed availability is influenced by lunar periodicity and during full moon and new moon, seed availability is more. However, there are higher number of collectors also during these periods.
- *Fishing labor*: Women have no access to the fishing vessel/boat due to social taboo. However, fisherwomen of Sundarbans often perform various off-shore activities as fishing labor like fish sorting and grading, while the men go for fishing in the sea for 8–10 days.
- *Dry fishing:* Fisher women of Sundarbans have major contribution in fish drying. The winter migratory bag net fishery contributes a lot in dry fish business. The temporary fishing camps establish for drying those fishes which are known as *Khuties*. The women of fisherfolk community are active in these camps for earning their livelihood. Washing and cleaning of the fishes are mainly done by the men and the women participate in fish grading and drying as well as watch and ward during the fish drying process (Roy et al. 2017). Sun drying is the simplest and most ancient method of preserving fish and is extensively practiced in lower Sundarbans area. This activity is more during October to middle of February and thus women fisherfolk contribute significantly to the family income (23 per cent annually).
- *Rural small scale aquaculture:* The geography of Sundarbans along with its brackish/saline watered creeks, channels, rivers is also dotted with numerous fresh water ponds. Most of the households have small ponds. Women are involved in aquaculture activities for food and livelihoods centered around these ponds.
- *Capturing/harvesting of fish:* Fisherwomen of Sundarbans often harvest fish from the water channels, canals, paddy fields, creeks with gill nets, scoop nets, and various indigenous traps like *Baki, Chokhia, Ghuni* etc. for household nutrition as well as for additional income.
- *Fish vending/marketing:* Fisherwomen are also involved in fish vending or fish marketing to earn their livelihoods, in spite of social problems and taboos.
- *Net mending:* It is another livelihood option of the fisherwomen of Indian Sundarbans. It also supports their men in fishing operation. Sometimes, they also earn by making nets.

- *Crab fishery:* Women are essential to the crab fishery in the Sundarbans, helping to catch and process crabs for profit. Women are often partner with their spouse or a small group to catch wild crabs for their livelihoods.

Besides these, fisherwomen also work as agricultural labour/ labour in non-fisheries sectors.

17.4 Seasonal Activity of Fisher Women

The life and livelihoods of the women of fisherfolk community in Sundarbans are closely connected with the seasons. Seasonal calendar, a participatory tool, was used to explore gender-specific workload, income etc. across seasons. The fisher women of Sundarbans collect fin and shell fish almost nine months in a year and generally involve in fish processing/drying during mid-October to end of January. They also harvest fishes from water channels during the monsoon (Roy et al. 2017). With the onset of monsoon, some fisherwomen also work as agricultural labour, particularly for transplanting of rice. In winter, they also collect timber from forests and also collect crabs from rivers and creeks. The daily routine of fisher women of Sundarbans shows three types of activities which are generally performed by them; those are "household work," "income generating activities" and "spare time for self." They spent 16 hrs a day in various household and income generating work and had very less time to spare for themselves. Women with dual working jobs routinely spend two or three hours more per day on work-related activities than men, according to household labor studies (Levine et al. 2001). Long work hours and malnutrition may affect women's sociology, economy, and health (IFPRI 1995).

17.5 Gender Participation Matrix in Fisheries in Sundarbans

Men and women in the fishing community in Sundarbans take up distinctly different sets of activities (Table 17.1). Capture of fish from rivers and estuary is done by the fishermen as there is social taboo on women going out for fishing in fishing boats. Fish seed collection is done by the fisher women and their children and fish seed segregation is also done by them. All female members are engaged in activities, consequent to fish landing either on a part time or full time basis. Women play an exclusive role in post-harvest operations, consisting of a series of activities. Fishermen are engaged in unloading of fish on shore, boat repairing and maintenance, and fish marketing particularly auctioning etc. Both fisher men and women are involved in net making and mending (Roy and Bhaumik 2013).

Table 17.1 Gender participation matrix

Sl. no.	Activities	Role distribution	
		Men	Women
1.	Fish capture	***	*
2.	Fish seed collection	O	***
3	Fish seed segregation	O	***
4.	Unloading on landing	***	O
5.	Sorting and grading	*	***
	in the river/sea,	***	O
	after landing,	*	**
6.	Icing of fresh fish	**	O
7.	Processing of dry fish	*	***
	Washing	***	*
	Cleaning (scaling and gutting)	***	*
	Transport from boat	***	o
	Fish grading	o	***
	Drying	o	***
	Watch and ward	o	***
	Purchase of packing materials	*	***
	Packaging	***	*
		**	**
8.	Marketing	***	*
	Auctioning	***	O
	Transporting	***	*
	Vending of fresh fish	**	***
	Vending of dry fish	*	***
9.	Boat repair and maintenance	***	O
10.	Net making	**	**
11.	Net care and maintenance	***	**
12.	Bringing boat to the shore	***	O

°No participation
* Low level of participation
** Moderate participation
*** High participation

17.6 Tribulations Faced by Fisherwomen in the Context of Climate Change

17.6.1 Loss of Livelihood

Gathering fish seeds is one of the main sources of income for Sundarbans fishermen. However, the availability of commercially significant shellfish and finfish has decreased as a result of climate change, including the variations in surface water salinity and sea level rise.

The risk of occupational hazards such as accidents has increased during the dragging of nets due to sea level rise. Due to extreme climate events, fishing boats are not getting good catches, and fisherwomen are losing their employment and thereby their income avenues. Drying fish is an organized sector in lower Indian Sundarbans mainly dependent on the winter bag net fishery, is also affected due to the changes of surface water salinity impacting the bag net fishery and eventually affecting dry fish processing. Prolonged dry spells have affected homestead pond aquaculture. Dry spells are also affecting harvesting of fish from the water channels, canals, paddy fields, and creeks due to drying up of the waters.

17.6.2 Food and Nutritional Insecurity

Food security is defined as having constant access to enough food for an active, healthy life (Gross et al. 2010). To fulfil their daily nutritional needs, people need have access to wholesome food and safe drinking water. Islands are sinking and nearly disappearing as a result of climate change. Displacement and a reduction in agricultural land are being seen. Access to clean drinking water is also being jeopardized in some areas due to seawater intrusion. The population of many nutritionally rich Small Indigenous Fishes (SIFs), such as Kholisa (*Trichogaster fasciata*), Nandos (*Nandus nandus*), Akash Tangra (*Mystus vittatus*), Pabda (*Ompok pabda*), and others, has drastically decreased as a result of saltwater intrusion caused by Aila in 2009.

Due to *Aila* in 2009, salt water ingress had destroyed the natural habitats of many nutritionally rich Small Indigenous Fishes (SIFs) like *Kholisa* (*Trichogaster fasciata*), *Nandos* (*Nandus nandus*), Akash Tangra (*Mystus vittatus*), Pabda (*Ompok pabda*), etc. and that has caused drastic reduction of the population of such fishes in the freshwater bodies in Sundarbans (Sinha et al. 2014; Roy et al. 2016). In the Indian Sundarbans, little indigenous fish have long been a staple of the household diet. Vitamins and minerals abound in this tiny fish. Since women of fisherfolk community are solely responsible for the household food and nutritional security as they collect, grow, purchase, store, and prepare, decline in SIFs in natural water bodies has lead to nutritional insecurity of the community (Roy et al. 2016). Additionally, kitchen gardening is hindered by land erosion, rising soil salinity, and water salinity. Women in the fishing community must therefore come up with different ways to ensure that the family is fed, which forces them to put in more effort and longer hours in order to make ends meet.

17.6.3 Drudgery of Women in Sundarbans

Drudgery is physical and mental strain, agony, and hardship experienced by human beings (Moharana et al. 2016). The women in Sundarbans live in abject poverty. Fisher women are compelled to work for long hours in saline waters for fish seed collection. There is added fear of being taken away by tidal waters or crocodiles. Most of the fisher women suffer from skin diseases due to the long time exposure to saline water. Segregation of the desired seeds, which are very thin and tiny, causes strain to eye. A year long study covering 600 respondents across various age groups revealed fever, followed by eye disease, lower back pain, scabies, and respiratory illness to be the most prevalent ailments recorded in the studied population. Morbidities were found to be more common in males (45%) as compared to females (29%). The overall pattern of illness is in line with the expected morbidity scenario of an area like Sundarbans with its extreme heat, stifling humidity and crippling poverty. Due to extreme climatic events, the shortage of drinking water has become a serious issue in Sundarbans. Women of fishermen community are largely responsible for water collection. They now have to walk long distance to collect drinking water.

17.6.4 Effects of the Migration of Men

In Sundarbans areas, after the fury of cyclonic storms like *Aila*, *Nigar,* and *Hudhud*, the men of the fisher community have begun to migrate to urban areas or to other states in search of work as these extreme events have washed away their assets like, land, fishing boats, fishing equipment, animals, and others. Most of them are jobless and have no alternative option. When the men migrate, the women stay back with children and elderly family members. All the responsibilities of the families are then handled by the women and the role of primary caregiver has increased.

17.6.5 Lack of Access and Control over Resources

In Sundarbans, fisherwomen are losing access and control over resources. Due to climate change, the physical stability of the environment and ecology of the Sundarbans is at stake. In many islands of Sundarbans, there is still no Primary Health Care Centers. Roads are not proper or have broken due to natural calamities like flood, cyclones etc. Even pregnant women have to cross rivers to reach the nearest hospital in emergency or during any complications. Fisherwomen don't have proper sanitation facilities and this is compounded when their homes and surroundings are submerged during floods. Access to general education is denied for children, especially for girl child from fishermen families, as they have to look after their other siblings when their mother goes out for fishing or other related activities. Along with this, issues like violence, abuse also sometimes restrict the womenfolk in Sundarbans to exercise their rights. Financial inclusion is almost nil in case of the fisher women. In Sundarbans areas, most of the families have lost their assets in natural fury, so, the financial institutions are not interested to extend credit, especially to the women. Moreover, the fisher women do not have absolute control over most of the productive resources, like land, capital, mechanized equipment etc.

17.7 Socio-political Dynamics vis-a-vis Climate Change

Even after severe climate shock, the fisherwomen felt that they were mostly disregarded in any government program and that they were not given any input facilities and factors hindering the socio-economic upliftment of fisherwomen are given in (Table 17.2). Disparities in input distribution are frequently caused by the dominance of individuals or a specific group. There is no insurance policy for the communities of fishermen in inland open water fishing. The fishermen's community is unable to obtain insurance coverage for natural disaster damages because of a lack of financial inclusion. Even after such incidents, they are unable to acquire institutional financial services.

The islands of Sundarbans are very much prone to be damaged by natural calamities and islands' fishermen community perceived that a long time is required for getting any benefit from government schemes as they belong to lower stratum of the society. Intra-village politics is also a constraint that often limits their socio-economic development. Male members generally migrate for jobs and the fisher women stay behind become more vulnerable to criminal attack or sexual violence. The health and hygiene of fisherwomen is often forfeited at the cost of their family responsibilities.

Table 17.2 Factors hindering the socio-economic upliftment of fisherwomen

Sl. no	Factors	Coping strategies
1.	No input facilities/ government benefits being provided	Shift from family occupation to wage labour.
2.	Domination of an individual or a particular group in the social system	Informal mutual support among the community members
3	Lack of credit/ loan facilities	Borrow from the fish wholesalers or take "*dadon*" (credit) from the businessmen
5.	No insurance facility for damages (assets, fishery requisites, home etc) in natural calamities	Shift to other non-fishing occupation
6.	Long time required to get any benefit of Govt. schemes	Male members migrate to urban area in search of job, fisher women adjust family labour by engaging their children in fishing related activities
7.	Insecurity of foods	Adjustment in food consumption, sometimes skipping meal
8.	Politics of village panchayats in distributing government benefits	Informal mutual support among the community members
9.	Sexual violence and criminal attack	Occasional migration

17.8 Strategies to Revamp the Situation

Gradual increase of surface water temperature, salinity variation over a span of 27 years in territorial waters confirms the foot prints of global warming in Indian Sundarbans and the results of different foot prints of climate change create severe troubles in living, landscape management, and biodiversity range of that area (Mitra et al. 2015). Due to climate change, the distinct demarcation of the seasons is less evident and creating stress to life and livelihoods. Women in Sundarbans who have been displaced or remain in home land are more susceptible to sexual violence. More women casualty happens in natural disaster or extreme events as their movement is limited due to traditional clothing. In spite of tremendous contributions of the fisher women in household income as well as in food and nutritional security, under changing climatic scenario, no attention is being accorded to the fisher women. Various national and international organizations are now focusing on Indian Sundarbans, to protect the unique gene pool of the globe which is being threatened due to climate change, but the everyday struggle is faced by the women; particularly the fisher women almost remain unnoticed. Interventions are needed to protect the fundamental rights of the fisher women living in Sundarbans viz. right to food, health care, education, and work.

17.9 Conclusion

The Sundarbans, though rich in biodiversity and natural resources, presents a paradox of abundance and adversity for its inhabitants, particularly women from the fishing communities. Climate change has exacerbated the region's ecological fragility, leading to declining fish stocks, loss of traditional livelihoods, increased salinity, and frequent natural disasters. These changes have disproportionately affected fisherwomen, who continue to shoulder the dual burden of household responsibilities and income generation under extremely challenging conditions. Despite their significant contributions to fisheries and food security, they remain largely invisible in policy frameworks and institutional support mechanisms. Addressing the multidimensional vulnerabilities of fisherwomen in the Sundarbans requires an integrated approach ensuring access to education, healthcare, credit, infrastructure, and social protection while also recognizing and institutionalizing their roles in climate resilience and community development.

Points to Remember

- Sundarbans' fragile ecosystem supports millions, but is highly vulnerable to climate change impacts like cyclones, sea-level rise, and salinity intrusion.
- Fisheries are the economic backbone, with women playing a vital yet underrecognized role in seed collection, fish drying, aquaculture, vending, and net mending.

- Climate change has reduced fish availability, increased livelihood risks, and caused male out-migration, leaving women to shoulder household and economic burdens.
- Nutritional insecurity is rising due to the decline of small indigenous fish species and loss of freshwater sources and kitchen gardens.
- Women face high drudgery and health hazards, including long working hours, exposure to saline water, lack of sanitation, and limited healthcare access.
- Systemic neglect and gender inequality limit women's access to resources, government schemes, credit, education, and participation in decision-making.

References

Bhaumik U, Saha SK, Mitra PM (1993) Scenario of brackish water fin fish and shell fish seed collection in the Sundarbans, West Bengal. J Inland Fish Soc India 25(1):44–50

Census (2001) Government of India National Portal. https://www.india.gov.in/census-2001-data-online

Census (2011) Government of India National Portal. https://censusindia.gov.in/nada/index.php/catalog/42611

Chandra G, Sagar RL (2003) Fisheries in Sundarbans: problems and prospects. zunia.org. Available at: http://works.bepress.com/ganesh_chandra/5. Accessed 4 Apr 2014

Gross R, Schoeneberger H, Pfeifer H, Preuss H (2010) The four dimensions of food and nutrition security: definitions and concepts

Hazra S, Ghosh T, Dasgupta R, Sen G (2002) Sea level and associated changes in the Sundarbans. J Sci Cult 68(9–12):309–321

Hazra S, Dasgupta R, Samanta K, Sen S (2004) A preparatory assessment of vulnerability of the ecologically sensitive Sundarban Island system, West Bengal. In: Gossain et al (eds) Proc vulnerability assessment and adaptation due to climate change on Indian water resources, coastal zones and human health. IIT Delhi (India, NATCOM), pp 66–82

Hossain N (2009) Accounts of crisis: poor people's experiences of the food, fuel and financial crises in five countries. Institute of Development Studies, London. Available at: http://www.ids.ac.uk/files/dmfile/AccountsofCrisisFINAL.pdf. Accessed 7 Apr 2014

IFPRI (1995) Feeding the world to 2020: what role will women play? News and Views, IFPRI, Washington, DC

Islam MM, Chuenpagdee R (2013) Negotiating risk and poverty in mangrove fishing communities of the Bangladesh Sundarbans. Marit Stud 12:7

Jha BC (2012) The Sundarbans: an ecological paradise including fisheries and biodiversity. In: CIFRI compendium on Sundarbans: retrospect and prospects. p 16

Katiha PK, Ekka A, Biswas DK, Sinha A, Pandit A, Das BN (2012) The goods and services of Gosaba Islands in Sundarbans. In: CIFRI compendium on Sundarbans: Retrospect and prospects. pp 137–144

Levine JA, Weisell R, Chevassus S, Martinez CD, Burlingame B, Coward WA (2001) The work burden of women. Science 294:812

Mandal B, Mukherjee A, Banerjee S (2013) A review on the ichthyofaunal diversity in mangrove based estuary of Sundarbans. Rev Fish Biology Fish 23:365–374

Manna RK, Roshith CM, Das SK, Suresh VR, Sharma AP (2014) Salinity regime and fish species distribution in the Hooghly-Matlah estuary. In: Sinha RK, Ahmed B (eds) Rivers for life. IUCN, Dhaka, pp 232–240

Mishra SS, Gopi KC (2017) Fish diversity of Indian Sundarbans and its resources and research prospects. Fauna of Sundarbans Biosphere Reserve. Zoological Survey of India, Kolkata, India

Mitra A (2015) Facts and figures of climate change in Sundarbans. https://www.researchgate.net/publication/386280566_Facts_and_figures_of_climate_change_in_Sundarbans/citations#fullTextFileContent
Moharana G, Nayak J, Rout PK (2016) Drudgery and ergonomics in animal farming: implications for women. In: Compendium ICAR sponsored short course on empowering farmwomen through livestock and poultry intervention. pp 56–65
Mukherjee N (1993) Participatory rural appraisal, vol 1. Concept Publishing Company, New Delhi
Nandi J (2010) Field survey: west Dwarikapur. In: Sundarban Basir Sathe: Kemon Ache Sundarbaner Manush: Aila Parabarti Ekti Sameeksha. Manthan Samayiki, Kolkata
Priyadarshini S (2015) Climate change pushing Sundarban farmers into awkward jobs. Nat India
Raha AK, Zaman S, Sengupta K, Bhattacharyya SB, Raha S, Banerjee K, Mitra A (2013) Climate change and sustainable livelihood programme: a case study from Indian Sundarbans. J Ecol 107(335348):64
Rajshekar SC (2011) Agricultural infrastructure development in Sundarbans. World Bank Technical Assistance Final Report, New Delhi
Roy A, Bhaumik U (2012) Women's role in fisherfolks' communities of Hooghly Estuary. J Inter Des 16(4a):1100–1106
Roy A, Bhaumik U (2013) Participation of women of Sundarbans in fisheries decision making. J Inland Fish Soc India 45(2):23–27
Roy A, Manna RK, Sinha A, Das SK, Majumder S, Biswas DK, Mukherjee CN, Mitra A, Sengupta A (2016) Conservation of small indigenous fish (SIF): a road map to ensure food and nutritional security in coastal Sundarbans. In: Mondal B et al (eds) Family farming challenges and opportunities. Renu Publishers, New Delhi, pp 218–228
Roy A, Sharma AP, Bhaumik U, Pandit A, Singh SRK, Saha S, Mitra A (2017) Socio-economic features of womenfolk of Indian Sundarbans involved in fish drying. Indian J Ext Educ 53(2):142–146
Sharma AP (2012) Ecosystem structure and functions of mangroves in fisheries perspectives of Sundarbans. In: CIFRI compendium on Sundarbans: retrospect and prospects. p 165
Singh SS (2017) Climate change impact: Sundarbans steadily losing its famed mangroves. The Hindu, Kolkata, 1 July 2017
Sinha A, Das SK, Roy A, Mitra A, Mukherjee CN, Sengupta A (2014) Small indigenous fish: an alternate source to improve rural livelihood and nutritional security in coastal Sundarbans. In: Natl Conf on Wetland Adaptation, Abstracts. pp 73–74
Williams M (2015) Women in today's fisheries economy. Yemaya, (50):2–4

Fish Farmer Producer Organization: A Potent Tool for Paradigm Shift in the Fisheries Sector, India

18

Piyashi DebRoy, D. C. Yateesh, Anurag Singh, Arun Pandit, Basanta Kumar Das, and Pranab Gogoi

Abstract

Fish Farmer Producer Organizations (FFPOs) represent a transformative institutional model for addressing systemic challenges faced by small-scale fishers and marginal fish farmers in India. This study examines the evolution, structure, and functions of FFPOs as a response to constraints such as limited market access, technology gaps, inadequate infrastructure, and weak bargaining power. Drawing upon national initiatives like the Pradhan Mantri Matsya Sampada Yojana (PMMSY) and supported by empirical success stories, FFPOs have emerged as effective vehicles for inclusive rural development, sustainable aquaculture, and value chain enhancement. The analysis highlights how FFPOs, through collective action, enable access to credit, technology adoption, and improved market linkages, ultimately leading to increased income, productivity, and resilience among fisher communities. However, persistent challenges in governance, financing, and policy implementation require coordinated efforts across government, private, and civil society sectors. The study emphasizes the need for robust capacity-building frameworks and a conducive policy environment to unlock the full potential of FFPOs in driving the blue economy in India.

Beyond mere production, we forge opportunities to enhance the value of our Catch- an anthem to empowering small-scale fishers/farmers through collective action for sustainable development. – Fish Farmers Producer Organisation

P. DebRoy (✉) · D. C. Yateesh · A. Singh · A. Pandit · B. K. Das · P. Gogoi
ICAR – Central Inland Fisheries Research Institute (ICAR-CIFRI),
Barrackpore, Kolkata, India

B. K. Das et al. (eds.), *Small-scale Inland Fisheries in India*, Sustainability Sciences in Asia and Africa, https://doi.org/10.1007/978-981-95-3849-2_18

Keywords

Fish Farmer Producer Organizations · Small-scale fisheries · Collective action · Rural livelihoods

18.1 Introduction

Since time immemorial, fishing has been one of the primary occupations of humankind, and major civilizations have flourished along the banks of water bodies. Moreover, it has long served as a vital source of livelihood for many, providing essential support to deprived fishing communities for nutritional security, otherwise unattainable. This sector's significance in our country's economic activity is undeniable, contributing significantly to national income, foreign exchange, food production, and employment opportunities. The fisheries sector has, over the years, emerged as a sunrise sector in the Indian economy. It has shown consistent growth, outpacing many other sectors, and has become a cornerstone for food and nutritional security, offering livelihoods to a staggering 28 million fisher population. Remarkably, the sector has maintained a robust average growth rate of 10.34% per year and is a substantial contributor to the Agriculture GDP, constituting approximately 6.72% of the total Gross Value Added with steady growth.

Small-scale fishers and marginal fish farmers, constituting a significant segment of the 28 million fisher population in India, play a pivotal role in the country's fisheries sector. Despite their numerical importance, these small-scale fisheries and marginal farmers face unique challenges that are integral to understanding the broader dynamics of the industry. Issues such as limited access to financing, constrained market access, ensuring the quality of fish products, diseconomies of scale, and the inability to add value to their catches hinder their progress. The technology gap that exists among small-scale producers in the fisheries industry is a notable obstacle (Kamala et al. 2018). The adoption of fisheries technologies remains limited due to a lack of knowledge and resources in these communities. Moreover, the scarcity of storage, transportation, and processing facilities poses significant challenges to the efficient handling and marketing of fisheries products (Noopur et al. 2023). Cultural and social factors, such as sociocultural practices and norms, can impede the adoption of new technologies and fishing practices, further impacting the livelihoods of small and marginal fishers in India (Morgan et al. 2017).

Historically, cooperative societies were established in the agriculture and fisheries sectors to tackle key challenges and support rural communities. However, this traditional model has encountered several limitations over time, such as restricted access to capital, bureaucratic and complex governance structures, and an overreliance on government assistance. Additionally, issues like unequal member participation, limited entrepreneurial drive, and susceptibility to external influences have been noted (Singh et al. 2003). To overcome these challenges and ensure their sustainability, robust management and support mechanisms are essential. In response to these challenges, a new approach has been adopted, focusing on overcoming political and bureaucratic hindrances that impede farmer cooperatives. A supportive legal and policy environment is crucial in fostering the growth of fisher collectives and bridging the gap between fishers and consumers. In this context, Fish Farmer Producer Organizations (FFPOs) have emerged as potential drivers. According to the Ministry of Agriculture and Farmers Welfare, an FFPO is a type of Producer Organization (PO) where fishers are its members. The FFPO, as a producer-run organization, encompasses various fishery products, including artisanal and non-farm items. This approach aims to create a more streamlined and effective system to address the unique challenges faced by fishers, promoting collective strength and sustainable fisheries practices. The impact of Producer Organizations on the fisheries sector extends beyond economic dimensions. By fostering collective action, these entities empower fishers to negotiate better prices, access credit facilities, and adopt modern fishing practices.

India has a rich history of collective institutional arrangements that paved the way for the emergence of Fish Farmer Producer Organizations (FFPOs). The cooperative movement, dating back to the early 1900s, was the first major attempt to organize producers, exemplified by successful models like Amul and fisheries cooperatives such as Matsyafed. In the 1990s, Self-Help Groups (SHGs) became prominent, particularly among women, fostering financial inclusion and collective action. Joint Forest Management Committees and Water User Associations also emerged as community-based natural resource management bodies. Watershed committees further strengthened participatory governance in rural areas. The introduction of Producer Companies in 2002 provided a hybrid model that combined cooperative principles with corporate efficiency, offering a new legal structure for farmer groups. These diverse institutional experiences highlighted both the potential and challenges of collective organization. FFPOs have built upon this legacy, offering a more business-oriented, market-linked approach to empower fishers and farmers. The aggregation of fishery products through FFPOs not only facilitates improved market access but also ensures economies of scale, making it advantageous for fishers to collaborate.

18.2 FFPOs Unveiled: Transforming Indian Fisheries

India's fisheries sector is witnessing unprecedented growth, fuelled by a rising demand for fish and seafood products. In recent years, the concept of Fish Farmer Producer Organizations (FFPOs) has emerged as a favored model, spearheading sustainable fish farming practices and elevating the socio-economic conditions of small-scale fish farmers. FFPOs stand as "collective organizations formed by fish farmers uniting to pursue shared objectives in production, marketing, and value addition of their fish and seafood products. The primary goal is to offer a platform to small-scale fish farmers and fishers in India, facilitating their access to technical, financial, and marketing support- resources that would be challenging to attain individually."

18.3 Objectives of Developing FFPOs in India

FFPOs have emerged as a transformative institutional model for improving the socio-economic conditions of fishers and fish farmers in India. Their core objectives span from economic empowerment to sustainable value chain development. The following table outlines these primary goals:

Objective	Description
Economic empowerment	Strengthening collective bargaining power through aggregation of resources and collective action.
Enhancing productivity	Promoting efficient, cost-effective, and sustainable use of aquatic resources and farming inputs.
Realizing higher returns	Ensuring better prices and income through market linkages, liquidity support, and fair pricing.
Building capacities	Developing entrepreneurial and technical skills among members to foster self-reliance and growth.
Developing value chains	Creating integrated and sustainable fisheries value chains for income generation and sectoral growth.

Specific functions and objectives drive Fish Farmer Producer Organisations (FPOs) intending to improve the socio-economic status of their members. The following factors help to clarify the role of FFPOs in empowering farmers.

18.4 Broad Services and Activities by Indian FFPOs

Production and Productivity FFPOs play a crucial role in the production and productivity aspect of the fisheries sector in India. They ensure a steady supply of quality inputs for fish production at affordable rates, engaging in various fish culture activities that employ diverse techniques. Moreover, FFPOs disseminate technology and innovations while maintaining strict quality control measures. By aggregating smaller lots of produce from farmer members in India, they contribute to efficient and sustainable fish farming practices.

Post-Harvest Management and Infrastructure In the realm of post-harvest management and infrastructure in India, FFPOs provide essential machinery and equipment for storage, transportation, and logistics. They offer value-added services such as cleaning, sorting, grading, and packing, contributing to the preservation and quality of the harvested fish. FFPOs also take a step further by establishing processing facilities at the fish farm level, promoting high-value addition. Activities like cold chain development and traceability interventions further enhance post-harvest management efforts.

Marketing and Branding FFPOs extend their services into marketing and branding initiatives in India. They play a pivotal role in creating brands, packaging, and standardizing fish products. By marketing the aggregated produce with enhanced negotiation strength, FFPOs secure better deals for their members in India. Additionally, they provide valuable market information to support informed decision-making among farmers. FFPOs also explore opportunities for both domestic and export sales, operating fish vending kiosks to increase accessibility and visibility in the Indian market.

Together, these activities play a pivotal role in fostering FFPOs in India, ensuring the adoption of sustainable fish farming practices (Fig. 18.1), improved market access, and increased income for fish farmers (Source: Guidelines on the Formation and Promotion of Fish Farmer Producer Organizations—FFPOs).

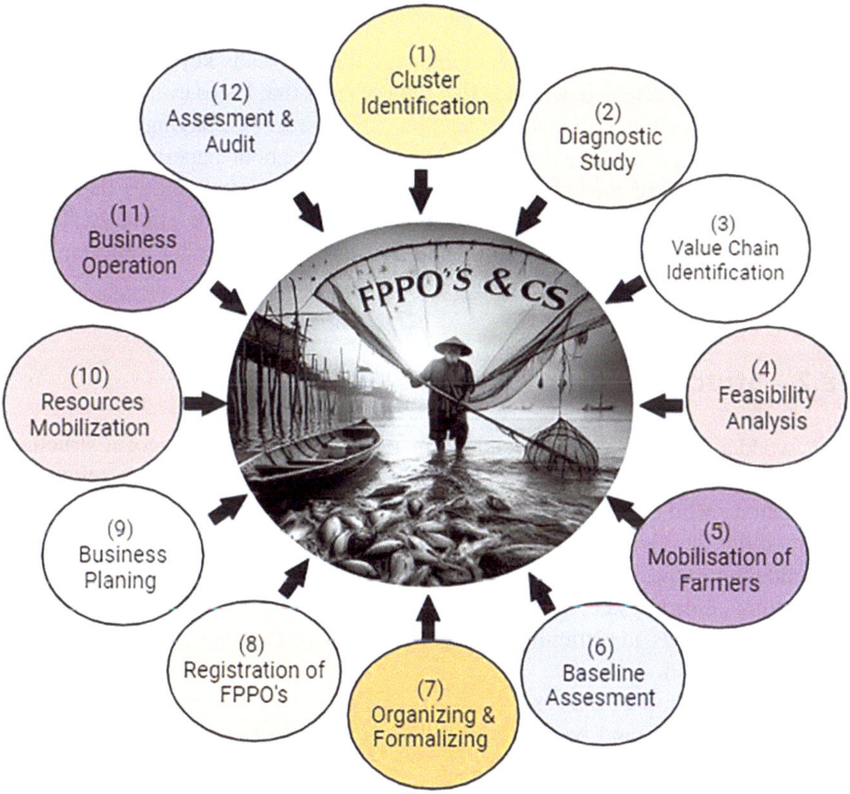

Fig. 18.1 Activities of FPPOs

18.5 FFPO Organizational Structure and Governance

The organizational structure and governance of Fish Farmer Producer Organizations (FFPOs) in the fisheries sector play a pivotal role in ensuring their effectiveness and impact.

When it comes to Fish Farmer Producer Organizations (FFPOs), how they're set up and run is a big deal. Think of it as the roadmap that guides their success in the fisheries sector. They usually work together in a cooperative way, teaming up in clusters or associations. This teamwork boosts their power to negotiate with suppliers and buyers because they pool their resources. This way of working, as pointed out by Yadav and Sharma in 2020, helps FFPOs benefit from doing things on a larger scale, like getting better access to markets and more services.

18.6 Key to Trust: How FFPOs Keep Things in Check

Imagine FFPOs as a team—how they manage themselves is key to building trust. They use good governance, which is like a set of rules that keeps everything running smoothly. This includes things like keeping proper records, having regular meetings, and being clear about finances. This isn't just about paperwork; it's about building trust among members and others involved. Research shows that FFPOs with clear and open rules tend to do better financially and operationally. They're more likely to pay back loans on time and use funds efficiently for important projects (NABARD 2021).

18.7 Structure and Governance

Ever wonder who makes the decisions in FFPOs? Well, they've got a structured team for that. It's like a well-organized team with different roles. There's the Board of Directors, who make big decisions, the Executive Committee, who handle day-to-day stuff, the General Body, where everyone gets a say, sector-specific subcommittees for specific tasks, and functional departments for different jobs. This team setup, guided by keeping things clear, fair, and using the right skills for the job, makes sure FFPOs in agriculture and fisheries stick together and work well for everyone involved.

In a nutshell, how FFPOs are set up and run is the backbone of their success. It's not just about rules and paperwork; it's about teamwork, trust, and making sure everyone benefits in the fisheries sector.

18.8 Evaluation of FFPO Initiatives Linked with Small Scale Fishers in India

In the global arena, numerous Fish Farmer Producer Organizations (FFPOs) have emerged, marking successful strides in uplifting the livelihoods of small-scale fish farmers. The Indian context provides a notable example, where the National Fisheries Development Board has established the FFPOs to catalyze socio-economic improvements for small-scale fish farmers.

Specifically, the Pradhan Mantri Matsya Sampada Yojana (PMMSY), slated for implementation from FY 2020–21 to FY 2024–25, stands as a pivotal initiative aimed at ensuring the comprehensive socio-economic development of fishers, fish farmers, and fish workers, with the ambitious goal of doubling their incomes. A cornerstone of the scheme involves the establishment of 500 Fish Farmers Producer Organizations/Companies (FFPOs/Cs) with focus on areas like West Bengal, Assam, Odisha and Uttar Pradesh, strategically designed to empower fishers and fish farmers, bolstering their bargaining power within the market and promoting inland aquaculture clusters. Within this framework, 300 FFPOs are slated to be set up under the direct auspices of PMMSY, while an additional 200 will be established through a synergistic convergence with the ongoing FPO Scheme of the Department of Agriculture, Cooperation, and Farmers Welfare.

FFPOs can procure inputs like fish seed, feed, nets, or ice in bulk at lower prices and distribute them to members. They can also set up common infrastructure like cold storage, hatcheries, or fish drying units. FFPOs help small-scale inland fishers bypass middlemen and connect directly with Wholesale markets, retail chains, exporters, and online platforms. This improves income by ensuring a larger share of profits remains with fishers. FFPOs receive training through government schemes like PMMSY and from institutions like NFDB or ICAR-CIFE. These trainings focus on the scientific fish farming practices, record keeping, quality standards, and financial literacy. These strategic alignment of the PMMSY, coupled with collaborative initiatives, reflects a concerted effort to transform the socio-economic landscape for fishers and fish farmers in India. As the nation endeavours to double the incomes of these crucial stakeholders, the establishment and evaluation of FFPOs emerge as instrumental components, serving as beacons of progress and empowerment within the fisheries sector.

18.9 Exploring the Growth Trajectory: FFPO Registrations in India

Embarking on an exploration of fish farming in India, our attention shifts to the captivating evolution of Fish Farmer Producer Organizations (FFPOs). Illustrated below (Fig. 18.2) is a comprehensive year-wise breakdown of FFPO registrations. The story unfolds with a gradual ascent from 2014 to 2017, laying the foundation for a more pronounced shift. Beyond 2018, the trend takes a notable upswing, marking a substantial surge in FFPO registrations. This uptick signifies an increasing

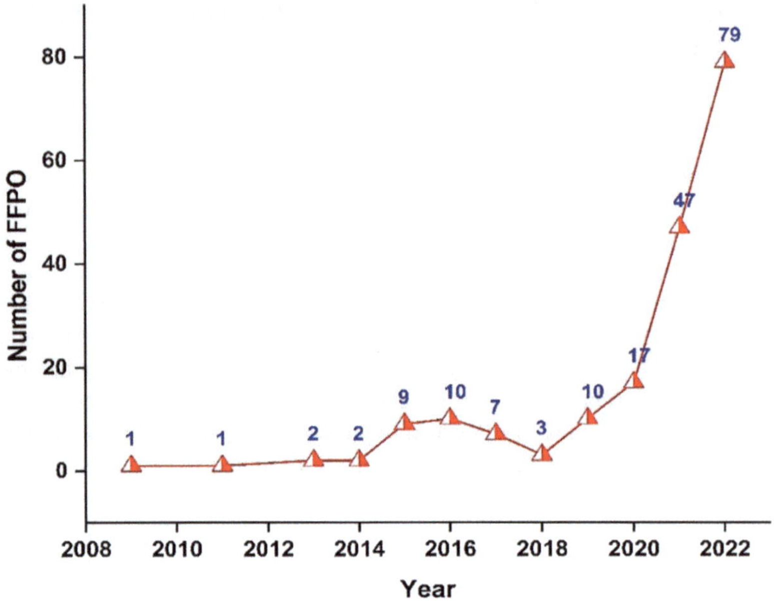

Fig. 18.2 Year wise number of registered FFPOs in India (FPO Platform for India, Tata Cornell Institute)

recognition of the indispensable role played by collective organizations in shaping the fish farming sector.

Moving forward to (Fig. 18.3), the geographical distribution of FFPOs across different states in India is unraveled. Karnataka emerges as a frontrunner, boasting the highest number of FFPOs. Following closely are states like Uttar Pradesh, Tamil Nadu, and Assam. Noteworthy contributions also come from Andhra Pradesh, Bihar, and Maharashtra. This distribution underscores the widespread popularity and adoption of the FFPO model, a driving force for the development and expansion of the fisheries sector across various regions of India.

18.10 Institutional Support for FFPOs

Network of central government institutions assumes pivotal roles in providing indispensable support to Fish Farmer Producer Organizations (FFPOs). Vital stakeholders such as the Small Farmers Agribusiness Consortium (SFAC), National Cooperative Development Corporation (NCDC), and the National Bank for Agriculture and Rural Development (NABARD) collaborate to fortify FFPOs on their developmental journey.

Complementing these central entities, state governments, international aid agencies, the corporate sector, and non-governmental organizations (NGOs) join forces to offer critical financial and technical assistance to Resource Institutions (RIs).

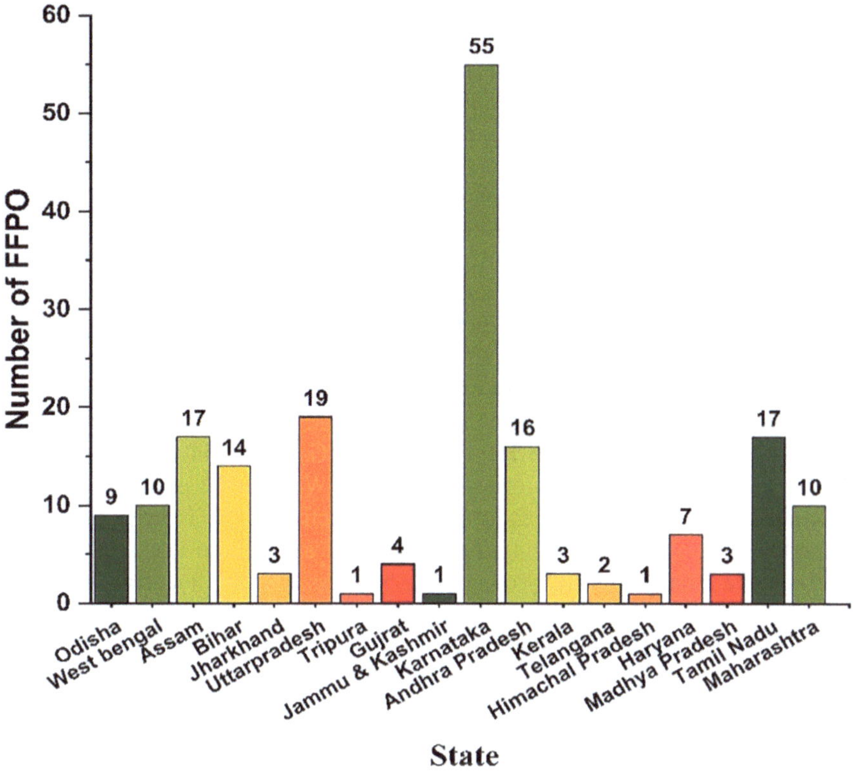

Fig. 18.3 State wise distribution of FFPOs in India (FPO Platform for India, Tata Cornell Institute)

These entities, identified by SFAC, submit project proposals directly for FPO promotion. The concerted efforts of SFAC and NABARD in conducting training programs for the Board of Directors (BoDs) and Chief Executive Officers (CEOs) of FPOs underscore a collective commitment to enhance the skills and knowledge base crucial for the effective functioning of these organizations. This collaborative approach, involving various governmental bodies and support organizations, exemplifies a holistic strategy to strengthen and empower FFPOs, ensuring their sustained growth.

18.11 Distinguishing Features of Fish Farmer Producer Organizations (FFPOs)

In the realm of agricultural organizations, Fish Farmer Producer Organizations (FFPOs) stand out with distinct features that set them apart from traditional counterparts. Unlike cooperative societies, FFPOs operate under the Indian Companies Act, granting them a broader geographical scope encompassing the entire Union of

India. This expansive reach allows FFPOs to engage with a diverse membership base, including individuals, groups, associations, and producers of goods or services. One noteworthy difference lies in the voting rights structure, where FFPOs adhere to the principle of "*one member, one vote*," with no external interference from government authorities. Additionally, FFPOs demonstrate a higher degree of autonomy compared to cooperative societies, operating autonomously within the provisions of the Act. The profit-sharing model in FFPOs is designed to be commensurate with the volume of business, fostering a more dynamic and market-oriented approach. Unlike cooperative societies, FFPOs enjoy enhanced financial autonomy with borrowing powers determined by a Special Resolution in general meetings. This flexibility allows FFPOs to establish more robust relationships with corporate entities, facilitating the joint creation of producer companies. This transaction-based partnership model fosters a collaborative approach between producers and corporate entities within the FFPO framework. In comparison to non-profit organizational models like Self-Help Groups (SHGs) and Joint Liability Groups (JLGs), FFPOs offer a business-oriented structure. While SHGs primarily focus on communal financial resilience and microfinance activities, JLGs serve as mechanisms for securing bank financing, particularly for agricultural loans. In contrast, FFPOs carve a unique niche by combining business acumen, financial autonomy, and a dynamic membership structure, distinguishing them as a distinct and innovative force in the agricultural landscape.

18.12 Success Stories in Fisheries Sector: Transformative Impact of FPOs on Livelihoods and Economic Prosperity

Fish Farmer Producer Organizations (FFPOs) have increasingly emerged as powerful institutional platforms to enhance the economic resilience, market access, and bargaining power of small-scale fishers across India. While the structural and operational features of FFPOs provide them with inherent advantages over traditional cooperative or non-profit models, the true value of these organizations is most vividly illustrated through their on-the-ground impact. The following case studies provide concrete evidence of how well-structured FFPOs can successfully transform fisheries-based livelihoods and contribute to community development.

Case Study 1: Bhavi Aqua and Fish Farmer Producer Company (BAFFPC)
Bhavi Aqua and Fish Farmer Producer Company (BAFFPC) stands as a shining example of FPO success, receiving recognition as the best FPO on World Fisheries Day. Established in April 2017 in Andhra Pradesh, BAFFPC is engaged in agriculture, allied activities, and fisheries. Under the leadership of CEO Shri. G. Manoj Kumar Reddy, this FPO, initiated by KPL MACTS, strives to enhance fishermen's income by addressing critical challenges, including securing better prices, engaging in value-added activities, and improving marketing facilities. Operating in floriculture, horticulture, and pisciculture, BAFFPC not only offers capacity-building to its fisher members but also witnessed a strategic increase in sales, rising from Rs. 1.35

lakhs in 2017–18 to an impressive Rs. 51.70 lakhs in 2019–20. By implementing value addition through products like fish pickles and chutneys, BAFFPC strengthens both backward and forward linkages in the supply chain. With a focused approach on branding ("Gunapatis") and establishing direct consumer connections, BAFFPC empowers farmers through digital marketing, ensuring sustainable growth and fetching better prices for their produce.

Case Study 2: MatsyaSahay FPO in Kerala
The MatsyaSahay FPO in Kerala serves as a compelling illustration of the transformative impact FPOs can have on the fisheries sector. Documented collaboratively by the National Fisheries Development Board (NFDB) and the Kerala State Fisheries Department, this success story vividly portrays how FPOs can rejuvenate the livelihoods of fisherfolk. Over a three-year period, MatsyaSahay FPO achieved an impressive 87% increase in revenue, averaging INR 280,000 annually. Strategic initiatives, such as establishing direct market connections and embracing technology, successfully eliminated intermediaries and ensured better pricing for the fish catch. The success of MatsyaSahay FPO can also be attributed to its unwavering focus on capacity building, leading to a remarkable 35% increase in the average fish harvest per member. The positive outcomes documented in this case study underscore the profound impact that well-structured and strategically operated FPOs can have on the economic prosperity and sustainability of communities within the fisheries sector.

These case studies strengthen that FFPOs, when effectively structured and strategically managed, can act as transformative agents for small-scale fishers. By integrating business acumen with community ownership and capacity building, FFPOs have the potential to reshape the inland and coastal fisheries sector, making it more equitable, sustainable, and profitable.

18.13 Challenges Faced by Fish Farmer Producer Organizations (FFPOs)

Fish Farmer Producer Organizations (FFPOs) face a myriad of challenges that impede their effectiveness and growth. Firstly, there exists a significant challenge in the form of limited awareness and understanding among fish farmers regarding FFPOs. This lack of awareness hampers the engagement and collaboration between fish farmers and FFPOs, restricting the realization of potential benefits and opportunities these organizations offer. Moreover, FFPOs encounter obstacles related to insufficient financial and technical resources required for their establishment and sustainability. Inadequate funding and a lack of technological support hinder FFPOs from implementing efficient practices and adopting modern technologies essential for the promotion of sustainable fish farming. Another critical challenge revolves around capacity gaps within FFPOs, particularly in governance, management, and leadership skills. These gaps hinder effective decision-making, strategic planning, and leadership within the organizations, impacting their overall efficiency and performance. Furthermore, FFPOs grapple with the lack of supportive policies, regulations, and institutional frameworks. The absence of a conducive environment for the development and growth of FFPOs restricts their potential impact.

18.14 Future Perspectives for Fish Farmer Producer Organization

The future of Fish Farmer Producer Organizations (FFPOs) appears promising as the demand for fish and fishery products is projected to rise in the upcoming years. FFPOs hold significant potential in addressing this increasing demand while simultaneously enhancing the socio-economic conditions of small-scale fish farmers. However, several challenges loom, including obstacles related to access to finance, technical expertise, and efficient marketing channels.

To surmount these challenges, prioritizing capacity-building initiatives is essential. Providing FFPOs with the necessary technical and financial support becomes crucial in fostering their growth. Additionally, creating an environment through favorable policies and regulations is imperative. The long-term success and sustainability of FFPOs rely heavily on coordinated efforts among government bodies, research institutions, development agencies, and private sector partners. Such collaboration is essential to effectively tackle the diverse challenges faced by FFPOs and to support the continued growth and resilience of this vital sector.

18.15 Conclusion

In the evolving landscape of India's fisheries sector, the emergence of Fish Farmer Producer Organizations (FFPOs) remains a transformative force, exemplified by their pivotal role in driving sustainable fish farming and fishing practices and improving the socio-economic conditions of small-scale fishers. The success stories of Bhavi Aqua and MatsyaSahay FPOs illustrate the potential of FFPOs in enhancing productivity, market access, and income for fisher communities. Despite the remarkable progress, challenges such as limited access to finance, technological gaps, and bureaucratic hurdles persist, necessitating targeted interventions. The future of FFPOs appears promising as they navigate these challenges and capitalize on the growing demand for fish and seafood products. The objectives of FFPOs, as outlined in the Pradhan Mantri Matsya Sampada Yojana (PMMSY), align with the broader vision of empowering fishers and fish farmers. However, addressing challenges exists in FFPOs requires a holistic approach, including simplifying registration processes, providing better training, improving access to funds, and fostering an environment of equality and fairness. Policymakers play a crucial role in creating an enabling environment, incorporating FFPOs into Sustainable Development Goals (SDGs), encouraging research and innovation, and allocating resources strategically. The trajectory of FFPO registrations in India reflects a positive trend, with increasing recognition and adoption across states. Karnataka leads the way, followed by states like Uttar Pradesh, Tamil Nadu, and Assam. This widespread popularity emphasizes the scalability and effectiveness of the FFPO model in diverse regions. As FFPOs continue to evolve, their success hinges on robust organizational structures, good governance, and a clear roadmap for collective action. The journey of FFPOs in India serves as a beacon for the fisheries sector, offering a glimpse into the potential for sustainable development, growth, and enhanced livelihoods within the fishers and fishing communities.

Looking forward, the growth of FFPOs requires concerted efforts from stakeholders, including government bodies, NGOs, and the private sector. Collaborative initiatives, aligned with the PMMSY and existing FPO schemes, can further accelerate the establishment of FFPOs, ensuring a comprehensive socio-economic transformation for fishers and fish farmers. The narrative of FFPOs in India, as presented in this chapter, lays the groundwork for continued exploration, research, and policy advocacy, with the ultimate goal of fostering a thriving and sustainable fisheries sector that benefits all stakeholders involved.

Things to Remember

- FFPOs are a proven success in promoting sustainable practices of fish farming and the upliftment of the socio-economic conditions of small-scale fish farmers.
- The government is actively endorsing FFPOs through initiatives such as Pradhan Mantri Matsya Sampada Yojana (PMMSY) and One District-One Product (ODOP).
- The trajectory of FFPO registrations in India reflects a positive trend, emphasizing the scalable and effective nature of the FFPO model across diverse regions.
- FFPOs, exemplified by success stories like Bhavi Aqua and MatsyaSahay, drive sustainable fish farming practices, enhancing productivity, market access, and income for small-scale fish farmers.
- Addressing challenges like a lack of awareness, insufficient resources, capacity gaps, and the absence of supportive policies is crucial for ensuring the long-term success of FFPOs.

References

Kamala IM, Kennedy JS, Devanand II (2018) Technology gaps analysis and prominence of two spotted mite Tetranychus urticae Koch of jasmine (Jasminum sambac L.) in Tamil Nadu. Indian Res J Ext Educ 18(2):93–96

Morgan M, Terry G, Rajaratnam S, Pant J (2017) Sociocultural dynamics shaping the potential of aquaculture to deliver development outcomes. Rev Aquac 9(4):317–325

NABARD (2021) Performance of farmer producer organizations in India. Retrieved from https://www.nabard.org/

Noopur K, Chauhan JK, Walia SS, Verma MR, Dhar U, Choudhary S, Chikkeri SS (2023) Constraints in vegetable production in India: a review. Indian Res J Ext Educ 23(3):14–19

Singh M, Dubey MK, Khare YR (2003) Constraints faced by the District Cooperative Society (DCS) members. Indian Res J Ext Educ 3(1):94–95

Gender Equity in Small-Scale Fisheries in Inland Sector of India

19

Piyashi DebRoy, Anurag Singh, Shreya Bhattacharya, Basanta Kumar Das, Pranaya Kumar Parida, and Prateep Kumar Nayak

Abstract

This chapter explores gender equity in India's small-scale fisheries, highlighting the essential yet undervalued role of women across the value chain from harvesting and processing to marketing and resource management. Despite comprising nearly half of the workforce, women face systemic barriers including limited access to resources, decision-making, education, and leadership roles. Beside this, the chapter also overviews gender roles across diverse aquatic systems rivers, reservoirs, wetlands, and estuaries revealing regional differences in women's participation and empowerment. Issues like gender-based violence, discrimination, and cultural norms further constrain their contributions. However, government interventions, policy reforms, and growing grassroots efforts are fostering inclusive development. Achieving gender equity is not only a matter of social justice but also key to enhancing food security, sustainable resource use, and economic resilience in India's fisheries sector.

Keywords

Gender equity · Small-scale fisheries · Women in fisheries · Empowerment · Policy

Recognizing women's role in small-scale fisheries goes beyond gender equity, as it highlights their key contributions to livelihoods, conservation, and community resilience.

P. DebRoy (✉) · A. Singh · S. Bhattacharya · B. K. Das · P. K. Parida
ICAR – Central Inland Fisheries Research Institute (ICAR-CIFRI), Barrackpore, Kolkata, India

P. K. Nayak
School of Environment, Enterprise and Development, Faculty of Environment, University of Waterloo, Waterloo, ON, Canada

B. K. Das et al. (eds.), *Small-scale Inland Fisheries in India*, Sustainability Sciences in Asia and Africa, https://doi.org/10.1007/978-981-95-3849-2_19

19.1 Introduction

19.1.1 Gender Equity in Small Scale Fisheries

Small-scale fisheries (SSF) are typically labor-intensive and involve fishing households operating with low investment, limited capital, and minimal energy inputs. SSF often rely on small boats or shore-based operations, with short fishing trips conducted near the coast. Generally, community-managed SSF play a vital role in supporting livelihoods and ensuring food security in both marine and inland environments (FAO 2005; World Bank 2012). Globally, small-scale fisheries are diverse and widespread, contributing significantly to employment and nutrition. Millions of people depend on these fisheries, which function across a variety of ecological settings (FAO 2016). Despite this global diversity, there is no universally accepted definition that fully captures the complexity of the sector. Women play a vital and often underrecognized role in SSF, accounting for an estimated 40% of those directly engaged in the sector whether through employment or subsistence activities (ICSF 2023). In India, women are actively involved throughout the fisheries value chain, particularly in post-harvest tasks such as processing, marketing, drying, salting, sorting, and trade. They also manage household finances and sometimes contribute to boat construction and repair (FAO 2020). In inland and open-water fisheries, women engage in activities like fish selling, grading, drying, prawn seed collection, subsistence fishing, shellfish gathering, and household-level fish processing (Kusakabe 2003). These roles are essential for maintaining the economic viability and food security of fishing communities. Nevertheless, women in SSF frequently encounter challenges such as unequal access to resources, limited opportunities, and a lack of formal recognition for their contributions. Gender equity in this context requires acknowledging the distinct roles, needs, and barriers faced by both men and women. It does not imply treating everyone identically, but rather ensuring fair outcomes through tailored support. According to FAO (2023), achieving gender equity and equality is fundamental to the sustainability of small-scale fisheries, necessitating structural and institutional reforms that eliminate discrimination and develop inclusive environments for all stakeholders in the sector.

19.1.2 Importance of Addressing Gender Equity in the Context of Indian Fisheries

India has a rich and diverse range of inland and coastal water resources that support fisheries and aquaculture. The country's inland lakes cover approximately 0.772 million hectares, while reservoirs span around 3.524 million hectares. Additionally, India has an extensive river network stretching across 45,000 km^2, along with floodplain wetlands totaling about 0.354 million hectares. The estuarine area is estimated at 0.3 million hectares, and backwaters and lagoons cover approximately 0.19 million hectares (ICAR 2017; NFDB 2022). This rich aquatic resource supports a significant population of inland fishers, totaling 23.12 million. Notably, 44% of these fishers, equating to 10.17 million, are women. (Handbook on Fisheries Statistics 2022). Despite constituting nearly half of the labor force in small-scale fisheries, women often face limitations in starting their own businesses due to a lack of access to technology, credit, and land. Granting them more authority can trigger economic multipliers that benefit not only individuals but entire communities, families, and the broader economy. In post-harvest operations, such as processing and marketing, women play a crucial role in ensuring household nutrition and food security, as their increased participation has the potential to enhance nutritional outcomes by diversifying the types of fish consumed.

In several inland regions of India, including West Bengal, Bihar, Odisha, and Assam, women play a significant role in subsistence fishing and prawn seed collection from rivers and wetlands. Despite their active involvement, structural barriers such as traditional gender norms, landlessness, and male-dominated cooperative systems continue to restrict their access to formal credit, fisheries technology, and participation in decision-making processes. In most fisheries cooperatives and panchayat-based waterbody leasing systems, ownership and lease rights are predominantly conferred upon male members, thereby limiting women's eligibility for policy-related benefits and government subsidies (Sugunan and Sinha 2001). Across various states, including Bihar, Manipur, Nagaland, Odisha, and Tamil Nadu, women are primarily engaged in supportive roles such as construction and repair of fishing gear, post-harvest processing (including drying), and fish marketing, indicating their multifaceted contributions to inland fisheries despite limited recognition and empowerment (Sinha et al. 1994).

The long-term sustainability of Inland ecosystems hinges on adopting equitable fisheries practices, a goal frequently advanced through the active involvement of women. This involvement leads to improved resource management, reduced overfishing, and a more sustainable utilization of marine ecosystems. Communities that embrace gender inclusivity tend to be more resilient, robust, and better equipped to address challenges like resource depletion and climate change. Empowering women in fishing communities not only contributes to economic progress but also strengthens the social fabric of these communities.

19.2 Gender Disparities in Small Scale Fisheries in India

19.2.1 Gender Roles and Division of Labor in Fishing Communities

In India's inland small-scale fisheries (SSF), women play a crucial yet often underappreciated role across the entire fisheries value chain, starting from seed collection and fish harvesting to processing, marketing, and financial management. Traditionally, gender roles in fishing communities assigned direct fishing activities such as harvesting, boat operation, and net mending to men, while women were mostly engaged in post-harvest processing and marketing. However, these distinctions are gradually becoming less rigid. In riverine fisheries, for example, women take on a wide range of responsibilities, including transporting the catch from the riverbank after the harvest, grading and sorting fish, drying fish, collecting prawn and fish seed, repairing nets, and even providing meals for family members working at the riverside (Chandra and Sharma 2015). Their participation in cleaning, sorting, processing, and selling fish is vital for market readiness and significantly contributes to household income and local economies.

Despite their extensive knowledge and generational experience, women's contributions are often undervalued and excluded from official fisheries data and policy frameworks (Bandyopadhyay and Joshi 2024). Persistent barriers including limited access to credit, technology, and formal training continue to reinforce traditional gender roles and restrict women's economic influence. Leadership and decision-making positions in fisheries cooperatives and organizations also remain largely male dominated, further limiting women's participation in governance.

However, recent developments indicate a shift toward more inclusive practices. In Odisha's Boudh district, for instance, Women's Self-Help Groups (WSHGs) that leased community tanks achieved a 60% increase in carp polyculture yields, improving both income and household nutrition (De et al. 2013). Across the state, more than 7800 WSHGs have adopted scientific polyculture techniques, including the cultivation of nutritionally rich mola fish, thereby enhancing food security and generating additional income (Fishery News 2023). These encouraging examples highlight the need for broader recognition of women's roles in inland SSF and underscore the importance of gender-sensitive policies and institutional reforms for a more inclusive and sustainable fisheries sector.

19.2.2 Limited Access to Resources and Decision-Making for Women

Women engaged in inland small-scale fisheries (SSF) across India continue to face significant systemic barriers in accessing productive resources and participating in decision-making processes. Land ownership, which is crucial for securing fishpond leases and fishing licenses, is predominantly male dominated. In Odisha, women were routinely excluded from the panchayat tank leasing system until the introduction of Mission Shakti, which facilitated Women's Self-Help Groups (SHGs) in accessing community tanks (Panemangalore 2021). Similar patterns persist across the sector, where women's access to credit, appropriate technologies, and fisheries related training remains severely limited. Deep-rooted socio-cultural norms further constrain their roles to less remunerative and informal tasks, while mobility restrictions often prevent them from participating in core fishing activities. Additionally, women are largely under represented in fisheries cooperatives and formal governance structures, marginalizing their voices in resource allocation and policy decisions. In Srikakulam, Andhra Pradesh, women fish vendors operating in local markets ("mandis") report the absence of basic sanitation facilities despite their substantial economic contributions (NCAER 2021). A study in Kerala highlights gender disparities in access to credit, transportation, ice, and storage infrastructures, those are key inputs that directly affect productivity and competitiveness (Rejula et al. 2023). Institutional analyses also reveal that women are often classified as informal or auxiliary workers, limiting their eligibility for formal welfare schemes and government support (ICSF 2023). These compounded inequalities not only suppress women's earning potential but also hinder broader objectives of food security, livelihood resilience, and sustainable fisheries governance.

Despite these challenges, there is emerging evidence of progress in select regions, suggesting that targeted interventions can enhance more inclusive outcomes. In West Bengal, cooperative support schemes have enabled the establishment of women-led fish vending units, offering access to hygienic infrastructure, cold storage, and better market linkages (National Platform for Small Scale Fish Workers (NPSSFW) 2025). Similarly, in Assam, women engaged in prawn seed collection have benefitted from training in sustainable harvesting practices under state and NGO led programs, resulting in improved productivity and income generation (FAO 2022). These examples reflect a mixed landscape, while the structural barriers remain pervasive, strategic investments and institutional support can significantly enhance women's roles in the fisheries value chain. However, without consistent policy efforts to address gendered disparities in resource access, decision-making, and institutional recognition, the long-term sustainability and equity of inland fisheries will remain at risk.

19.2.3 Gender-Based Violence and Discrimination in the Sector

Gender-based violence (GBV) and discrimination remain critical yet under-addressed issues in India's inland small-scale fisheries (SSF) sector. Women in these systems frequently operate within informal, insecure environments that expose them to various forms of violence and discrimination that range from verbal abuse and harassment to exclusion from economic opportunities and decision-making spaces. The highly gendered division of labor, wherein men dominate capture and production while women are often confined to post-harvest roles, that entrenches unequal power dynamics. These dynamics make women more vulnerable to economic marginalization and social exploitation. Structural gaps, including the lack of legal protection, inadequate grievance redressal mechanisms, and absence of female representation in cooperatives or unions, further suppress women's ability to report abuse or assert their rights (ICSF 2021; FAO 2022). Discriminatory practices such as unequal wages, restricted access to market spaces, and denial of formal licenses compound these challenges, perpetuating a cycle of invisibility and dependence.

A global review by Williams et al. (2022) confirms that GBV in fisheries is not only prevalent but often normalized within sectoral institutions and practices. They identify five interrelated types of GBV such as physical, sexual, psychological, economic, and cultural those persist across various fisheries systems, including inland sectors. A clear example of such violence was seen in Kerala where municipal officials reportedly harassed a woman fish vendor by discarding her catch onto the street, a public act of humiliation that highlights the power imbalances and lack of accountability in local fisheries governance (The Indian Express 2021). These harms are frequently reinforced by policy gaps, labor violations, and social norms that fail to recognize the risks faced by women fish workers. Importantly, the study argues that GBV is inseparable from broader fisheries governance and equity goals, and addressing it must be central to sustainable management strategies. Although state-level initiatives such as Telangana's SHE Teams have begun to create safer environments for women vendors and fish workers, their reach remains limited and uneven (Government of Telangana 2022). Unless gender-sensitive laws, strong institutional protections, and empowerment programs for both women and men are systematically implemented, gender-based violence will continue to weaken the social fabric and economic stability of inland fisheries.

19.3 Initiatives and Interventions Promoting Gender Equity in Small Scale Fisheries

19.3.1 Government Policies and Programs Supporting Women in the Sector

In the heart of rural India, various efforts and initiatives are gradually shaping the growth of the small-scale fisheries sector (Table 19.1). These programs act as guiding lights, helping to promote gender equality and empower women across different parts of the industry. As these initiatives gain momentum, they begin to address long-standing gender gaps by highlighting the valuable roles women play. The aim is clear, to give more recognition and voice to women involved in small-scale fisheries, like skilled artisans, these programs build opportunities for women to lead, participate, and succeed. Through training, women gain the skills needed to handle a range of activities from net making and fishing to processing and marketing. Access to resources like loans, modern tools, and knowledge becomes a key part of their empowerment.

Supporting these efforts, several national policies and frameworks promote gender inclusion. The *National Policy on Marine Fisheries* (Department of Fisheries 2017) and the *Draft National Inland Fisheries and Aquaculture Policy* (Department of Fisheries 2020) both highlight the importance of including women across the fisheries value chain. In addition, broader frameworks like the *National Policy for Women* (Ministry of Women and Child Development 2016) and the *Gender Budgeting Guidelines* (Ministry of Finance 2021) provide a supportive policy environment that encourages planning and action for gender equality. In line with these policies, various programs have been launched that directly promote gender equity in the fisheries sector.

Table 19.1 Government programs supporting women in small-scale fisheries

Program/scheme	Key provisions	Contribution to gender equity
NFDB Scheme for Women in Fisheries	Financial and technical support to women's SHGs and cooperatives in fish processing, marketing, fish vending units, and hygienic retail infrastructure	Promotes entrepreneurship among women in post-harvest fisheries, enables capacity building and infrastructure access
Pradhan Mantri Matsya Sampada Yojana (PMMSY)	Subsidies (30–60%) for women beneficiaries for value chain components (cold chain, fish kiosks, solar drying, etc.)	Encourages inclusivity through targeted financial assistance and support for women-led activities
National Livestock Mission (Fisheries Component)	Productivity enhancement and livelihood support in aquaculture including women groups	Empowers women to participate in aquaculture and increase household incomes
National Scheme of Welfare of Fishermen	Group accident insurance, savings-cum-relief, housing, community halls, and training	Benefits extended to women fishers involved in the trade and improves social security
Mahila Kisan Sashaktikaran Pariyojana (MKSP)	Promotes women's collectives in agriculture and allied sectors	Indirect support to women in fisheries through SHGs engaged in fish farming and fish-based livelihoods
Rashtriya Krishi Vikas Yojana (RKVY)	Financial assistance and convergence with other schemes for integrated development of allied sectors	Provides access to fisheries 1 related components for women SHGs under state-level projects
NABARD Fisheries Schemes	Credit access, entrepreneurship loans, and support for rural fishery value chains	Facilitates credit empowerment of fisherwomen and encourages enterprise-based livelihoods
Swachh Action Plan-Fisheries Component	Focus on hygiene and sanitation in fish markets, landing centers	Enhances health and working conditions for women handling fish in unhygienic environments
ICAR-CIFE Waste to Wealth Model	Conversion of fish waste into manure and other by products	Introduces innovative livelihood options for women in fish market waste management and circular economy

Source: Dof, Govt. of India

19.3.2 State-Specific Initiatives

Many states in India may have their own schemes and initiatives to support women in fisheries (Table 19.2). State fisheries departments often implement programs that target the specific needs of women engaged in small-scale fisheries.

In a significant stride towards enhancing the lives of women fish vendors, a collaborative effort between the Kerala State Road Transport Corporation (KSRTC) and the Department of Fisheries, marked the commencement of a free bus service tailored specifically for women involved in the fish trade, to address transportation challenges faced by women fish vendors, ensuring accessibility and convenience in

Table 19.2 State level programs supporting women in fisheries

State and program	Key provisions	Contribution to gender equity
Kerala—Mahila Kisan Sashaktikaran Pariyojana (MKSP)	Skill development, capacity building, financial assistance for women in agriculture and fisheries	Enhances women's skills and economic empowerment in fisheries
Andhra Pradesh—Matsyakara Bharosa	Financial aid during non-fishing seasons to fishermen families including women	Provides economic stability and social security for women fishers' families
West Bengal—Matshya Sampad Bikash Yojana (MSBY)	Fisheries sector development including women's skill training and participation	Promotes women's engagement and capacity building in fisheries
Odisha—Fish Pond Yojana for Women	Financial support for constructing and managing fish ponds	Encourages active participation of women in aquaculture
Tamil Nadu—Women Component Plan (WCP) for Fisheries	Dedicated budget allocation for women-centric fisheries schemes	Supports holistic development and empowerment of women fishers
Maharashtra—Stree Shakti Puraskar	Recognition and awards for women's achievements in various sectors, including fisheries	Motivates and publicly acknowledges women's contributions
Gujarat—Sakhi Mandal Scheme	Promotion of women's SHGs with financial assistance for income activities	Strengthens women's self-help groups, including fisheries-related enterprises
Telangana—SHE Teams in Fisheries	Safety, health, and environment teams focusing on women's well-being	Creates safer work environments and addresses harassment concerns
Odisha—Mamata Scheme	Financial aid to women's SHGs engaged in fisheries activities	Empowers women through group-based economic initiatives
Andhra Pradesh—Financial Assistance for Women SHGs	Support for women's SHGs to expand fisheries enterprises	Facilitates entrepreneurial growth among women fishers
Kerala—Subhiksha Keralam Project	Skill training, financial aid, market linkages for women in fisheries	Comprehensive support enhancing women's economic participation
Assam—Pradhan Mantri Matru Vandana Yojana (PMMVY)	Maternity benefit scheme for pregnant and lactating women	Improves health and well-being of women including those in fisheries
West Bengal—Women-Operated Fish Vending Outlets	Establishment and promotion of women-run fish vending platforms	Promotes economic independence and market access for women vendors
Tamil Nadu—Fishermen's Welfare Board Initiatives	Skill development, financial aid, healthcare services for women fishers	Enhances welfare and capacity of women in fishing communities

Source: State Government websites

their daily activities. This collaborative governance, where public transport and fisheries departments join hands to empower and uplift the women contributing to the vibrant fisheries sector.

In the vast landscape of India's inland fisheries sector, wide range of schemes has emerged, each designed to bring meaningful change to the lives of women engaged in small-scale fisheries. These diverse initiatives, covering multiple aspects of the sector, are united by a common goal, to address the various challenges faced by women and to empower them in economic, social, and holistic ways. From the safety provided by SHE Teams in Telangana's fish markets to Kerala's Mamata Scheme promoting financial independence, Assam's efforts to include women under the PMMVY program, West Bengal's support for women-managed fish vending outlets, and Tamil Nadu's Fishermen's Welfare Board offering special programs, all these schemes together form a powerful narrative of empowerment, inclusion, and progress. They reflect a determined commitment to improving the status of women in small-scale fisheries across India's diverse regions.

19.4 Gender Equity in Small-Scale Fisheries Across Rivers, Reservoirs, Wetlands, and Estuaries

Gender equity in small-scale fisheries across rivers, reservoirs, wetlands, and estuaries is vital for the sustainability and resilience of these aquatic ecosystems and the communities dependent on them. Women play diverse and crucial roles in fishing activities, from harvesting and processing to marketing and resource management, yet their contributions often remain underrecognized and unsupported. Ensuring equal access to resources, technology, training, and decision-making not only empowers women but also strengthens the social and economic fabric of fishing communities. The data presented here has been collected primarily under the institutional project conducted from 2018 to 2021.

19.4.1 Rivers

In the inland riverine fisheries along the Ganga River in India, women play a significant yet often underrepresented role in sustaining livelihoods. In the middle stretch of the Ganga covering districts such as Chiraigaon (Varanasi), Jamarwa (Fatehpur), Doganpur (Kanpur), and Ballipatti (Farrukhabad) in Uttar Pradesh women actively participate in fishing activities, particularly using gill nets and bag nets in open-access waters. Most of these women are between the ages of 35–55, with approximately 75% over the age of 50, reflecting a long-standing connection with the river and its resources. These women, typically lacking formal education, have acquired traditional knowledge and skills through decades of fishing experience, often spanning 27–33 years. They primarily belong to joint families, where fishing is both a source of income and a way of life. For 75% of these women, fishing is the main occupation, while the remainder supplement their income through agricultural labor. In contrast,

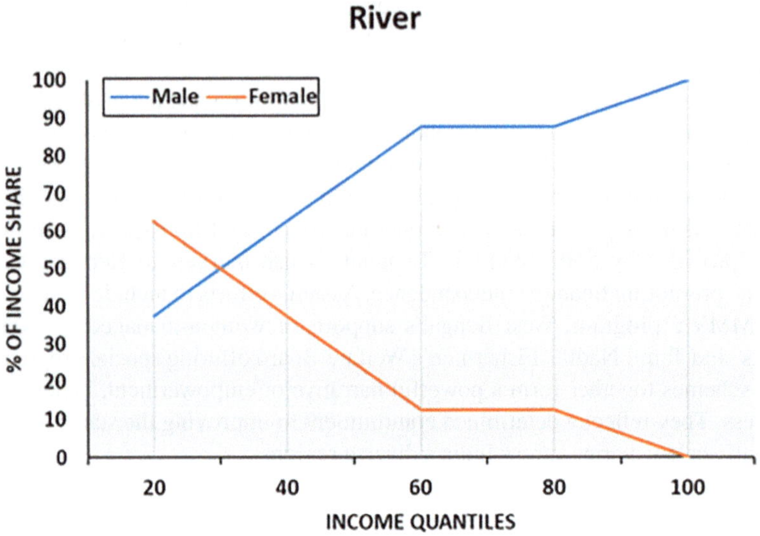

Fig. 19.1 Income distribution among male and female fishers in river Ganga. (Source: National Mission for Clean Ganga Project (2017–2020))

the lower stretch of the Ganga, encompassing Balagarh (Hooghly), and Godakhali, Diamond Harbour, and Fraserganj in South 24 Parganas, West Bengal—also shows significant participation by women in fisheries. Here, women are generally above 45 years of age and share similar levels of experience as those upstream. However, some socio-economic differences are observed. Women in the lower stretch have an average of 3 years of formal education and typically live in nuclear families (75%). Fishing remains their primary occupation. Despite the shared dependence on open-access riverine resources, income levels differ between the two regions. Women in the lower stretch earn approximately ₹5750 (about $70) per month solely from fishing, while those in the middle stretch earn around ₹4210 (about $51), with ₹3250 (about $40) coming directly from fishing activities (Fig. 19.1). These variations reflect broader socio-economic differences across regions. Overall, women in both stretches of the Ganga continue to play a vital role in maintaining household livelihoods through their sustained engagement in riverine fisheries, despite limited recognition, resources, and institutional support.

19.4.2 Reservoirs

The role of women in reservoir fisheries across India varies significantly depending on regional contexts and management systems. A study conducted on six reservoirs reveals (Fig. 19.2) differing levels of women's engagement, shaped by cultural practices, livelihood structures, and socio-economic conditions. These case studies

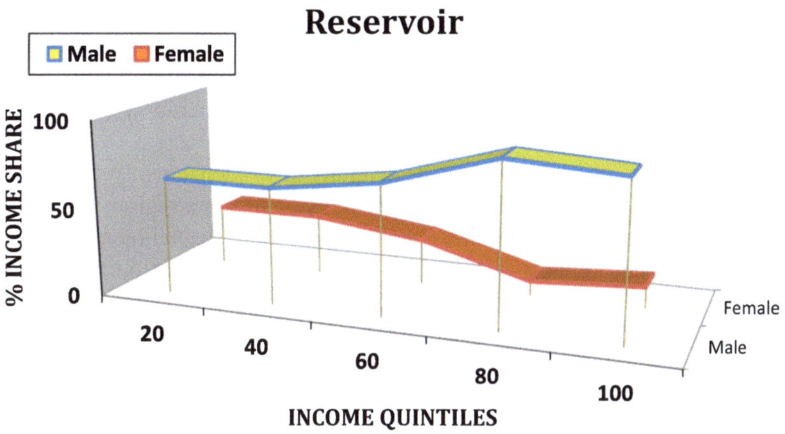

Fig. 19.2 Income distribution among male and female fishers of reservoirs in India. (Source: ICAR-CIFRI (2017–2020) Annual report)

provide important insights into the status of gender equity in India's inland small-scale fisheries sector.

In the Derjang Reservoir in Odisha's Angul district, fishing is primarily a male dominated activity. Women are largely confined to household responsibilities and reproductive roles. Their involvement in fisheries is limited to occasional tasks such as sorting and cleaning after harvests. Some women also contribute to household income through bidi (hand-rolled cigarette) making. Educational disparities are evident, with a higher proportion of illiterate women compared to men, and average monthly incomes of male fishers are significantly higher.

In Jharkhand's Patratu Reservoir, women mainly handle domestic responsibilities and have limited involvement in fisheries. Educational levels among fisherwomen are generally low, and while some women engage in fish vending, their participation remains minimal. However, the introduction of cage culture in the reservoir has led to improvements in children's education among fisher families, indicating a positive indirect impact.

In contrast, the Jargo Reservoir in Uttar Pradesh reflects a setting where women are not involved in fisheries activities. Most women are homemakers, and a large number are illiterate, highlighting restricted participation and educational barriers in this region.

In Tamil Nadu's Krishnagiri Reservoir, women actively participate in fishing activities alongside their husbands, and sometimes with their children. Despite their involvement, they face health issues such as skin rashes due to poor water quality, indicating occupational health risks associated with their roles.

The Nagarjunasagar Reservoir, spanning Telangana and Andhra Pradesh, presents a different scenario where women accompany their husbands on coracle fishing trips. In several households, women have higher education levels than men, which challenges traditional gender roles and offers potential for enhanced decision-making participation.

In Karnataka's Harangi Reservoir, women participate directly in coracle fishing with their spouses. Approximately 54.55% of the women engage in fishing, contributing to a combined monthly household income of ₹5405 (about $65). This joint effort reflects a collaborative model of income generation and household responsibility.

These cases collectively illustrate the diversity of women's involvement in reservoir fisheries across India. From complete exclusion in some regions to active participation and joint ventures in others, women's roles are shaped by a complex mix of cultural norms, education, health conditions, and institutional support. Understanding these patterns is essential for designing gender-inclusive policies and interventions that enhance equity and sustainability in inland fisheries.

19.4.3 Wetlands

In the wetlands of Nayacharra (North Bengal) and Katiganga (Murshidabad district, West Bengal), surveys have highlighted the gendered dynamics of fisheries-based livelihoods (Fig. 19.3). In these regions, men are primarily engaged in fishing, while women mostly perform domestic duties, constituting 81.48% of the female population. However, some women contribute economically through alternative sources such as domestic work, cooking, or bidi-making. Though their direct participation in fishing activities is limited, their contributions play an important role in

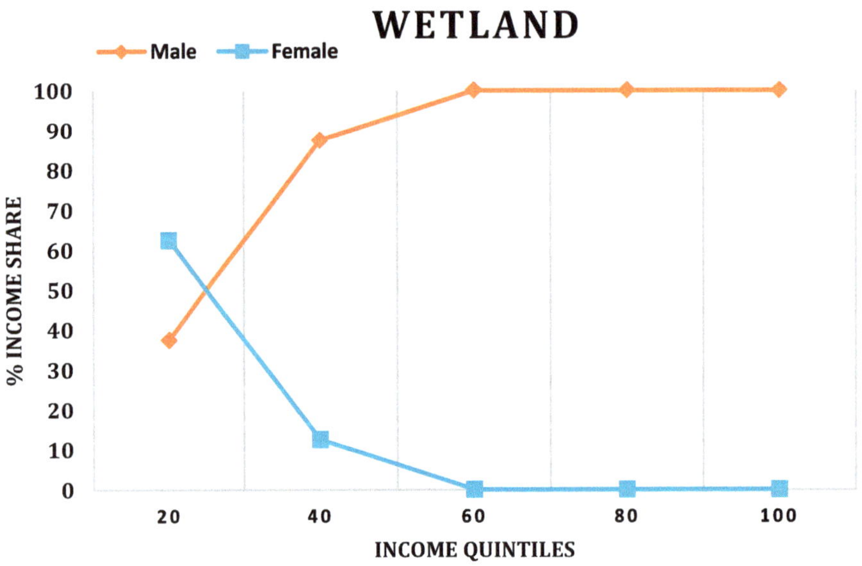

Fig. 19.3 Income distribution among male and female fishers of wetlands in India. (Source: ICAR-CIFRI (2017–2020) Annual report)

supporting household incomes and maintaining livelihood security, reflecting resilience and adaptability within these wetland-based communities.

In contrast, the Sundarbans region of West Bengal, particularly in the Sandeshkhali block, presents a different picture. A significant number of women are actively involved in estuarine fisheries. Their responsibilities extend beyond domestic roles and include feeding and monitoring aquaculture units, participating in fish farming, and taking part in decision-making processes regarding the distribution and utilization of profits from aquaculture. In several cases, homestead ponds are registered in women's names, either as part of inheritance or upon becoming household heads following the death of a spouse. These instances reflect gradual shifts in gender roles, moving towards enhanced agency and ownership for women in the fisheries sector.

19.4.4 Estuaries

In Odisha's estuarine systems, particularly in Khordha and Puri districts, women fishers report improved participation in recent years (Fig. 19.4). Their involvement spans various post-harvest activities, including fish drying, grading, sorting, marketing, and processing. Nevertheless, gender disparities persist. While women are recognized as active members of Primary Fishermen Cooperative Societies (PFCS), none hold positions on the Boards of Management, indicating limited representation in leadership roles. About 42.86% of surveyed women acknowledged ongoing gender bias that favors men in fisheries operations and governance. Despite these challenges, many women have begun to engage independently in fisheries-related

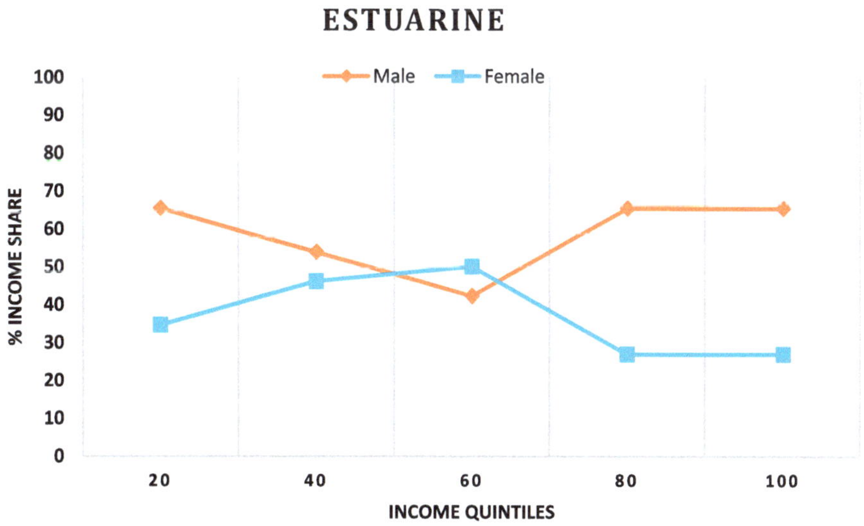

Fig. 19.4 Income distribution among male and female fishers of estuarine areas in India. (Source: National Mission for Clean Ganga Project (2017–2020))

businesses, reducing economic dependence on their spouses and asserting greater financial autonomy.

A similar pattern is observed in Odisha's Chilika lagoon, where women are engaged in fish vending and marketing, and are often registered with cooperatives, yet like in other regions, their representation at decision-making levels remains absent. These findings highlight both progress and persistent structural limitations in achieving gender equality in inland fisheries.

Across the aquatic landscapes of the Ganga River, reservoirs, wetlands, and estuarine systems, gendered income disparities are evident. Men continue to earn higher incomes and maintain greater involvement in core fisheries activities across most sites. However, in the lower income categories, particularly among the poorest households where women contribute at comparable levels to men, challenging conventional gender roles. In estuarine fisheries, women's income distribution appears more balanced, especially in the middle-income segment, indicating positive trends toward gender inclusion. These observations underscore both the entrenched inequalities and the emerging shifts that define gender dynamics in small-scale inland fisheries across India.

19.5 Challenges and Opportunities in Achieving Gender Equity in Small Scale Fisheries in Inland Sector of India

The inland small-scale fishing sector in India has a complex interplay of opportunities and difficulties in its quest of gender parity. Deeply ingrained cultural conventions that frequently limit women to conventional roles within fishing communities provide challenges. Women's ability to lead and actively participate in fishery activities is hampered by their limited access to resources, including technology, credit, and training, which further perpetuates gender inequities. One major obstacle that prevents women from having as much influence in establishing policies that impact their communities is their unequal access to leadership positions and decision-making forums. Cultural conventions and limitations on mobility can also prevent women from participating in some types of fishing, but these difficulties also present chances for revolutionary transformation. Initiatives focused on education and skill development can enable women to take on a wider range of responsibilities and make a bigger impact on the value chain of the fishing industry. Women can assume leadership roles and break down barriers by promoting fair access to economic and educational resources. Policies and initiatives that are gender-sensitive can help to build an atmosphere that is supportive of inclusivity and equitable involvement.

Inspite the challenges, there are significant economic opportunities waiting to be tapped. Placing greater emphasis on the involvement of women at every stage of the fisheries value chain, from marketing to post-harvest operations, has the potential to yield substantial financial benefits for families and communities. Additionally, by addressing gender-based violence and ensuring safe working conditions, the industry can create a more inclusive environment that attracts and retains women. This two-fold approach, harnessing economic opportunities while developing a secure and supportive workplace, can contribute not only to the empowerment of women but also to the overall growth and sustainability of the small-scale fisheries sector in the inland regions of India.

So achieving gender parity in India's inland small-scale fishing industry requires a coordinated effort to break down social barriers. This should be coupled with focused programs that empower women with increased economic and social clout also by identifying and resolving obstacles while simultaneously seizing opportunities, the industry can pave the way for a more inclusive and equitable future for women. The transformative journey towards gender equity necessitates a comprehensive approach, acknowledging the challenges and capitalizing on the potential for positive change.

19.6 Conclusion

Gender equity in India's inland small-scale fisheries is both a pressing necessity and a promising opportunity. Women play critical roles throughout the fisheries value chain, initiating from harvesting and seed collection to processing, marketing, and household management, yet their contributions often remain invisible in official statistics, undervalued in policy frameworks, and unsupported by institutions. Across diverse ecosystems such as rivers, reservoirs, wetlands, and estuaries, regional and socio-economic differences influence the degree of women's participation, income levels, and access to resources.

Despite forming nearly half of the labor force in this sector, women continue to face deeply entrenched gender-based challenges, including restricted access to land, credit, technology, and formal training, as well as limited representation in cooperatives and decision-making bodies. These inequalities are further compounded by gender-based violence, discriminatory norms, and systemic barriers that marginalize women from economic opportunities and governance structures. However, several promising interventions have begun to shift this narrative. Government programs such as Mission Shakti, PMMSY, National Policy for Women, and state-level schemes including SHE Teams in Telangana, fish vending support in West Bengal, and the free bus services in Kerala, demonstrate that targeted and inclusive strategies can improve women's participation, income, safety, and decision-making power. Case studies across inland waterbodies also highlight growing instances of women-led cooperatives, improved access to aquaculture, and increased involvement in sustainable practices, particularly in regions such as Odisha, Assam, and the Sundarbans. The economic and ecological sustainability of India's inland fisheries depends heavily on recognizing and empowering the women who form its backbone. When women have access to resources, education, and safe working conditions, they not only contribute to household and community well-being but also play a key role in improving nutrition, resource management, and resilience to environmental change. Achieving gender equity in small-scale fisheries requires a multidimensional approach, addressing structural barriers, reforming institutional frameworks, promoting inclusive policies, and investing in women's skills, leadership, and safety. This transformation will not only uplift women's status within fisheries but also strengthen the sustainability, productivity, and social fabric of inland fishing communities. A gender-inclusive inland fisheries sector is not only a matter of social justice, it is essential for the long-term success and resilience of the sector itself.

Points to Remember

- Small-scale fisheries in India employ 23.12 million inland fishers, with women comprising 44% of the workforce.
- Women's contributions to small-scale fisheries include post-harvest activities like processing, marketing, and financial management, crucial for household nutrition and food security.
- Despite their significant role, women often face limitations in accessing resources and decision-making opportunities within the sector.
- Gender-based violence and discrimination persist in small-scale fisheries, hindering women's engagement and well-being.
- Initiatives promoting gender equity offer training, access to resources, and leadership opportunities for women in fishing communities.
- In various aquatic ecosystems, including rivers, reservoirs, wetlands, and estuaries, women's roles in fisheries vary, impacting income distribution and participation.
- Cultural norms and limited access to resources present challenges to achieving gender parity in the fishing industry.
- Addressing gender disparities and enhancing women's involvement in small-scale fisheries can yield economic benefits and contribute to sustainable development in India's inland regions.

References

Bandyopadhyay S, Joshi L (2024) Raw deal for women in the fisheries sector. The Hindu Business Line. https://www.thehindubusinessline.com/opinion/raw-deal-for-women-in-the-fisheries-sector/article67970693.ece. Accessed 20 June 2025

Chandra G, Sharma AP (2015) Gender roles in inland fisheries of India: a cross country study [Abstract]. In: Proceedings of the 5th global symposium on gender in aquaculture and fisheries, November 2015. https://www.genderaquafish.org/wp-content/uploads/2018/04/gaf5-abs-26-ganesh-chandra.pdf

De HK, Chattopadhyay DN, Radheyshyam, Saha GS, Dash AK, Satpati TS, Pal S (2013) Grow out carp polyculture by women: a case study from Odisha. Indian Res J Ext Educ 49(1&2):78–82

Department of Fisheries, Ministry of Agriculture & Farmers Welfare (2017) National policy on marine fisheries, 2017. Government of India. https://dof.gov.in/sites/default/files/2020-08/National_Policy_on_Marine_Fisheries_2017_0.pdf. Accessed 16 June 2025

Department of Fisheries, Ministry of Fisheries, Animal Husbandry & Dairying (2020) Draft national inland fisheries and aquaculture policy, 2020. Government of India. https://dof.gov.in/

sites/default/files/2020-07/Draft_National_Inland_Fisheries_and_Aquaculture_Policy_2020.pdf. Accessed 16 June 2025

FAO (2005) Increasing the contribution of small scale fisheries to poverty alleviation and food security. FAO Technical Guidelines for Responsible Fisheries No. 10. FAO, Rome

FAO (2016) The state of world fisheries and aquaculture 2016: contributing to food security and nutrition for all. Food and Agriculture Organization of the United Nations, Rome

FAO (2020) The state of world fisheries and aquaculture 2020: sustainability in action. FAO, Rome. https://www.fao.org/documents/card/en/c/ca9229en. Accessed 16 June 2025

FAO (2022) The role of women in small scale fisheries: a snapshot from India. FAO, Rome

FAO (2023) Voluntary guidelines for securing sustainable small scale fisheries: gender equity and equality. FAO, Rome. https://www.fao.org/voluntary-guidelines-small-scale-fisheries/key-thematic-areas/gender-equity-and-equality/en. Accessed 10 June 2025

Fishery News (2023) Nourishing communities, empowering women: the odyssey of nutrition-sensitive aquaculture in Odisha. Fishery News, February. https://www.fishery.news/nourishing-communities-empowering-women-the-odyssey-of-nutrition-sensitive-aquaculture-in-odisha. Accessed 1 June 2025

Government of Telangana (2022) SHE Teams initiative for women's safety. Telangana State Police Department. https://sheteams.tspolice.gov.in. Accessed 19 June 2025

Handbook on Fisheries Statistics (2022) Department of Fisheries, Ministry of Fisheries, Animal Husbandry & Dairying, Government of India, New Delhi

ICAR (2017) Handbook of fisheries and aquaculture. Indian Council of Agricultural Research (ICAR), New Delhi

ICAR-CIFRI (2017–2020) Annual Report (2017–2020). ICAR-Central Inland Fisheries Research Institute, Barrackpore, West Bengal

ICSF (2021) Women in fisheries in India: a brief status report. International Collective in Support of Fishworkers, Chennai

ICSF (2023) Women in fisheries: a collection of articles from Samudra Report. International Collective in Support of Fishworkers, Chennai. https://www.icsf.net/resources/women-in-fisheries-2023. Accessed 16 June 2025

Kusakabe K (2003) Gender issues in small scale inland fisheries in Asia: women as an important source of information. FAO, Rome. https://www.fao.org/3/ad070e/ad070e08.html

Ministry of Finance (2021) Gender budgeting handbook for government ministries and departments. Government of India. https://finmin.nic.in/sites/default/files/Gender_Budgeting_Handbook_2021.pdf. Accessed 16 June 2025

Ministry of Women and Child Development (2016) National policy for women 2016 – articulating a vision for empowerment of women. Government of India. https://wcd.nic.in/sites/default/files/draft%20national%20policy%20for%20women%202016_0.pdf. Accessed 16 June 2025

National Mission for Clean Ganga (NMCG) (2017–2020) Annual report 2019. ICAR-CIFRI, Ministry of Jal Shakti, Government of India

National Platform for Small Scale Fish Workers (NPSSFW) (2025) Home. Retrieved October 26, 2025, from https://smallscalefishworkers.org/)

NCAER (2021) Socio-economic impact of COVID-19 on women fish vendors and informal sector workers. National Council of Applied Economic Research, New Delhi

NFDB (2022) National Fisheries Development Board annual report 2021–22. Government of India, Hyderabad

Panemangalore A (2021) Nutrition sensitive carp–mola polyculture through women's groups: a successful policy intervention by the government of Odisha, India. WorldFish

Rejula K, Chauhan JK, Walia SS, Verma MR, Dhar U, Choudhary S, Chikkeri SS (2023) Constraints in vegetable production in India: a review. Indian Res J Ext Educ 23(3):14–19

Sinha F, Srinivasan KA, Singh RK, Srinivasan V (1994) The blue revolution: case study of women in the inland fisheries sector. CABI, New Delhi, 127 p

Sugunan VV, Sinha M (2001) Sustainable fisheries development in reservoirs of India. CIFRI, Barrackpore

The Indian Express (2021) Fish vendor harassment case: rights panel books Attingal municipal staff. https://www.newindianexpress.com/cities/thiruvananthapuram/2021/Aug/13/fish-vendor-harassment-case-rights-panel-books-attingal-municipal-staff-2343966.html. Accessed 12 June 2025

Williams LA, Harper S, Kleiber D, Franz N, Cohen P (2022) Gender based violence in the fisheries sector: understanding the hidden harms. Fish Fish 23(6):1383–1398

World Bank (2012) Hidden harvest: the global contribution of capture fisheries. The World Bank, Washington, DC. https://documents.worldbank.org

MIX
Papier aus verantwortungsvollen Quellen
Paper from responsible sources
FSC® C105338

If you have any concerns about our products,
you can contact us on
ProductSafety@springernature.com

In case Publisher is established outside the EU,
the EU authorized representative is:
Springer Nature Customer Service Center GmbH
Europaplatz 3, 69115 Heidelberg, Germany

Printed by Libri Plureos GmbH
in Hamburg, Germany